P McGregor

A System of Logic

Comprising a discussion of the various means of acquiring and retaining knowledge,

and avoiding error

P McGregor

A System of Logic
Comprising a discussion of the various means of acquiring and retaining knowledge, and avoiding error

ISBN/EAN: 9783337248321

Printed in Europe, USA, Canada, Australia, Japan

Cover: Foto ©berggeist007 / pixelio.de

More available books at **www.hansebooks.com**

A

SYSTEM OF LOGIC,

COMPRISING

A DISCUSSION OF THE VARIOUS MEANS OF ACQUIRING AND
RETAINING KNOWLEDGE, AND AVOIDING ERROR.

BY P. McGREGOR, A.M.

NEW YORK:
HARPER & BROTHERS, PUBLISHERS,
FRANKLIN SQUARE.
1862.

Entered, according to Act of Congress, in the year one thousand
eight hundred and sixty-two, by

HARPER & BROTHERS,

In the Clerk's Office of the District Court of the Southern District
of New York.

PREFACE.

The following treatise is the result of an attempt to comprise within moderate limits everything of general interest which properly belongs to Logic, free from prolixity, obscurity or misrepresentation. Much that occurs in other works on the same subject, has been rejected as useless, irrelevant or erroneous, while I have endeavored to supply numerous deficiencies, and to exhibit a clear and accurate view of the principles and processes of logical thought, divested of scholastic figments, which only perplex and mislead the student.

CONTENTS.

INTRODUCTION.

		PAGE
§ 1.	Nature, Foundations, and Limits of Logic	13
§ 2.	Objects, Uses, and Study of Logic	14
§ 3.	Advantages of Knowledge, and Evils of Ignorance	15

PART I.

OF THE ULTIMATE SOURCES AND ELEMENTS OF KNOWLEDGE, AND THE PRIMARY PROCESSES BY WHICH IT IS ACQUIRED AND RETAINED.

CHAPTER I.

A GENERAL VIEW OF THE LIMITS, DIVISIONS, AND IMMEDIATE SOURCES OF KNOWLEDGE.

§ 1.	Necessary Limits and principal Divisions of Knowledge	29
§ 2.	Of the various Faculties by which Knowledge is acquired and retained	32
§ 3.	Of Propositions	37
§ 4.	Of Probability	40
§ 5.	General Criterion of Truth, and immediate Source of Error	45

CHAPTER II.

OF REASONING.

§ 1.	Nature, General Principle, and Expression of Reasoning	49
§ 2.	Special Principles of Reasoning	52
§ 3.	Processes and Criterions of Reasoning	65

CHAPTER III.

OF THE PRIMARY MEANS OF ACQUIRING CONTINGENT KNOWLEDGE.

§ 1.	Reality of Apprehensions, and Means of avoiding the primary Errors which they directly occasion	71
§ 2.	Primary mental Processes by which contingent Knowledge may be acquired	75
§ 3.	Primary external Processes by which contingent Knowledge may be acquired	86

CHAPTER IV.

OF THE PRIMARY MEANS OF RETAINING KNOWLEDGE.

	PAGE
§ 1. Reliability of Memory, and Means of avoiding the primary Errors which it tends to produce.............................	93
§ 2. Primary Processes by which Knowledge is retained..........	96

CHAPTER V.

OF GENERALIZATION.

§ 1. Nature of Generalization..	98
§ 2. Principal Processes of Generalization.............................	99
§ 3. Extension and Uses of Generalization...........................	107

CHAPTER VI.

OF HYPOTHESES.

§ 1. Nature and Uses of Hypotheses...................................	109
§ 2. Methods of testing Hypotheses.....................................	113

PART II.

OF THE PRINCIPLES AND METHODS OF INVESTIGATION.

CHAPTER VII.

OF INVESTIGATION IN GENERAL.

§ 1. Of Dispositions affecting Investigation...........................	119
§ 2. Of Habits affecting Investigation...................................	121
§ 3. Of Things which require no Proof.................................	127
§ 4. Of Things which may generally be admitted as proved	129
§ 5. General Modes of determining the Validity of Proofs	134

CHAPTER VIII.

OF STUDY.

§ 1. Nature and Uses of Study..	139
§ 2. Subjects, Modes, and General Rules of Study	140
§ 3. Selection and Study of Books.......................................	148

CHAPTER IX.

OF ORIGINAL INVESTIGATION.

§ 1. General Character, Uses, Prerequisites, and Methods of Original Investigation	151
§ 2. Of Direct Discovery	155
§ 3. Of Indirect Discovery	156
§ 4. Of Invention...	162

CHAPTER X.
OF CAUSES AND EFFECTS.

§ 1. Sources and Applications of the Knowledge of Causes and Effects 165
§ 2. Various Kinds of Causes 169
§ 3. Methods of determining Causes and Effects 171

CHAPTER XI.
OF LANGUAGE.

§ 1. Origin and Progress of Language 182
§ 2. Uses of Language 189
§ 3. Imperfections and Abuses of Language 194
§ 4. Interpretation of Language 200

CHAPTER XII.
OF EVIDENCE.

§ 1. General Principles of Evidence 209
§ 2. Criterions of Testimony 212
§ 3. Various Kinds of Testimony, and Peculiarities of each 226
§ 4. Means of ascertaining the Origin and Character of Written Testimony 228

CHAPTER XIII.
OF CLASSIFICATION.

§ 1. Nature and Uses of Classification 238
§ 2. Principles and Methods of Classification 240

CHAPTER XIV.

Tabular View of the Means of acquiring Knowledge 246

PART III.
OF FALLACIES.

CHAPTER XV.
NATURE AND CLASSIFICATION OF FALLACIES.

§ 1. Nature of Fallacies 249
§ 2. Classification of Fallacies 250

CHAPTER XVI.
SOURCES OF FALLACIES, AND MEANS OF GUARDING AGAINST THEM.

§ 1. Sources of Fallacies 252
§ 2. Of Prejudices 254
§ 3. Means of guarding against Fallacies 260

CHAPTER XVII.

OF PARALOGISMS, OR FALLACIES OF PRIMARY ASSUMPTION.

	PAGE
§ 1. Paralogisms of Intuition	266
§ 2. " " assuming what is attempted to be proved	267
§ 3. " " Comprehension	269
§ 4. " " Signs	271
§ 5. " " Memory	274
§ 6. Intrinsic Paralogisms of Testimony	275
§ 7. Extrinsic " "	280
§ 8. Paralogisms of Misinterpretation of Language	283

CHAPTER XVIII.

OF SOPHISMS, OR FALLACIES OF INTERMEDIATE REASONING.

§ 1. Sophisms of Confusion	287
§ 2. " " Generalization	289
§ 3. " " Causation	292
§ 4. " " Probability	298

CHAPTER XIX.

OF ABERRANCIES, OR FALLACIES OF IRRELEVANCY.

§ 1. Aberrancies of Confusion	306
§ 2. " " Appeals to Authority	316
§ 3. " " Appeals to Desires	320

CHAPTER XX.

Table of Fallacies 325

PART IV.

A SPECIAL SURVEY OF THE PRINCIPAL BRANCHES OF KNOWLEDGE.

CHAPTER XXI.

CLASSIFICATION OF KNOWLEDGE, ACCORDING TO ITS SUBJECTS.

§ 1. Scientific Knowledge	331
§ 2. Mixed "	339
§ 3. Particular "	340
§ 4. Tabular View of Knowledge	341

CHAPTER XXII.

OF MATHEMATICS.

§ 1. Peculiarities of Mathematics	344
§ 2. Uses " "	346
§ 3. Study " "	347

CHAPTER XXIII.

OF THE PHYSICAL SCIENCES.

		PAGE
§ 1. Of the Physical Sciences in general		351
§ 2. " Mechanical Sciences		352
§ 3. " Ethereal "		358
§ 4. " Organical "		362
§ 5. " Geographical "		367

CHAPTER XXIV.

OF THE MENTAL SCIENCES.

§ 1. Of the Mental Sciences in general.................. 374
§ 2. Of Logic and Psychology.............................. 376
§ 3. Of Theology.. 377
§ 4. Of Morality, or Ethical Science..................... 381
§ 5. Of Jurisprudence.. 383

CHAPTER XXV.

OF MIXED KNOWLEDGE.

§ 1. Of Philology... 385
§ 2. Of Ethnography... 391
§ 3. Of Technology.. 392

CHAPTER XXVI.

OF PARTICULAR KNOWLEDGE.

§ 1. Of History... 396
§ 2. Of Chronology... 402
§ 3. Of Biography.. 403

CHAPTER XXVII.

OF THE KNOWLEDGE OF FUTURITY.

§ 1. Of the Knowledge of Futurity in general....... 405
§ 2. Sources of our Knowledge of Futurity........... 406

PART V.

OF THE RETENTION OF KNOWLEDGE.

CHAPTER XXVIII.

OF THE RETENTION OF KNOWLEDGE BY SIMPLE REMEMBRANCE.

§ 1. General Laws and Rules of Remembrance..... 413
§ 2. Of the Relations of Thoughts......................... 419

CHAPTER XXIX.

OF THE RETENTION OF KNOWLEDGE BY MEANS OF EXTERNAL SIGNS.

	PAGE
§ 1. Of External Signs in general	425
§ 2. Of the Retention of Knowledge by Writing	427

CHAPTER XXX.

OF THE MEANS OF POSSESSING A READY COMMAND OF OUR KNOWLEDGE.

§ 1. Requisites to possessing a ready Command of our Knowledge 429
§ 2. Means of acquiring and employing the preceding Requisites 431

CHAPTER XXXI.

Tabular View of the Means of retaining Knowledge 434

NOTES 435
INDEX 449

INTRODUCTION.

SYSTEM OF LOGIC.

INTRODUCTION.

§ 1. NATURE, FOUNDATIONS, AND LIMITS OF LOGIC.—Subjects and Definition of Logic.—On what founded.—Advantages of uniting its three principal subjects in one Science.—Why the communication of Knowledge is excluded.

LOGIC is the science which exhibits the foundations and primary elements of knowledge, the proper means of investigating truth, the nature and sources of erroneous opinions, the modes in which we must proceed in order to secure the former and avoid the latter, and the best methods of retaining knowledge after it has been acquired. It may, therefore, be defined *the science of the acquisition and retention of knowledge, and the means of avoiding error.* It is founded on the following principles, the truth of which becomes evident from a slight consideration (1).*

1. *We have only one set of intellectual faculties, the laws of whose proper exercise are identical, throughout the numerous fields of human inquiry.* Thus we have not one faculty of vision for Botany and another for Chemistry, nor one faculty of reasoning for Morality and another for Geology; and, in order to sound reasoning, the premises must necessarily imply the conclusion professedly inferred from them, whatever be the subject of consideration.

2. *Those faculties operate uniformly, and are liable to mislead us only in certain ways.* Thus the senses always operate by impressions on the nerves, through which corresponding impressions are produced in the seat of consciousness; and they are liable to mislead us only by presenting something which appears like a different thing. So Reason constantly operates by showing that one thing necessarily implies another; and it

* The figures at the ends of paragraphs refer to the notes, which precede the Index.

can occasion error only by leading us to believe that this is the case, when in reality it is not.

The acquisition and retention of knowledge, and the means of avoiding error, form subjects sufficiently concise and connected to be discussed as one science; and we throw unnecessary obstacles and dangers in the way of the inquirer, if we separate them, and, after furnishing him with a part, either leave him to think he has mastered the whole, when in reality he has not, or tacitly refer him to some unknown quarters, for a knowledge of several of its most important parts, of which he is still ignorant.

The propriety of including the retention of knowledge will be readily perceived, by observing that it is not sufficiently extensive to form a separate science, while it is essential to render knowledge available. In order to be of any value, truth must not only be discovered, but secured in such a manner that we can bring it before the attention at pleasure. It is further to be observed that many truths can be discovered only by retaining in the memory many others previously acquired.

The subject of the communication of knowledge should be excluded from Logic, on account of its great extent and its distinct nature.

§ 2. OBJECTS, USES, AND STUDY OF LOGIC.—General and special objects of Logic.—Its Utility.—Study of Logic.—Who may study it successfully.

Logic is designed to aid us in every inquiry, and not to dispense with any other science. Its general object is, to show the capacity of our intellectual faculties, and the modes in which they must be employed, in order to acquire and retain knowledge, and avoid error. Its principal special objects are—(1) to assist us in determining the truth of any given proposition—(2) to guard us against the errors which we are liable to adopt, in the various departments of investigation—(3) to furnish all the other aids which general discussions and directions can supply, in the pursuit of knowledge—(4) to point out the best means of retaining our intellectual acquisitions, so that we may use them at pleasure—and (5) to give us the proper degree of confidence in our intellects, so that we may avoid both dogmatism and skepticism.

A man who has ascertained the laws of proper investigation and the sources of error, will evidently deviate from the paths of truth much less frequently than one who differs from him only in having paid no attention to those subjects, and who may consequently be led unawares, by some common prejudice or illusion, into a wrong path, from which he would have been restrained by a knowledge of Logic. To investigate some points rightly, and to know the nature and conditions of proper investigation, are very different things; the latter of which is never acquired without study, and can rarely be acquired at all without the aid of Logic, while it is of great value in all the most important fields of human inquiry.

The utility of Logic appears from the fact that we can never adopt an erroneous opinion without first violating one or more of its principles, just as a person cannot commit a solecism in language without first violating some rule of Grammar.

In order that Logic should answer its objects, its principles and rules must be well understood and remembered: for otherwise they will be overlooked, and consequently violated, at the very time when their aid is most requisite. Its various parts should, therefore, be studied with such care that there will be no danger of misunderstanding them; and the more important parts should be repeatedly reviewed, until they are permanently impressed on the memory. The student should particularly beware of adopting false views of those rules and principles: for, as they are applicable to all investigations, he will thus lay the foundations of error on every other subject.

Logic may be mastered without any previous preparation or extraordinary abilities: and, therefore, it may be studied successfully by any person who will bestow on it a little care and labor, while it requires much less of either than some other subjects of comparatively little importance.

§ 3. ADVANTAGES OF KNOWLEDGE, AND EVILS OF IGNORANCE.—
Benefit of understanding the value of Knowledge.—Its various Advantages: (1) Its effects on the main pursuits of our lives.—Evils arising from mistaken views.—How Knowledge would prevent them. —(2) Its effects on the Emotions.—(3) Pleasures derived from it.— (4) Its influence on evil Habits—(5) on physical Welfare and Safety —(6) on Morality—(7) on Impositions—(8) on Superstition—and (9) on mental Discipline.—Evils of Ignorance.—Threefold Benefit

of the proper acquisition of Knowledge.—Bearing of this section on Logic.

The pursuit of knowledge is often much less pleasant, for the time being, than that of sensual pleasure, gain, fame, or amusement; and, even when we have engaged in it, we are liable to be led astray by doing that which is easiest and most pleasant at the moment. Hence it is necessary that we should clearly see the benefits which result from advancing actively and circumspectly in the right course, in order that the subject may receive proper attention. The following are the principal advantages of knowledge.

1. *Knowledge is indispensable to prevent us from being fatally mistaken, regarding the main pursuits of our lives.* For, in order to this, we must choose proper ends, and right and judicious means of accomplishing them, while the ignorant cannot know what ends are proper, or what are the best means of securing them. The ends at which he aims are what particularly distinguishes a wise from a crafty man. The latter often exhibits much ingenuity and activity in effecting his ends: but, as he never sufficiently considers the end, he only secures and accelerates his own ruin; and, the more power he possesses, the worse for himself and those connected with him.

Striking instances of the evils which result from ignorance on this subject, are furnished by the innumerable votaries of sensuality, avarice, vanity, and ambition, who have formed a great majority of mankind, up to this day. They have all thought themselves on the highway of happiness, while they were treading the paths of lasting misery. They have erred, not only in expecting too much from their favorite objects, but in overlooking others, of much more consequence. Even where the general object of their pursuits was proper, they have erred egregiously regarding its comparative importance. The accumulation of money, for example, may be properly sought by right means and to a reasonable extent; but the case is greatly altered when it is made the paramount object of life, and pursued through right and wrong, by men who have paid no attention to much more important matters, of which they are profoundly ignorant.

No person knowingly blasts his own permanent welfare; and hence this is always done ignorantly, and if the individual knew more, he would act differently. Some frequently confess that they are acting foolishly: but all they mean, is, that such conduct is deemed foolish by others, or that it may possibly lead to some disagreeable consequences, the real nature and extent of which, however, they do not consider, and consequently do not know, while they think that these will be more than counterbalanced by the benefits which they expect from the course pursued. Every one necessarily does what he deems best at the time, upon the whole; and hence those follies are inevitably accompanied by gross ignorance on many important subjects, although this is generally culpable, and, therefore aggravates, instead of palliating, the guilt of their conduct.

In order to answer the purpose, knowledge must relate to the particular subject in hand: a knowledge of Mathematics will not supply the place of an acquaintance with Physics, and much less with Psychology or Morality, any more than a superabundance of water will supply the place of solid food. In order to permanent happiness, which must be the main object of every enlightened mind, we must know where it lies, and the course which we must adopt in order to secure it. But when we have learned the real nature and sure tendency of different pursuits and practices, we can choose proper ends; and a knowledge of our duties, and the ways in which these ends can be rightly effected, will prevent us from erring fatally or seriously in the pursuit.

2. *Knowledge is requisite in order to the due exercise and regulation of the emotions, on which happiness mainly depends.* The emotions are not directly under the control of the Will, but are excited by the contemplation of their respective objects; and, consequently, these must be perceived by the mind before the emotion can be excited, while the mind can never perceive anything of which it is totally ignorant. Thus we cannot sympathize with the joys or sorrows of others unless we know what they are; and we cannot feel affection and reverence for the Eternal, unless we learn those attributes of his character which alone excite these emotions towards him.

That happiness depends mainly on the due exercise of benevolent and sympathetic emotions, and the suppression or eradication of those of a contrary kind, is proved by observation, and by the known force of such emotions. Let a man surrounded with all physical comforts and enjoyments only have some strong malevolent emotion excited, and he immediately feels unhappy. On the other hand, let one destitute of many of those advantages have his mind filled with strong pleasant emotions—regarding the past, the present or the future—and he is happy while under their influence. Now there are always objects within the reach of our mental vision which excite such emotions: but we must diligently search for them and keep them in view, in order to benefit by them; and this requires a knowledge of the nature and importance of those emotions, which are undervalued by the ignorant, because they do not strike the attention like objects of sense.

3. *Knowledge furnishes various direct pleasures which cannot be enjoyed by the ignorant.* The absorbing interest which the Mathematician and the Philologist frequently feel in their studies, is a striking instance of the direct pleasure derivable from knowledge, even in its most abstract form: and although these are studies in which the majority of mankind cannot be expected to feel a deep interest, yet the case is otherwise with various departments of knowledge. If we except the few who are insane, or only a little above idiocy, all mankind delight in observing the beauties and wonders of nature. Many, also, feel much interested in witnessing great and stirring historical scenes: and although these may be placed beyond the reach of observation, yet the pages of History disclose them, in countless numbers. To persons who prefer tracing the lives of distinguished or remarkable individuals, Biography offers an extensive field of similar enjoyment.

The sources of these pleasures are as varied as the subjects of thought. For, besides the mathematical and physical sciences, Ethnography, History and Biography, there is the wide field of the mental sciences, which will ever possess the strongest attractions for all who desire to penetrate to the causes of observed phenomena, and trace the ultimate laws by which they are regulated.

There are not only different fields of enjoyment, but also various subdivisions of the same field, so that every individual's precise taste may be gratified. He who disrelishes the stormy scenes of politics or war, may trace the progress of religion, science and literature, or study the history of manners and social life; he who delights in contemplating external nature, as it is presented to our immediate view, can study Geography, while those who prefer to analyse its materials, may study Chemistry and Geology: one who desires to contemplate vast objects, may have recourse to Astronomy; and he who would inspect the minute works of the Creator, can study Physiology and Entomology, while those who wish to examine the inventions of man, are furnished with an extensive field, in the various processes and results of Art.

Nor do those pleasures, by any means, end with the first acquisition: for they may be renewed whenever we choose to recollect the objects which first excited them; and although much may have been forgotten, yet the most striking and impressive parts will generally be remembered, on account of the strong attention and feeling which they originally excited.

4. *Knowledge is necessary to prevent mankind from addicting themselves to evil practices which mar their happiness.* The desire of enjoyment exists constantly in every mind: and hence those who are unacquainted with the pleasures which accompany knowledge, and are never obtained by the ignorant, devote themselves to the only enjoyments of which they know. From this source have sprung the various forms of sensuality, with the fearful evils which they have inflicted on the human race, and many pastimes which exert a most disastrous influence on their devotees. Yet such enjoyments will ever be eagerly sought by those who have found nothing better; and this can be done only by acquainting ourselves with the mighty and wonderful works of God, the treasures of Science and Art, the records of History and Biography, and the numerous objects which exercise our sympathies, and require our active efforts, throughout the world.

Those various subjects are much more than sufficient to occupy all the time that can be spared from important duties; and they furnish a field of harmless and exalted enjoyment which the longest life and most diligent study can never exhaust. When we learn to enjoy such pleasures, and at the same time know the great evils that result from those practices, it will not be a difficult matter to discontinue them forever. For such knowledge discloses to us those objects which excite our strongest and purest emotions, as well as the worthlessness of those practices as means of happiness, and the numerous privations and sufferings which they entail on their votaries.

5. *Knowledge is requisite to our physical welfare and safety.* The value of the useful arts, for these purposes, is obvious; and although the dependence of many of them on sciences apparently of no practical application is not so easily seen, it is not the less real. A knowledge of the properties of abstract quantity seems, at first sight, to be utterly removed from the business of life; yet it forms the foundations of many important arts, and of the science of Astronomy, whose aid is requisite to enable the navigator to cross the ocean in safety, and convey the superabundant food of one hemisphere to the famishing millions of another. So the discoveries of the Chemist, the Botanist, and the Physiologist, improve the art of Agriculture, and promote the preservation and restoration of health.

Without the aids furnished by superior intelligence, man would be in a worse condition than the lower animals: for in childhood he is helpless, and he comes to maturity very slowly; he is destitute of natural clothing or means of defence; he is much inferior to many of the brutes in bodily strength; and the spontaneous productions of the earth do not suffice for his sustenance. Hence not only his welfare, but his very existence is wholly dependent on his superior knowledge.

We are incessantly surrounded by agencies, and tempted to yield to certain allurements, which tend to injure health and produce premature death; and, in order to escape the bad effects of exposure to their influence, we require a knowledge of their nature, and of the proper means of guarding against them, which is often unattainable without extensive and careful investigation.

It is only by the aid of knowledge that men will industriously follow proper methods for supplying their physical wants, or secure the fruits of their labor, and make a right use of them after they have been acquired. But a knowledge of God, of man, and of external nature, produces industry, justice, abundance of everything requisite to supply our physical wants, temperance in the use of them, a proper degree of care against external dangers, and a general observance of the laws of health.

The physical welfare of a community also requires many conveniences which can be furnished only where a dense population admits of a great division of labor, while such a population cannot exist happily without an extensive and accurate knowledge of the art of cultivating the soil, and of the proper modes of regulating the distribution of its productions. For otherwise the larger portion of the community will inevitably suffer from want, and drag on a wretched existence, surrounded by the strongest temptations to vice.

6. *Knowledge is necessary to secure common morality, and render ordinary business safe and agreeable.* We are beset by various inducements to act immorally towards others, so strong that we shall frequently yield to their influence, unless we are fortified against it by a knowledge of the motives to virtue, and the sure consequences of vice. Hence serious offences of this kind will always abound among a people ignorant on the subject of their duties, and an ignorant community is addicted to several vices, against which he who deals with its members must incessantly guard, in order to avoid serious pecuniary loss, and other great evils incident to such intercourse.

The prevalent vices of nations and individuals widely vary; and as every one is apt to look at the bright side of his own and the dark side of his neighbours' character, he readily concludes that he is, upon the whole, tolerably virtuous, while an impartial observer might find that both are equally immoral, their respective failings differing only in kind, and not in degree.

The general connection between ignorance and crime is shown by the fact that, in every intelligent community, the majority of criminals belongs to the small fraction of society which is illiterate, and that the slaves of vice are uniformly found to be grossly ignorant regarding the nature and sanctions of morality. They frequently know something on these subjects; but their views of them are radically erroneous. Knowledge removes such evils, by imparting proper affections towards others and steady moral principles.

7. *Knowledge guards us against the numberless impositions that are practised on the ignorant, by the designing and unprincipled.* Impositions of this kind often enable men to obtain others' property without giving a

fair equivalent in return, while they escape the penalties attached to robbery or theft; and hence the extreme prevalence of such frauds, which have produced much evil. Ignorance and its ordinary concomitant, credulity, may be said to be the capital on which the various classes of deceivers and impostors have traded in all ages: and they disappear only where men have become too enlightened to be deceived; for the same bad training that makes one man a credulous dupe, will make a person of a different disposition a cheat.

Frauds and impositions have been extremely common, on account of the wide field presented by ignorant credulity, and because there are as many temptations to such deceptions as there are evil passions or depraved appetites. Yet a knowledge of the devices and falsehoods of the unprincipled, the criterions of truths, and the laws of nature, would banish all these evils from society. A very ordinary knowledge of Physiology and Pathology, for example, would enable us to detect the impositions of a quack, who professed to cure all diseases with a single nostrum: a knowledge of the character of God and the nature and condition of man, would banish religious deceptions: the young and unwary would escape the snare, if they knew the character and objects of the insnarer: and men would very rarely believe any false assertions, if they were well acquainted with the requisites of credible testimony and the sources of error.

8. *Knowledge is requisite to free the mind from superstition.* History abounds with instances of the dismal effects which have flowed from this source. Under its influence, the most civilized and enlightened nations of antiquity became addicted to the vilest and most cruel practices, even to murdering their own offspring and immolating themselves, in order to appease the supposed wrath or procure the favor of deities that existed only in their own benighted imaginations. And even where those more revolting superstitions passed away, the puerilities and disgusting practices of succeeding times debased the mind, and shut out the light of truth. Superstition not only produces particular evils, but also substitutes its own worthless forms and pernicious doctrines in the place of truth, and generally surrounds the minds of its victims with a web of prejudices and errors, which renders them satisfied with fatal ignorance.

Wherever true religion is absent, superstition inevitably appears: for the phenomena of nature lead us irresistibly to a superior power, of which our conceptions must be either accurate or the reverse; and the belief in a future state is too deeply rooted in our nature to be eradicated by any sophistry, however subtle. Hence some form of superstitious belief and practice exists wherever people's views of the superior power and a future state are radically erroneous. But a knowledge of the character and government of the only true God, of the relation in which we stand to him, and of the laws of nature, banishes superstitious opinions and practices, as the effects disappear with their cause.

9. *The acquisition of knowledge is requisite to discipline the intellect, and fit it for a proper performance of duty.* It is a matter of daily observation that the characters of persons and the amount of good which they effect, depend much more on their training than on the native force of their understandings. Many men of ordinary abilities become, by proper mental discipline and instruction, happy and useful members of society, while others, of great native talents, have often, for want of these advantages, spent miserable lives, and were justly regarded as public pests.

The intellectual, like the corporal, faculties require suitable exercise, in order to the proper performance of those functions for which they were conferred, while such exercise is found only in a proper course of study and observation, for the attainment of knowledge. As a man cannot be made a good sailor by following the plough, so a person cannot be fitted for properly discharging his various duties, unless his intellect has been exercised on those very subjects with which he is to be conversant in after life. In order to discharge our duties towards God or man, we must possess a correct knowledge of the divine character; and this is unattainable unless our minds are properly trained in investigating his works and words.

Ignorance inevitably leads to innumerable pernicious errors, both of opinion and practice. We must think and act; and unless we are guided by knowledge, we shall both think and act in such a manner as not only to miss the great object of all our pursuits, but also to inflict many serious and permanent evils upon others. On many subjects, ignorance is evidently so dangerous and disreputable that men are not satisfied without possessing something which will pass for knowledge, while they may be, in reality, so ignorant that they readily adopt for truth errors recommended by their own prejudices or the authority of persons in whom they confide, without ever mistrusting that there is anything wrong.

Besides religion, we meet with opinions on many other subjects, firmly believed by the unenlightened portion of mankind, which men acquainted with those subjects know to be totally false. Such are, the opinion that some persons are lucky and others unlucky, independently of character, conduct and circumstances; that, by means of certain simple manipulations, a person can be made to see better without than with the use of his eyes; that men can perform miracles by means of satanic agency; that certain diseases can be cured by rubbing the affected part to a corpse—and so forth.

Thus ignorance not only excludes knowledge, but substitutes in its place a spurious belief, much worse than none.

Some have maintained that ignorance is favorable to happiness, and knowledge dangerous and pernicious. But happy ignorance exists only in the realms of pure imagination. We have the highest authority for asserting that, in ancient times, the people perished for want of knowledge;* and the need of it is equally great in every age. We are also told that ignorance is an evil, and knowledge a great good,† statements which are confirmed by all the annals of our race. Certain kinds of knowledge are liable to be abused: but this cannot occur with a man who knows his duties, and is at the same time influenced by those emotions and moral principles which uniformly accompany an accurate and extensive acquaintance with the most important subjects of human investigation: and hence the evils attributed to knowledge are, in reality, the effects of ignorance.

From the preceding survey, we see that the proper acquisition of knowledge furnishes three distinct advantages: 1. It supplies the information requisite to right conduct, and to avoid pernicious courses. 2. It trains

---

* Prophecies of Hosea, chapter iv., verse 6.

† Proverbs of Solomon, chapter xviii., verse 15; chapter xix., verse 27; and Ecclesiastes, chapter vii., verse 12.

the faculties, and renders them able to perform their functions properly. 3. It affords various direct enjoyments, which can be repeated indefinitely.

The degree of knowledge requisite for the purposes mentioned, is not attainable except by means of investigations conducted in accordance with the principles of Logic; and these require to be studied, in order to be known. This is proved by the grave errors and mistakes committed by many investigators, and the ignorance or erroneous opinions of a great majority of mankind, on all the most important subjects, from the earliest times to this day.

B

PART I.

OF THE ULTIMATE SOURCES AND ELEMENTS OF
KNOWLEDGE, AND THE PRIMARY PROCESSES
BY WHICH IT IS ACQUIRED AND
RETAINED.

CHAPTER I.

A GENERAL VIEW OF THE LIMITS, DIVISIONS, AND IMMEDIATE SOURCES OF KNOWLEDGE.

§ 1. NECESSARY LIMITS AND PRINCIPAL DIVISIONS OF KNOWLEDGE.—(1) Knowledge limited to Intuitions, Comprehensions, and Inferences.—Peculiarities of these several classes.—Boundaries of the Knowable and of the Known.—Distinction between Knowledge and Belief.—(2) Knowledge either Mediate or Immediate.—Definition of Cognition, Consciousness, and Discernment.—Truths known by the latter.—(3) Knowledge consists of Necessary, Contingent, and Hypothetical Cognitions.—Branches belonging to each.—Common properties of all.

1. EVERYTHING which we can know, must belong to one or other of the three following classes of truths.

(1) Those which are self-evident, or which we know must be such, and cannot possibly be otherwise, independently of anything made known to us by our senses. Such are, the existence and essential nature of time and space—that contradictories cannot co-exist, and that a thing is equivalent to itself. These we term *intuitions*, and the faculty or power by which we know them *Intuition*.

(2) Truths not necessarily such, but made known to us directly by *Comprehension*, the faculty by which we directly know truths which are not self-evident. Thus, when we behold the sky, we certainly see a blue expanse; when we smell a rose, we feel a particular odor; when we have succeeded in effecting a difficult object which we deem important, we feel a pleasant emotion; and when we think of a tree which we have often seen, we have an idea of its appearance. Truths of this class we term *comprehensions*.

(3) Inferential truths, or those which are necessarily implied in intuitions, comprehensions, or suppositions, and which we term *inferences*. By *necessary implication* is meant, such a connection that the inferences must be true, and cannot possibly be false, if the things from which they are inferred are true.

These three classes of truths include everything that

can possibly be known: for it is evident that we cannot possibly know a thing if it is not knowable by any of the faculties by which we obtain direct knowledge, nor susceptible of being found to be implied in anything which we can either know or suppose. Hence—(1) any investigation which lies beyond these limits is fruitless, such as an inquiry into the origin of life, or the atomical structure of matter—(2) everything within these limits is knowable—and (3) no statement is entitled to be classed with known truths, till it has been clearly ascertained to be an intuition, a comprehension, or an inference.

It is also evident that everything which we actually know, must be known directly or indirectly, and that whatever is known directly must be so because it is known to be either self-evident or presented by some faculty that we possess of directly knowing truths which are not of the former class. It is equally evident that a truth can be known indirectly only in consequence of its being found to be necessarily connected with something that is known directly. Hence the intuitions, comprehensions and inferences which a person knows to be such, form all his actual knowledge. Intuitions and inferences not known to be such, things once comprehended but afterwards totally forgotten, and things never comprehended, evidently form no part of our actual knowledge.

Knowledge can be only of truths: for although we may *believe* error we cannot *know* it, since this implies that it is either a truth known directly to be such, or that it is sustained by conclusive proof, which error cannot possibly be. Knowledge also implies belief; and therefore a truth which a man rejects, or does not believe, is not known to him, however well it may be known to others, and even if he formerly believed it himself, on good grounds.

2. We distinguish unknown truths from the known by terming the latter *cognitions*, which may be defined—truths known to be such. Intuitions and comprehensions may be termed *immediate* or *direct* knowledge, as we know them directly, without the intervention of any proof or process. The two faculties of Intuition and Comprehension may be designated by the common term *Consciousness* or *Discernment;* and truths known by it may be called *discernments.* These are accompanied

with a direct knowledge of their certainty, so that they require no proof; and they may be said to be *discerned*. Inferences, being made known only by intermediate proofs and processes, may be termed *mediate* or *indirect* knowledge.

Consciousness includes the knowledge which we necessarily have of the reality of all our present thoughts and their immediate objects, or those things of which we think. It is self-evident that we cannot think without knowing that we think, and that we cannot know unless there is something that we know. When we see, for example, there must be something that we see, and we necessarily know that we see. So when we feel, we necessarily know that we feel, and that there is something which we do feel.

3. There is no necessity for our possessing comprehending faculties, such as we actually possess, nor for the existence of the things comprehended. Thus some men cannot see, and the things which we see might have had no existence. Comprehensions may, therefore, be termed *contingent* truths: and those which are necessarily implied in them are properly classed under the same term. As these depend on contingencies, they are not necessarily true; for a necessary connection between two things does not imply the existence of either, but only that if one exists, the other must exist.

Another class of cognitions is, that which expresses certain properties or relations of things merely supposed or assumed, and having possibly no actual existence. These may be called *hypothetical* truths. Such are, the properties of a machine which has yet no actual existence, but is merely planned by the inventor, and consequences which have been proved to follow a certain course of conduct, which is only supposed, and has not been actually adopted.

As inferences from intuitions are necessary truths as much as the original intuitions, both classes fall under that class of cognitions. Hence, *necessary*, *contingent* and *hypothetical* cognitions include all human knowledge. The first class comprises what must be; the second, what actually was, is, or will be, though not necessarily; and the third, what will be, if certain things are assumed or pre-supposed.

These three classes of truths do not differ in respect

to certainty: all are equally certain, where they agree with the proper criterions, and we are liable to err in regard to each. Statements have been believed to be self-evident which are untrue, and fallacious mathematical demonstrations are by no means unknown, while, at the same time, contingent and hypothetical cognitions may be established conclusively, although this is frequently more difficult than the demonstration of a mathematical theorem.

§ 2. OF THE VARIOUS FACULTIES BY WHICH KNOWLEDGE IS ACQUIRED AND RETAINED.—Definition of Faculty.—Nature of Apprehension. —Knowledge acquired through Apprehensions.—Reasoning.—Nature of Remembrance.—On what dependent.—Similitudes, Ideas, Phantasms, and Prototypes. — Comprehensions. — Knowledge dependent on Remembrance. — Nature of Emotions, and what we know directly by them.—Nature and general laws of Attention.—Nature of Abstraction, and of Conception.—Notions and Imaginations.—Nature of Generalization.—Six things which necessarily exist in all Thinking.—Frequent Errors.

By a *Faculty* is understood a power, capacity or susceptibility of thinking, feeling or acting. Hence whatever we do, we must have a faculty of doing, and whatever we feel, we must have a faculty of feeling. We cannot, for instance, see without the power of seeing; we cannot reason without the power of reasoning; and we cannot feel, unless we are susceptible of feeling.

That we see colors, hear sounds, and smell odors, are sure truths, whatever difference of opinion may prevail regarding the origin of such cognitions. The faculties by which we thus obtain immediate knowledge, through the influence of things external to the mind, are termed *apprehending* faculties or *Apprehension*, and the cognitions obtained *apprehensions*. They are all properly classed together here, since they resemble each other in depending on some impression made on a nerve, by means of which a corresponding impression is made on the mind.

The knowledge thus obtained consists of *sensations*, or the pleasant and painful feelings which we experience, and *perceptions*, or that of which we are conscious, besides what we feel. Thus, when we view a green field, the green expanse which we behold is quite distinct from the pleasing or painful feelings which accompany this perception.

As we know intuitively that certain truths necessarily imply others, we learn the existence and observable properties of material and living beings by applying the faculty of Intuition to our apprehensions. Those truths are not apprehended directly; but they are implied in our apprehensions, and learned by means of *reasoning* or drawing necessary inferences, the faculty of Intuition, when thus applied, being termed *Reason*. Thus, when we view a tree, the eye perceives only extended colors: but, by observing all the phenomena, and drawing the necessary inferences, we find that there is a solid body without us, of a particular form, size, and color.

In drawing inferences, we are not confined to what we have personally apprehended: for, by means of testimony, we learn the apprehensions of others; and we can draw inferences from these as we do from things primarily apprehended by ourselves. Certain sounds and visible characters are found to denote certain things, by means of which we learn the thoughts of others. We determine what those sounds and characters imply, by the proper application of our faculties, as in other investigations.

While tracing inferences, we are frequently obliged to reason about things not apprehended at the time. Thus, in proving conclusions, it is generally necessary to refer to things which are not present to our senses, such as things previously seen, heard or proved. This is done by means of *Remembrance*, the faculty by which we know our former thoughts.

If we carefully consider the phenomena of Remembrance, we shall find that it is not a simple faculty, but that it depends on two things. If we view a tree, and then close our eyes, we may still discern a faint and fleeting likeness of it; and we find that the same is true of all our apprehensions: and so all other thoughts, also, have their likenesses, after the originals have disappeared. These likenesses I term *similitudes*. They may be subdivided into two classes — *ideas*, or similitudes of apprehensions — and *phantasms*, or similitudes of other thoughts, besides apprehensions. The original of a similitude I term its *prototype*. Trains of similitudes pass spontaneously through the mind, according to certain laws; and the simple faculty of discerning them I call *Memory*. Apprehensions, similitudes and all other things directly discerned, except intuitions, are included under the general term *comprehensions*.

When we consider the phenomena of similitudes, Intuition leads us to the conclusion that they all had their prototypes: for all other suppositions involve an absurdity, as we shall see hereafter. Thus, by means of Memory and Intuition, we can know and reason about the past as if it were the present. The last alone is known immediately, in respect of all contingent truths, whose past and future are known only by their being necessarily connected with something present.

On comparing the peculiarities of the various objects of thoughts, we find a large class which, unlike apprehensions, is immediately independent of anything beyond the mind. These consist of similitudes and of feelings which differ widely from those of Apprehension, although remotely dependent on them. These are generally termed *emotions;* and they resemble apprehensions in being known solely by Consciousness. We know the existence and character of our feelings solely from experiencing them, and neither by intuition, by reasoning, nor by the testimony of others, although we may investigate their origin, laws and component elements, as in the case of other thoughts. They also directly teach us nothing but their own existence and nature. We feel nothing but our feelings: we cannot feel the truth of any assertion. Thus we cannot feel that the sky is blue, that Alexander the Great died at Babylon, that things equal to the same thing are equal to each other, nor that the three angles of a triangle are equal to two right angles.

We *attend* to that of which we think; and the power of doing so is termed *attention.* This is of two kinds. *Spontaneous* attention is that which is produced by some present feeling, without desire or effort; and it is generally proportional to the strength and vividness of the feeling; but where these become intense, they nearly absorb the attention, or make us overlook other present objects of thought. *Voluntary* attention is that which is produced by a desire or effort of the Will, directing it to something which we deem of consequence; and it is generally proportional to the importance of the subject, according to our belief; but where we consider this very great, the subject may nearly absorb the attention, like strong feelings. The actual degree of attention is frequently the result of the two kinds operating simultaneously, and proportional to their united amount: but it often depends on one alone.

Attention is evidently not a distinct power of the mind, but merely a general name, to denote the exercise of our faculties with reference to particular objects of thought: and it is usually exerted with reference to several objects at the same instant. During our waking state, we generally attend simultaneously to our apprehensions and ideas. Thus, when we walk with a friend, we attend to the organs of motion and speech, hear his conversation, observe various objects around us, and attend to the subjects of conversation, so that we collect the sense of what he says, and form our own replies. But it is observable that the force of attention diminishes as the number of objects simultaneously considered increases.

When we discern different things at the same time, we have the power of concentrating the attention on some and overlooking the rest. Thus, while viewing a tree, we may confine our attention to the trunk, the branches, the leaves, or the blossoms; and, with respect to the latter, we may consider their form, their odor, their color, or their position and arrangement. This faculty is termed *Abstraction*, which may be defined as that by which we fix the attention on particular objects of thought, and withdraw it from others, at pleasure, or as long as the object is present. In the case of apprehensions, the phenomena of course vanish when their causes are withdrawn; but, with regard to similitudes, we may retain and consider them till mere weariness or exhaustion induces us to turn to something else.

Abstraction is *spontaneous* where it results simply from the pleasure or pain of a particular feeling, and *voluntary* where we voluntarily limit the attention, for some purpose, which is discovered by the aid of Reason. Hence the exercise of voluntary abstraction is dependent on the latter faculty. Yet it is a distinct power, which is requisite in order to discover the peculiar and general properties of the various objects of human research.

Besides observing the phenomena of Comprehension, and drawing inferences from them, we can modify the simple elements, and combine them into a new whole, different from any which we ever comprehended. Thus, we can think of a small horse from having seen a large one, or of a circle being elongated into an ellipse, or of the half of it, without the rest, or of a green horse with an ox's head. The combination may be either of the comprehensions or of the modifications previously thought of, or of the two blended, as in the case of the last instance just mentioned, in which case there is a double exercise of the same faculty. We may afterwards modify and recombine the combination, and so on, without any definite limit, except what arises from the greater difficulty of the new combinations, on account of their greater complexity.

This faculty we term *Conception*, and the thing thought of, a *conception*. If this be of something not described nor known, we term it an *imagination:* if it be of something described or represented to us by others, we term it a *notion*. Thus, an inventor imagines something new, and a person forms a notion of a plant or an animal from a description. Conception, when employed in imagining, is termed *Imagination:* but it is essentially the same throughout, both in its nature and its processes.

When we compare several objects, we often perceive that, although they greatly differ in some respects, yet they have certain properties in common. Thus, the sky and the ocean are both *blue;* flint, iron, and diamond are all *hard;* water, oil of turpentine, and alcohol are all *liquid;* and oxygen, hydrogen, and nitrogen are all *gaseous* bodies; the planet Venus *revolves round the Sun,* and so does Mars, &c. When we thus note or ascertain a common property of several things, we are said to *generalize;* and the knowledge obtained is rendered permanent and available by means of *general terms,* or words that express the common property, wherever it exists. In this process there is no new faculty employed, but only several of those already described.

In all thinking there necessarily exist six distinct things—(1) that which thinks, or the mind—(2) the faculty by which it thinks—(3) the thought, or some act or exercise of the faculty—(4) the object of the act, or that which is discerned, known, believed or supposed—(5) a knowledge of the reality both of the act and of its immediate object — and (6) something which originates or causes the act, which may be either in the mind or without. Thus, when we see, there are—(1) the mind, that sees—(2) the faculty of vision—(3) an exercise of this faculty—(4) the thing seen, or the colors—(5) a knowledge that we think and see these colors—and, (6) the cause of our seeing them, which is, the action of rays of light on the retina.

Various errors have arisen from confounding several of those things with each other, which happens the more readily because they are designated by the same term. A common instance is, confounding apprehensions with those qualities of substances which cause them, as when it is said "I feel the heat of the fire." The confusion becomes evident from the difficulty which is generally experienced in distinguishing the perception of color from the objective reality which causes it. We do not readily believe that there can be nothing either in or on the colored substance which in the least resembles its color, any more than there can be anything in a bell resembling its sounds. A similar, but much less common, error is, confounding thought with the thinker. The former is only an act of the latter, and totally different from its essence or substance, of which it cannot possibly form a part.

§ 3. OF PROPOSITIONS.—Definition of Propositions.—Subject and Predicate.—Expression of a Proposition.—Converse, contrary, and contradictory of a Proposition.—Identical Propositions.—Important property of these.—Simple, Alternative or Disjunctive, and Complex or Compound Propositions.—Absolute and Conditional or Hypothetical Propositions.—Affirmative and Negative Propositions.—Universal, General, Particular or Indefinite, and Individual or Singular Propositions.—Frequent ambiguities.—Various forms of Propositions.—Caution.—Combinations.

Everything affirmed or denied is expressed by a *proposition*, which is, an assertion of a truth, assumption, supposition, belief or opinion: and it is either expressed in words or simply declared by the mind. It may refer to the past, the present, or the future, or to any two of them, or to all time.

Every proposition necessarily consists of at least two parts, the one relating to the thing of which something is said, or the *subject*, and the other, to what is said of it, which is termed the *predicate*. Thus, in the proposition "just men abhor deception" the first two words are the subject, and the latter, the predicate. These are essential parts of every proposition, since, in every assertion, there must be something of which we assert, and something which is asserted regarding it. The subject and predicate may each consist of a single term, as "John sleeps," or of a long clause, as "every one who desires the welfare of his country, will cheerfully submit to privations, for the public good, during times of general distress;" or each part may consist of several connected clauses, as "the true patriot, and the wise and upright statesman, will not be turned from the path of duty, either by the threats of the powerful or the clamor of the multitude."(2)

The proper expression of a proposition requires at least two words, one denoting the thing spoken of, and the other, what is said of it. The common idiom of a language may, indeed, require only one word, as in the Latin expression *pluit*, (which is equivalent to "it rains:") but, in all such cases, some second word is understood. Thus the preceding verb has some nominative understood, such as *Jupiter*, or *Deus* (God). So the English expressions *yes* and *no* are only abbreviations for a responsive repetition of the terms of the question.

The *converse* of a proposition is one in which the subject is asserted of the thing predicated, so that subject and predicate change places. Thus—"those who abhor deception are just men," is the converse of the first example given above. The *contrary* of a proposition is one which predicates the contrary attribute of the same subject. Thus—*"John is weak"* is the contrary of *"John is strong."* By *contrary* attributes are meant those which are most unlike, of the same class, as *good* and *bad*, *wise* and *foolish*, *hard* and *soft*, *high* and *low*, *black* and *white*, *light* and *heavy*. The *contradictory* of a proposition is one which denies of the subject the attribute which the former asserted. Thus—"John is not strong" is the contradictory of "John is strong."

An *identical* proposition is one which predicates the subject of itself, or whose subject and predicate are identical, as "a man is a man"—"azote is another name for nitrogen"—"Philip was the father of Alexander"—"London is the capital of the British Empire." To this class of propositions belong all verbal definitions, or those which explain the signification of terms, provided that they are accurate—and it is an evident property of the whole class that, if the original proposition is true, so is the converse.

A *simple* proposition is one which attributes a single property to a single thing, as "John died." One which attributes one or other of several properties to a subject, is termed *alternative* or *disjunctive*, as "John is either in London or in Paris or in New York." A *complex* or *compound* proposition attributes various properties to the same or to different things, as "man is mortal, and yet frequently forgets his mortality"—"John died yesterday, and James died to-day." Propositions of this kind consist of several simple propositions united, into which they may be resolved. On the other hand, where the different things contained in a compound proposition form one whole, and the same thing is attributed to every one of them, the compound proposition may be expressed simply. Thus, the compound proposition "John is a descendant of Adam; Mary is a descendant of Adam, etc." is tantamount to "All mankind are descendants of Adam."

An *absolute* or *unconditional* proposition affirms the predicate absolutely, without any condition, as "All men are mortal." A *conditional* or *hypothetical* proposition predicates only upon some condition or supposition, as "If report be true, all men are mortal"—"Although he should do that, he would gain nothing by it."

An *affirmative* proposition asserts the predicate of the subject, as "John is dead." A *negative* proposition denies it, as "John is not dead"—"No man is mortal."

A *universal* proposition predicates of all the individuals of a class or all the parts of a whole, as "All men are mortal"—"No matter is unextended." A *general* proposition predicates of most of a class or whole, as "Most men are rational"—"Carbonic acid is generally gaseous." A *particular* or *indefinite* proposition affirms or denies of a small or indefinite part, as "Some men are wise." A *singular* or *individual* proposition predicates of a single individual or part of a whole, as "John thinks" —"This piece of wood is brittle."

The classes of propositions defined in the preceding paragraph are frequently expressed in such a manner as to render the extent of the subject ambiguous or doubtful. Thus, in the proposition "Man is mortal," the subject may mean either "every man" or only "Most men." So, "Men say so"—may mean "All" or "Most" or only "Some men." Such ambiguities have been a frequent occasion of error.

Propositions may frequently be varied in the form of expression, so as to bring them under a different class, without in the least changing their signification. Thus, the conditional proposition "If report be true, all men are rational," is equivalent to the absolute proposition "It is reported that all men are rational," or, "The rationality of all men is reported." So the affirmative proposition "All men are mortal" is tantamount to the negative proposition "No man is immortal;" and "John is not dead" is equivalent to "John is alive." As the negative of an attribute is tantamount to the affirmative of its contradictory, every negative proposition may be converted into an equivalent affirmative. Hence it appears that, in examining propositions, we should regard their real signification or import, rather than their form.

The several kinds of propositions may be combined with each other indefinitely. Thus, the proposition "If John did that, he is either a knave or a fool," combines the conditional and alternative forms. So we may combine the conditional with an affirmative or negative, simple or compound, universal or particular, and so forth.

§ 4. OF PROBABILITY.—Definition of *Probability.*—(1) Probabilities founded on previous experience regarding the concomitance of certain properties.—What these imply.—When they become certainties.—Principle of Reasoning.—(2) Probabilities founded on what we know has happened in cases apparently similar.—Principle of Reasoning. — Why we often err. — Connection between Agencies and their Results. — (3) Probabilities based solely on what must happen.—Distinction between these and the preceding classes.— Source of frequent error, and mode of avoiding it.—(4) Probabilities founded on the known connection between Causes and Effects. —Use of Experience.—Common error.—(5) Probabilities based on actual investigation of proof.—Distinction between Probability and Certainty. Principles of Reasoning in all cases of Probability.— Influence of individual Experience.—Uses of Probability.—Circumstances in which it exists, and to what generally proportional. —Resultant Probabilities.—Means of ascertaining their value.

A *probability* is, a proposition implying facts which tend to prove, but which do not absolutely prove, that it is a truth. Probabilities are of various kinds, the most common of which are included under one or other of the five following classes.

1. In comparing two things, we frequently observe that they possess many obvious properties in common; and although there is no proof that they possess unob-

served properties in common, yet experience informs us that this has been found to hold true, in similar instances; and this we indicate by saying that such is *probably* the case. Here the probability implies, not only that a proposition may be true, but that it has actually been found true, in similar cases; and the probability is greater or less, according as the cases in which the unobserved attributes were afterwards found to be common, are more or less numerous, or as the resemblance is more or less extensive.

Of this kind is the probability that a certain effect will follow from an agency similar to one whose effects are known, that an effect has been produced by a cause known to have produced similar effects, and that two or more similar phenomena have similar causes, antecedents, concomitants or effects. Other cases are, that the testimony of a person of doubtful veracity or a stranger, regarding some unknown subject, is true, that a man will continue to act as he has hitherto done, in apparently similar circumstances, that a certain phenomenon has been preceded by its usual antecedent, or the reverse, and that a newly discovered species will be found similar, in unobserved attributes, to known species to which it bears a general resemblance.

This class of probabilities is based on the self evident principle that where the determining conditions or agencies are the same, the results will be the same. Whenever we ascertain that the conditions or agencies are actually the same, in two or more cases, the results must be the same, and probability gives place to certainty, as the former exists only where we know but of a partial similarity of the determining conditions.

2. Results are found to vary, in cases apparently similar, while the previous variations are known. Thus, if a person has succeeded in effecting a certain result, by the same apparent means, in seven cases out of ten, we say the probability, or chance of his succeeding, in the next attempt, is seven tenths, and of his failing, three tenths, the cases being all alike, so far as is known. We reason on the self-evident principle that results must follow as they have done, in the same circumstances. But there is frequently no means of ascertaining that the several cases are, in reality, perfectly alike; and hence the future results often turn out differently from what the probabil-

ity indicated. Such probabilities are evidently proportional to the ratio which the favorable cases bear to the whole.

Whatever affects the agencies concerned in producing results, will affect these in a corresponding degree. Thus, greater attention to the laws of health, in a community, increases the probability that a person of a certain age, taken at random, will live so many years; and greater temperance among seamen will lessen the chances that a ship will be lost at sea. The probability varies, in such cases, as the agencies which determine the results vary.

3. One or other of several results must happen, and we know no reason why one should happen more than another. Thus, as a certain day must be fair or foul, we say it is an even chance that it will be fair. Here the probability varies inversely according to the number of possible results, being greater as these are fewer, and conversely. These differ widely from the preceding kinds of probability: for here our expectations are based solely on what must happen, independently of any knowledge as to what has happened, and, unless experience prove the contrary, there may be unknown agencies which will interfere with the expected results. Thus, if a certain time of the year is generally fair, there is more than an even chance that the day will be fair, and conversely. It is, therefore, very unsafe to act on the assumption that certain results will follow, where we are ignorant of what has happened, in circumstances similar in reality, and not merely in appearance.

4. A certain change must have resulted from one or other of several causes, or a cause must have produced one or other of several results, and we have no reason to decide in favor of one agency or result more than another. Thus, a disease must have arisen from some violation of the laws of health, or from some original unsoundness of constitution, or from both combined; and, therefore, we might say, where we know nothing regarding the person affected, the chances are one third, or one to three, that it has proceeded solely from an unsound constitution. The degree of probability varies with the number of causes or results. But experience is here peculiarly requisite, since we cannot generally decide, with certainty, as to the number of causes which may have op-

erated, and we are liable to substitute mere suppositions instead of the real agencies concerned in producing the results. Many baseless scientific theories have originated in this manner. Experience frequently alters abstract probabilities of this kind, so that the actual probability is the result of a combination of various elements. Thus, in the case just mentioned, when we come to know the extreme frequency of violation of the laws of health, and the comparatively rare cases in which original constitutional defects are the sole cause of disease, we shall find that the chance of such a defect being the sole cause of a disease, instead of being one to three, is not one to three hundred.

With regard to the results which a certain agent may be expected to produce, we generally reason more on our previous experience than on necessity; and we are apt to assume a degree of similarity, in two cases, which does not actually exist, as when it is assumed that a medicine will produce one or other of two results on a patient, because it has done so in other cases apparently similar, although the former may differ in several important peculiarities.

5. The proofs which support a certain proposition have been examined, and we infer that it is true; but the examination has not been so close or thorough as to exclude the possibility of mistake. Here the probability varies with the degree of care employed in the examination. If it has been conducted by others, and not by ourselves, the degree of probability will vary with their character, being higher or lower, according as they are careful investigators and faithful relators or the reverse.

When the subject has been examined in such a manner as to leave no possible room for any error, the conclusion is certain, and not merely probable. But we can never safely assume this regarding the conclusions of others, unless we have properly tested them, as we cannot know that their examination is of that character. They furnish only probabilities, which vary with the character and circumstances of our informants.

In this class of cases, we reason on the self-evident principle that the probability of a proposition varies according to the nature of the proof by which it is sustained.

From its nature, no degree of probability can amount

to a certainty; nor does the one pass by insensible gradations into the other. When a proposition has been found to be certainly true, this immediately places it in a class widely removed from probabilities. We must beware, however, of adopting any probability as a cognition until we have obtained conclusive proof.(3)

In every case of probability we reason upon self-evident principles, as in cases where we arrive at certainty; but, in the former, we reason from evidences which are incomplete or inconclusive, and consequently the results partake of the nature of their sources, whereas, in the latter, the foundations are known truths, and, therefore, if our arguments or proofs are legitimate, our conclusions are absolutely true.

Men are apt to base their views of probability mainly on their individual views and experience, while they either reject or do not know the experience or knowledge of others: and hence the same proposition often appears very probable to one and extremely improbable to another. Thus, Herodotus disbelieved the statement of the Phœnician mariners, who had sailed round Africa, that, during part of their voyage, the sun rose on the right hand when they faced its position at noon, whereas a person who possessed a more extensive knowledge of Astronomy would consider such a statement highly probable, if not a matter of course.

Although probability never amounts to knowledge, yet it is often of great consequence, both in scientific investigations and in the ordinary business of life. In the former case it acts as a guide and stimulant, and many truths would never have been discovered, had there been no previous indications pointing to a certain conclusion. It guides us in establishing the proposition in question, by pointing out what is wanting, in order to that end; and it stimulates to investigation by showing that more or less of the task is already accomplished, and thus promising important results for comparatively little toil. It is also of great value in many of the ordinary affairs of life; for it enables us to determine the most eligible course, where certainty is unattainable.

Probability exists only where we have partial knowledge, and are at the same time ignorant on some points which we require to know, in order to possess certainty. Where we are totally ignorant, we know of no probability; and where our knowledge is sufficiently extensive, the probability gives place to certainty. The degree of probability is generally proportional to our knowledge of what is requisite to be known, in order to arrive at certainty.

The probability of the truth of a certain proposition is often the resultant of several probabilities combined, every one diminishing or increasing the preceding, according as it is discordant or accordant. Thus, the probability that a certain testimony is true, may be the resultant of the probability that the witness knows the truth, and that he asserts what he believes. The probability of his knowing the truth, again, may be the resultant of several remoter probabilities, such as his means of ascertaining the truth, the use he has made of them, the tenacity of his memory, and the relation between his statements and those of others who possessed the means of forming a right judgment on the subject.

The resultant of several accordant and independent probabilities is generally proportional to their total amount; and where there are discordant probabilities, the resultant generally varies with the difference between the amount of the former and that of the latter kind.

The resultant of connected probabilities, or probabilities of probabilities, generally varies as the product of the factors expressing the value of each. Thus, if it is an even chance that a witness correctly knows what he relates, and a similar chance that he reports correctly what he knows, the probability that his statement is true, is only as one to four. In such cases every new probability introduced may diminish the resultant.

§ 5. GENERAL CRITERION OF TRUTH, AND IMMEDIATE SOURCE OF ERROR.—Apprehensions real, and why.—Only possible sources of Error.—How these may be avoided.—Distinctions between Apprehensions and Ideas.—All other Comprehensions like the former.—How we ascertain whether a proposition is self-evident.—Requisites to the validity of inferences.—How these are distinguishable from Intuitions.—How inattention occasions error.—Expression of the general criterion of Truth, and the immediate source of Error.—How erroneous Belief may be avoided.—Its uniform concomitant. Belief includes Knowledge and Opinion.

Not only are we directly conscious of our apprehensions, but we know intuitively that it is impossible for any being to believe that he thus apprehends when, in reality, he does not. Thus, when I view the clear sky, I know immediately that I perceive a blue expanse; and when I feel pain, I have a direct and unerring knowledge of the reality of my sensation. This holds equally true of every sensation and perception.

It is self-evident that no apprehension can exist without attention. Thus, when the attention is completely absorbed by other thoughts, a person exposed to a freezing temperature can have no sensation of cold, and if a clock should strike near him, he cannot hear it. It is common to have apprehensions which are immediately forgotten, because they excite very little attention: but it is impossible for us to apprehend without any attention whatever to the thing apprehended; otherwise we should be conscious of that of which we were not conscious, or think what we did not think.

The only possible sources of error, on this point, are, mistaking one apprehension for another, or for a similitude: and it requires only a moderate degree of attention to avoid such errors, since the various kinds of apprehensions are palpably different from each other, and similitudes are much fainter and under the control of the Will. Thus, there is no danger of mistaking a sound for a color, or even the smell of a rose for that of tobacco. So it is very easy to distinguish the mere idea of the Moon from actually seeing it: for not only are our organs of vision very differently affected in the two cases, but it requires no effort of the Will to continue the apprehension, whereas the mere idea speedily gives place to some other thought, unless we retain it by a conscious effort of the Will. On the other hand, we may discern the idea at pleasure, whereas the apprehension is dependent on the presence of its cause, and ceases when that disappears.

The reality of other comprehensions admits of as little rational doubt as that of apprehensions. Thus, when I discern the idea of a tree, I am as conscious of the reality of the discernment as if I viewed the tree. Attention is here even more requisite than in the former case, as there is no external object to excite the comprehension.

We are liable to mistake an apprehension for its idea; but the peculiarities which distinguish the former are so obvious that it requires only a little attention to avoid this error. The same remark applies to the distinctions between the other kinds of comprehensions. Thus, an idea is distinguished from a conception by the former being spontaneous, while the latter is the result of a voluntary effort, and less vivid; and it is distinguished from a phantasm by our recollecting the latter's origin, and its exhibiting the peculiarities of its prototype. We avoid error here precisely as in the case of apprehensions and ideas.

Hence we may see that attention will enable us to avoid errors in regard to all comprehensions, and that these must arise from inattention, as in the case of apprehensions.

In order to determine whether a proposition is an intuition or self-evident truth, we have only to consider it with attention: for the very nature of an intuition is, that the attentive mind discerns its necessary truth, and its falsity to be an impossibility. We may first consider whether it is discerned to be self-evidently and necessarily true; and if we have any hesitation, it may be removed by considering whether its contradictory, or the proposition denying its truth, is possible: for if we have the evidence of consciousness either way, that suffices. A proposition which must necessarily be true, and one which cannot possibly be false, are evidently both entitled to equal and perfect credence, and free from the possibility of a rational doubt regarding their truth.

Our comprehensions and intuitions are often intuitively known to imply cognitions entirely different from themselves; these may be known, in the same way, to imply others, and so on, without any definite limit. The existence of A, for example, may necessarily imply that of B, which may prove that of C, and so on. The number of these intermediate cognitions is a matter of no essential importance. All that is necessary, in order to establish the truth of the last inference, is, that the original proposition be true, and that there be a self-evident connection between every cognition and the proposition which immediately follows. This can be ascertained by attentively considering whether one proposition necessarily implies that which is immediately founded on it, or deduced from it; for if it does, the connection becomes self-evident upon an attentive consideration and comparison of both.

Owing to the extreme rapidity of thought, we are liable to mistake inferences for intuitions or comprehensions, since we may overlook the processes by which we arrived at them: but a proper degree of attention will always show their true character; for the distinction between discernments and inferences is as easily discovered as that between intuitions and apprehensions.

Inattention occasions error by its leading us to overlook differences, or to draw immaterial distinctions, or by concealing something from view altogether. There is no other possible way in which we can be led into error: for we can mistake one thing for another only by overlooking differences or drawing unimportant or irrelevant distinctions; and if everything which concerns the subject is clearly before the attentive mind, there can be no false or unwarrantable inferences.

Thus it appears that *the general criterion of truth is, the evidence of attentive Consciousness, either direct or indirect.* The former tests comprehensions and intuitions, and the latter, inferences. It also appears that *inattention is the sole immediate source of erroneous opinions,* which all spring from our not being sufficiently attentive to those things which must be carefully considered, in order to attain to truth.

That we often believe what is false, does not in the least prove that we can never certainly know whether any proposition is true. Wherever our belief is erroneous, there must have been a want of attentive consideration of one or more of the circumstances necessary to form a correct opinion: for it is obvious that, if the original assumption cannot be false, and that it necessarily implies a certain inference, the latter must be true. There is a certain number of steps in every process, every one of which we can consider with the greatest attention; and hence we can know whether we have given the proper degree of attention to every point which ought to be considered.

Where there is any room for doubt or uncertainty, as to whether we have sufficiently attended to every point, we are not warranted in assuming the proposition in question as established. But wherever we know that this has been done, the proposition cannot be false. In every case of erroneous belief, we may discover a want of due attention on one or more points. Thus, in dreaming, insanity and delirium, similitudes are taken for prototypes, because characteristic peculiarities are overlooked. In such cases, indeed, there is hardly an attentive consideration of anything, but merely a train of ideas or apprehensions passing through the mind, accompanied with those thoughts which they immediately excite. So, when we mistake a false inference for an intuition, we cannot have carefully considered whether the proposition is really intuitive.

Belief includes all that we believe, or take to be true, whether true or false. It may be subdivided into *knowledge* and *opinion*. Knowledge is, belief based on attentive consideration and the evidence of consciousness at every step, so that there is no room for error, and no reasonable ground for doubt or disbelief. Opinion includes all other belief, whether correct or not.

CHAPTER II.

OF REASONING.

§ 1. NATURE, GENERAL PRINCIPLE, AND EXPRESSION OF REASONING.—Two kinds of Knowledge obtained by means of Intuition.—Nature and definition of Reasoning.—Identity of the process, in every case.—Its importance.—General principle of Reasoning.—Extent of its application.—Definitions of Syllogism and its parts.—Different modes of expressing Syllogisms.—Best mode.—Part frequently suppressed.—When it ought to be expressed.—Reason for a thing.

THE knowledge obtained by means of Intuition is of two kinds, *immediate* and *mediate*. The former regards the self-evident and necessary properties and relations of things: the latter consists of inferences which are necessarily implied in other propositions.

We frequently know by Intuition that if one thing is true, another thing must be equally true. Thus, if I know that it is full moon, I know, with perfect certainty, that it is not new moon. Here is an act of *Reasoning*, which is, simply drawing necessary inferences, or finding out, by means of self-evident truths, propositions that are necessarily implied in others. To *reason* is, to discern intuitively that one thing necessarily implies another, or that, if the former is true, the other is necessarily and inevitably true, and cannot by possibility be false. That such is the nature of all valid reasoning appears, not only from an analysis of the process, but from the fact that it cannot possibly be otherwise. There is no possible means of knowing anything indirectly, except by discerning that it is necessarily implied in something which we know directly. (4)

In all sound reasoning the thing implying is compared with the thing implied; and the necessary implication is discerned intuitively. The process may, indeed, be fallacious, and the supposed inference may not, in reality, be necessarily implied: but this is never the case in valid or legitimate reasoning, which is what we mean when we speak generally; and, even in fallacious reasoning, the inference is supposed or professed to be necessarily implied.

Thus it appears that *Reason*, or the faculty of reasoning, is only Intuition, applied to discover mediate knowledge, or to find out necessary consequences, by means of self-evident truths, and that the cognitions implied in our apprehensions are as dependent on Intuition as on Apprehension.

Whatever may be the subject, the reasoning process is identical in all cases: it is, always, simply discovering that one thing necessarily implies another. But it is of the utmost importance, because it is requisite to the acquisition and retention of all mediate knowledge.

The general principle of all reasoning is, that *wherever one thing necessarily implies another, the existence of the former conclusively proves that of the latter*, the necessary implication being always discerned by Intuition. It may be otherwise expressed thus: *wherever one proposition necessarily implies another, the latter is true if the former be true.* The conclusiveness of such inferences is evidently independent of the actual truth of the implying propositions: we may draw necessary inferences from a proposition that is merely assumed or supposed to be true, as well as from one which actually is so. This proceeding is frequently of the utmost importance, although, in order to avoid error, we must distinguish it from those cases in which the implying propositions should be actual truths.

A single act of reasoning is called a *syllogism*. Every syllogism necessarily consists of three parts, and no more, which may be designated as follows—(1) the *premise*, or implying proposition—(2) the *inference*, or proposition necessarily implied—and (3) the *connective*, or self-evident truth by which we know that the inference is necessarily implied in the premise. (5)

A syllogism may be expressed in various ways, as its parts may be differently expressed or arranged. The connective may be expressed in a more or less general, or in a particular form; and the other parts are, of course, susceptible of the usual variety of expression. The following may serve as an example of these variations.

1. Everything that thinks, exists: (*Connective*)
I think; (*Premise*)
Therefore I exist. (*Inference*)
2. I think; therefore I exist: for whatever thinks exists.
3. I know intuitively that if I think I exist: but that I think is a matter of immediate consciousness; therefore I exist.
4. I exist: for I think; and whatever thinks exists.
5. It is self-evident that whatever thinks, exists: therefore I exist; for I am directly conscious of thinking.
6. I am immediately conscious of thinking; and whatever thinks must exist: therefore I exist.
7. I certainly exist: for every being that thinks exists; and I know, by direct consciousness, that I think.

In stating a syllogism, the best arrangement is that which exhibits most clearly the necessary connection between the premise and the inference; and there is often little to choose, between several modes. The connective is frequently suppressed in discourse, because the mind can generally supply it, as it is suggested or rendered obvious by comparing the premise and the inferences. Thus the above syllogism may be expressed "I think: therefore I exist." The premise, also, is, in many cases, not distinctly expressed, as it is presumed to be well known to the party addressed, or to be apparent from the preceding assumptions. So the inference is often omitted, where it is presumed to be well known, and the question only regards the proof. But wherever there is any difficulty or doubt, every part of the syllogism should be expressed.

Where a premise is acknowledged to be true, it may be assigned as a proof of the inference; and, in such cases, it is frequently termed the *reason* for it.

§ 2. SPECIAL PRINCIPLES OF REASONING.—Definition and Nature of these Principles.—Their number.—(A) Definition and two general expressions of Truisms.—Eight principal modifications of these. —(B) Seventeen principles relating to duration and extension.— (C) Forty-seven principles relating to abstract quantity.—(D) Definition of Being, and of various kinds of Beings.—Necessary and contingent properties.—Seventeen principles regarding the necessary properties of substances in general.—Motion, Force, and Change. —Necessary property of Inanimate Beings.—Six principles regarding the necessary properties of Living Beings.—Thought.—Belief. —(E) Twenty principles regarding Determining Conditions and Agencies in general.—Causes and Effects.—How these are distinguished from uniform Antecedents and Consequents.—Nine principles regarding Desires and Volitions.—(F) Eighteen principles of Inclusion and Exclusion.

The special principles of reasoning are, those self-evident and universal truths which are essential to reasoning, and which we may rightly employ either as premises or connectives, in any argument or investigation. As all reasoning is only an application of Intuition, and intuitions are the only self-evident and universal truths that we can know, it follows that these are the only legitimate principles of reasoning. A complete enumeration of them is neither practicable nor desirable, since their number is indefinite, and many of them are of very little use: but the more important of them are stated under one or other of the following heads.

A. Truisms.

A *truism* is a proposition in which the predicate or a part of it, is asserted of itself, as—" a man is a man," and —" a Frenchman is a man."

The general formulas, or expressions, of all truisms are, that a thing is equivalent to itself, and that every part of a thing belongs to it. These are susceptible of innumerable modifications, of which the following are among the principal:

1. A thing is not different from itself.
2. Different things are not the same thing.
3. A whole is equivalent to all its parts. As "all the parts" are the same as "the whole," this expression is the same as saying that a whole is equivalent to itself.
4. Incompatibilities cannot coexist. This is only saying in other words that things which cannot coexist cannot do so. What these are, must be learned from other sources.

5. Propositions which have the same amount and kind of proof, are equally credible. Hence, if one of such propositions is unproved, all are unproved; if one is established, all are established; and if one is believed, all are equally entitled to belief.

6. The character of a proposition is identical with that of others tantamount to itself. Therefore the former is true, false or doubtful, according as the latter have been found to possess any of these properties. This principle enables us to vary the forms of expressions without altering the signification, and to combine many propositions into one, which is equivalent to the whole of them taken together. For it applies equally to cases where one proposition is tantamount to another, and to those where it is equivalent to several, which it comprises.

7. Every proposition whose predicate correctly expresses the thing denoted by its subject, is necessarily true; and so is its converse. If azote denote the same substance which is otherwise termed nitrogen, it must be true that "azote is nitrogen," and equally true that "nitrogen is azote." So if it be true that "London is the capital of the British empire," it is equally true that "the capital of the British empire is London."

8. Every proposition which declares that a part of the subject belongs to it, is necessarily true. Thus, "a Frenchman is a man"—"a horse is a quadruped," and "an eagle is a bird."

B. Principles relating to Duration and Extension, or Abstract Time and Space.

1. *Duration* and *extension* necessarily exist.
2. They are immutable, and independent of every other thing.
3. Every change occurs in time and space.
4. All their attributes are unaffected by any other thing, or by any change. The parts of duration succeed each other uniformly, beyond the power of control, and can neither be accelerated nor retarded. So, extension is unaffected by the bodies that may occupy portions of it, or pass through it.
5. Both are incapable of either motion or thought.
6. They cannot originate any change or motion.
7. They are imponderable, or without weight.
8. They cannot affect any of our senses. We cannot feel, taste, smell, see, or hear either time or space.

9. All the parts of each are homogeneous, or exactly alike in kind.

10. Every part of the one coexists with every part of the other. Every part of time exists everywhere, and every part of space exists always.

11. Their natures are essentially different; and one cannot possibly pass into the other. Time possesses no extension, and the parts of space never come into existence, nor cease to exist.

12. Every part of *duration* is preceded and followed by other parts of it; and therefore it is eternal, or without beginning or end.

13. Every part of *extension* is surrounded by other parts of it; and, therefore, it is infinite, or without any bounds.(6)

14. *Duration* consists of parts which come into existence successively, and then cease to be forever; and hence no two of them can coexist, nor can any of them which has ceased to exist, return, or exist again.

15. The parts of *extension* all lie without each other; and, therefore, no two of them can coincide.

16. *Extension* has position, but not form; and duration has neither position nor form.

17. *Extension* possesses mathematical solidity, or length, breadth and thickness; but it does not possess resistive solidity, or, in other words, it can offer no resistance to the motions of bodies in any direction; duration possesses no solidity of any kind.

C. Principles relating to Abstract Quantity.

1. *Abstract numbers* are illimitable in amount, and homogeneous in kind.

2. *Abstract magnitudes* are endlessly diversified in form and size; and they consist of at least four essentially different kinds—solids, surfaces, lines, and angles.

3. A *solid* has three dimensions—length, breadth, and thickness.

4. A *surface* has only two dimensions—length and breadth; and it may be considered the boundary of a solid.

5. A *line* has only one dimension—length; and it may be considered the boundary of a surface.

6. A *point* has no magnitude, but only position; and it may be considered the extremity or end of a line.

SEC. 2.] SPECIAL PRINCIPLES OF REASONING. 55

7. A *plane angle*, or one formed by two straight lines meeting in a point, indicates only the difference of direction of its two bounding lines, which is the sole measure of its dimension.

8. The dimensions of a plane angle may be determined by the number of aliquot parts of a circular circumference intervening between its two bounding lines, the center of the circle being at the angular point.

9. A *solid angle*, or one formed by three or more plane surfaces which meet in a point, termed its apex, indicates only the differences of direction of its bounding planes, which are the sole measure of its dimension. Planes, or plane surfaces, are those which are quite straight, and free from any curve or bend, in every direction.

10. The dimension of a solid angle may be determined by the number of aliquot parts of a spherical surface included between its bounding planes, the center of the sphere being at its apex.

11. *Number* and *magnitude* are essentially different; but the latter may be measured by the former.

12. *Abstract quantity* has no substantial existence; and, therefore, its nature and primary attributes can be known only by Intuition.

13. *Magnitudes* are extended, and have form, size, and position. *Numbers* possess none of these; they are only so many, and can occupy no part of extension.

14. The *whole* includes all its parts; and the latter are included in the former.

15. The *whole* is greater than a part; and a part is less than the whole.

16. *Magnitudes* which can be made to coincide, are equal.

17. If one of two *homogeneous quantities* is neither greater nor less than the other, it is equal to it.

18. Things *equal* to the same thing, or to equals, are equal.

19. If *equals be affected by equals*, the results will be equal. This principle comprises many others less general, of which the five following are the most important.

20. If *equals be added to equals*, the sums will be equal.

21. If *equals be subtracted from equals*, the remainders will be equal.

22. If *equals be multiplied by equals*, the products will be equal.

23. If *equals be divided by equals*, the quotients will be equal.

24. *Like powers or roots of equals* are equal.

25. If *unequals be affected by equals*, the results will be unequal. This principle comprises the five following ones.

26. If *equals be added to unequals*, the sums will be unequal; and that which includes the greater element will be the greater sum.

27. If *equals be subtracted from unequals*, the remainders will be unequal; and the greater quantity will leave the greater remainder.

28. If *unequals be multiplied by equals*, the products will be unequal; and the greater multiplicand will give the greater product.

29. If *unequals be divided by equals*, the quotients are unequal; and the greater dividend gives the greater quotient.

30. *Like roots or powers of unequals* are unequal; and the greater quantity gives the greater root or power.

31. If the same operations be performed on *several quantities*, the results will bear to each other the same relations as the original quantities. This principle comprises the 19th and 25th, but it is less explicit.

32. If *equals be affected by unequals*, the results will be unequal. This principle is the converse of the 25th, and comprises the five following subordinates.

33. If *unequals be added to equals*, the sums will be unequal; and the greater is that which includes the greater element.

34. If *unequals be subtracted from equals*, the remainders will be unequal; and the greater remainder is that which is left by subtracting the smaller quantity.

35. If *equals be multiplied by unequals*, the products will be unequal; and the greater multiplier gives the greater product.

36. If *equals be divided by unequals*, the quotients will be unequal; and the greater divisor will give the smaller quotient.

37. *Unlike roots or powers of equals* are unequal; the lower root of every number greater than unity is greater than the higher; and the higher power of every such number is greater than the lower.

38. A *straight line* is wholly in one direction. Hence,

SEC. 2.] SPECIAL PRINCIPLES OF REASONING. 57

it has two extremities or ends; it can never return into itself, however far produced or extended; nor can it inclose a space; and only one straight line can be drawn from one point to another.

39. *A straight line* may be indefinitely produced, or lengthened, from either extremity.

40. *Two straight lines* cannot coincide in two points without forming one straight line. Hence two straight lines cannot inclose a space: otherwise one straight line might do so.

41. *A straight line* is the shortest that can be drawn from one point to another. Hence any two sides of a triangle are together greater than the third.

42. *Straight lines* which intersect or cut one another, lie in different directions, and diverge indefinitely, as they are produced. Hence straight lines which lie in the same direction will never meet, however far produced; for othwise they would lie in different directions.

43. The *position of a straight line* is wholly determined by any two points in it, or, in other words, the position of a straight line cannot be in the least changed, as long as any two points in it continue in the same position.

44. The *direction of a straight line* in a plane is wholly determined by the angle which it makes with another fixed straight line in that plane; and, conversely, the size of the angle is wholly determined by the direction of the two straight lines. Hence straight lines which are in a plane, and make equal angles with another straight line in that plane, lie in the same direction, and, therefore, can never meet, though produced ever so far; and straight lines in a plane which never meet, make equal angles with another straight line which they intersect.

45. *Two straight lines which lie in different directions,* or make unequal angles with a third straight line, and are in the same plane, will meet, if produced indefinitely.

46. *Two straight lines in the same plane* which are inclined to each other, or nearer at one point than at another, will meet, if indefinitely produced. Hence straight lines in a plane which never meet, however far produced, are parallel, or everywhere equidistant.

47. *Two straight lines in a plane* which do not intersect each other, and are equidistant at any two points, lie in the same direction; and, therefore, will never meet, how-

C 2

ever far produced, and, therefore, are everywhere equidistant.

D. *Principles regarding the necessary properties of Substantial Beings.*

A *Being* is, whatever is, was, or will be. Beings are *unsubstantial* and *substantial*. The former consist of those which are incapable of action or passion, including duration and extension, and the states, properties or relations of substantial beings: the latter comprise all other beings. Substances are either *material* or *immaterial*. The former are, those which directly affect one or more of our senses, such as iron, water and air: the latter comprise all those which do not so affect any of our senses, such as the minds or souls of men. Substances are also divided into *inanimate* and *animate*, or living. The former are incapable of thought: the latter either think or have the power of doing so.

The attributes or properties of substances are of two kinds:—The *necessary*, which every substance must possess, and which are known only by means of Intuition;— and the *contingent*, which a substance may or may not possess, and which are learned by means of Comprehension.

The following are the principal necessary properties of substances.

1. Every substance possesses resistive *solidity*, the nature of which is discerned intuitively.

2. Every substance is *extended in space*, and *exists in time*, past, present or future; or, in other words, it must occupy some portion of space, and exist during some portion of time.

3. Every substance is *numerically one*, and can occupy only one part of space, at any instant of time: in other words, it cannot be two substances, nor be in two distinct places at once.

4. Every substance possesses *form* and *magnitude*, or it must have some shape and size; and it can have only one form and size at any instant.

5. The *solid particles* of substances are *incompressible* and *inextensible*, or, in other words, their mass or quantity of solid substance can neither be increased nor diminished, and much less annihilated. Bodies apparently solid are frequently compressed; but this arises from their being porous.

6. Substances are *impenetrable*, or, in other words, two substances cannot occupy the same part of space at the same instant.

7. Every substance must be either *at rest* or *in motion*, although it may be relatively at rest, and absolutely in motion, like a mountain, which partakes of the motions of the earth, though it is relatively at rest.

8. Every substance is *mobile*, or susceptible of motion.

9. All motion consists of a substance's *passing* from one part of space to another;

10. Every motion is performed *in time*, or, in other words, no motion is instantaneous.

11. Every motion of a substance is *from that part of space where it is to that which is immediately contiguous*. Consequently every substance which is found in a place different from one which it formerly occupied, must have passed continuously through every part of space along some course joining the two positions.

12. A substance impelled by a force moves in the *direction* in which it is thus impelled, or, in other words, it moves in the direction in which it is made to move. A *force* is anything which changes, or tends to change, a substance's previous state of rest or motion.

13. *Force* is requisite to change a substance's state of rest or motion.

14. The *velocity*, or *speed*, of a substance is proportional to the force which generates it, if there is no counter action or interference of any other force.

15. A substance affected by *several forces* which do not counteract each other, moves as if it were affected by one force equivalent to them all.

16. Substantial beings alone are *susceptible of change*, this term being understood in its ordinary signification; and every change consists of the motion of a substance from one part of space to another.

17. *Substantial beings* alone are capable of *thinking*, or originating motion, although all beings of that class may not possess that power.

18. *Inanimate substances* can have no *desire* or *aversion*. Hence they cannot be influenced by motives, or originate any change, and they can produce or undergo a change only where they are affected by some other agent, unless we assume that they were in motion from all eternity.

Of the necessary properties of living beings, the following are the principal.

1. Every *living being* is at least capable of *thought*, although he may possibly never have actually thought.

2. Every thought is *real*, and the act of a living being.

3. Every thought is precisely what it *appears to be* to the thinker. We must distinguish the thought from its object or cause, which may be different from what we believe it to be; but this does not in the least affect the character of the thought.

4. The *presence* or *absence* of thought is precisely as the thinker believes. We cannot believe that we do not think when we do, nor that we think when we do not.

5. *Belief without grounds* satisfactory to the believer, is impossible, real and not professed belief being, of course, understood. The grounds of belief may, in reality, be conclusive; yet if they do not appear to the individual of any weight, he will inevitably disbelieve.

6. *Belief against grounds* satisfactory to the believer is impossible. The proofs may be really worthless; yet if they appear satisfactory to the believer, he will necessarily believe.

E. Principles relating to determining Conditions and Agencies.

1. Where the *conditions* which determine results are the same, the *results* will be the same, however much other things may vary, the word "same" being taken to mean either absolute identity or perfect similarity. The determining conditions are, those things which determine, limit or fix a thing to be what it is, and not different.

2. Where the *determining conditions are different*, the results will be different, although other things should be the same.

3. Any *proposition established in one case*, independently of particular circumstances, holds true of all cases which differ only in those circumstances.

4. *What has once happened*, must always happen, wherever all the circumstances which influence the result are the same.

5. Every *change* is preceded or accompanied by *motion* of one or more beings; and change without motion is impossible.

6. Every *change* must have its *ultimate origin* either

SEC. 2.] SPECIAL PRINCIPLES OF REASONING. 61

in the volition of a thinking being, or in a substance which was previously in motion from all eternity.

7. Every *change* is necessarily preceded by something which produces it, or on which it is dependent, called its *Cause*, and with reference to which it is termed an *Effect;* or, in other words, everything which begins to be, is preceded by something which makes it be, and failing which it would not be. The effect being dependent on the cause, distinguishes the latter from a uniform antecedent, and the former from a uniform consequent.(7)

8. *Nothing can produce only nothing*, which may be otherwise expressed, something cannot spring from nothing.

9. In order to *produce a change*, a cause must be adequate to produce it: but what are adequate causes must generally be learned from other intuitions, and from experience.

10. Where a *cause known to operate*, is found inadequate to produce the whole of an effect, the residue is owing to some other cause.

11. *An agent cannot act where it is not;* or, in other words, an agent can produce no effect where it is not, except by some medium. The Sun produces effects on the Earth by the mediums of light and heat, without which we could neither see it nor feel its warmth.(8)

12. When *two sets of agencies or forces* operate, which tend to produce precisely similar effects, but directly counteract each other, the one set nullifies the other, and no such effect ensues, which may be more briefly expressed thus: equal and contrary forces neutralize each other.

13. The *same cause*, operating in the same circumstances, always produces the same effect.

14. An *effect* which depends solely *on a particular cause*, varies in proportion to the changes in the cause, and conversely; or, in other words, effects are proportional to their causes, and the intensity of a cause is measured by the effect. Hence, where the result is apparently otherwise, one or more additional agencies must be concerned in its production; and this is so frequently the case that an appeal to experience is generally requisite, before we are warranted in assuming that no such second agency is concerned.

15. Where *several agencies* or *forces* exactly co-operate or combine, to produce a particular effect, the result is

the same as would be produced by a single agency equal to the sum of those agencies.

16. A body, or a connected system of bodies, is in *equilibrium* when the forces which tend to move it in a certain direction are equal to those which tend to move it in the contrary direction. By a "body" is understood a material substance.

17. Where a body is urged in contrary directions by *unequal forces*, it will move in the direction in which it is impelled by the stronger force, as if it were affected by a single force equal to their difference.

18. When *the cause ceases to act*, its effect ceases to be produced. This principle must not be confounded with the very different proposition that an effect already produced will cease to exist when the producing cause ceases to operate; a proposition which is not even true. A change already produced may be permanent, or produce farther changes, although its cause may become inoperative, or even cease to exist. Thus, if a man ruins his health by dissipation, it is not immediately restored when he ceases to dissipate, and perhaps never; yet it is self-evident that, when he ceases from his bad habits, he will suffer no further injury directly from that cause.

19. The *presence or absence of the cause* proves the presence or absence of the effect, provided there is no counteracting or interfering agency.

20. The *presence or absence of the effect* proves the presence or absence of the cause, if the former has not been removed by some extraneous agency.

The following are the most important principles regarding desires and volitions:

1. A regard to *good and evil* is the only cause of the free or voluntary acts of living beings.

2. Every volition, or act of will, has for its object the *greatest apparent good;* or, in other words, the greatest apparent good is always willed. By the greatest apparent good is meant that which the individual believes to be such, although it is frequently by no means the greatest real good.

3. That which will apparently secure *more good*, is willed before that which is believed to secure *less;* and where one of *two evils* is deemed inevitable, that which appears the less is willed.

4. Every *apparent good* is an object of desire; and every *apparent evil* is an object of aversion.

5. Where there is no *knowledge or belief regarding good or evil,* there can be no volition or voluntary action.
6. Where the *apparent good and evil* are precisely equal, there can be no volition or voluntary action.
7. Where two objects appear, in every respect, *equally good,* there can be no choice.
8. Every *voluntary action* conforms to the decision of the Judgement which pronounces it a means towards the greatest good.
9. A *change of the Judgement* or *of belief,* regarding good and evil, is followed by a corresponding change of volition.

F. Principles of Inclusion and Exclusion.

1. A thing cannot *be,* and yet at the same time and in the same sense, *not be.*
2. An attribute cannot be truly *affirmed* and *denied* of a thing, at the same instant.
3. No being can possess either *contradictory* or *incompatible* attributes or properties.
4. Of two *contradictory propositions,* one must be *true* and the other *false.* Hence, if one is true, the other is false, and conversely.
5. Of two *contrary propositions,* both cannot be *true,* but both may be *false.* It is not generally necessary that a being should possess either of two contrary attributes, but it cannot possess both. Thus, the sky is neither white nor black; but if it were the one, it could not be the other.
6. A proposition cannot be *both true and false,* in the same sense. Hence, if one of several propositions must be true, and all but one are found to be false, this one is true. So, every proposition is false which necessarily implies something incompatible with truth; else the proposition implied would be both true and false.
7. If one thing *includes* a second, this second, a third, and so on, then the first includes the last. This principle applies both to physical objects and to inclusion by necessary inference. Thus, if London is in England, and England, in Europe, London is in Europe. So, if all kings are men, and all men are mortal, then all kings are mortal.
8. If one thing is *equivalent* to a second, this second, to a third, and so on, any one of the series is equivalent

to any other. As equals are included under equivalents, this principle includes the former.

9. A being possessing the *characteristic* or *peculiar marks of a class*, is one of that class; and a being which wants these marks, does not belong to the class. The "characteristic" or "peculiar marks" of a class are, those properties which belong to every individual of a class, and to no other being.

10. Every being possesses the *essential properties* of its class. Thus, a triangle must have three angles, and an animal must be alive. The "essential properties" of a class of beings are those which distinguish it from all others, and the want of which excludes a being from the class.

11. The *universal includes the particular ;* or, in other words, whatever belongs to a class individually, belongs to every one of that class; and whatever is wanting in a class individually, is wanting in every one of that class. Thus, if John is one of a class of beings individually rational, he is rational; and if he is one of a class of animals individually wanting fins, he is finless. This principle enables us to express a universal truth in an indefinite variety of less comprehensive forms.

12. Whatever *belongs to every individual of a class*, belongs to the class; and whatever is wanting in every individual of a class, is wanting in the class.

13. If some property of a thing is *proved*, all incompatible properties are *disproved*. Thus, if it is proved that a certain man is virtuous, it is disproved that he is vicious; or, if it is proved that he is a cheat, it is disproved that he is honest.

14. An *exclusive and uniform consequent, concomitant* or *antecedent* of a phenomenon, proves its past, present or future existence. This principle applies only where the connection of the two things is already known, which is generally ascertained either by their being necessarily connected as cause and effect, or by their being the effects of a common cause.

15. A *common consequent, concomitant* or *antecedent* of a phenomenon proves its probable past, present, or future existence: and the probability varies according to the extent of the connection, which is learned chiefly by previous experience of similar cases.

16. Every *intuition is self-evident*, whether expressed

universally, generally, or particularly: and hence it may assume an indefinite number of forms, all of which are self-evident, if properly expressed.

17. A thing either *is* or *is not*. This is the formula of alternation; and it is extensively employed to determine all the possible conditions of a problem.

18. The existence of a *property* necessarily implies the existence of something of which it is the property; and a property cannot inhere in anything with which it is incompatible. This principle is extensively applied, in its special forms, in ascertaining the existence and qualities of objects.

§ 3. PROCESSES AND CRITERIONS OF REASONING.—Method of establishing a Conclusion.—Primary Premises, Conclusion, and Chain of Reasoning.—Requisites to render a chain of reasoning valid.—Testing Syllogisms.—Parts of it often overlooked.—Ultimate foundations of all valid reasoning.—Two modes of testing a chain of reasoning.—Things to be examined, in all cases.—Effect of Ambiguity or Obscurity.—Two modes of proceeding.—Where both should be adopted.—Advantages of considering connectives generally.—Allowable course where any of the conditions of valid reasoning are wanting.—Nature and various kinds of Arguments.—Collateral chains of Reasoning.—Nature and use of Combination.—Representation of a complex Argument.—Proving too much.—Why the faculty of reasoning cannot be successfully impugned.—Relation of Reasoning to Language.—Evils of not looking beyond words, in reasoning.—How Memory and Language aid reasoning.—Other aids.—Judgement, and Intellect.

In establishing the proposition in question, it is generally requisite to proceed by degrees: the first proposition, which we term the *primary* premise, is discerned to imply a second, this second, a third, and so on, to the last inference, or *conclusion*, the premise of one syllogism being the inference of the preceding. The whole series is termed a *chain* of reasoning, the syllogisms being compared to links; and it may consist of an indefinite number of these. The conclusion is always connected with the primary premise by means of the seventh principle stated in the last subdivision of the preceding section, although it is so obvious that it is rarely expressed.

The nature of a chain of reasoning may be exhibited to the eye by the following diagram:

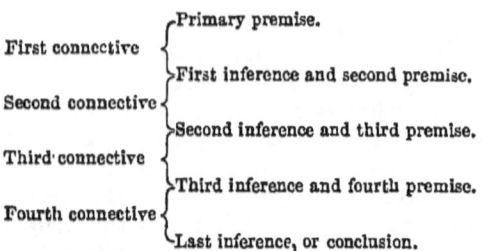

To render a chain of reasoning valid, it must possess the four following characteristics.

1. *The primary premise must be all that it is rightly assumed to be.* If it is assumed to be a discernment, it must not be an inference or a proposition founded on testimony; and if it is assumed to be a supposition which is possibly true, it must not be an impossibility. So, if it is assumed to be proved by testimony, this must be conclusive; or if it is assumed to have been demonstrated by a process of reasoning, this must be valid.

2. *Every successive premise must be necessarily implied in that which immediately precedes it.*

3. *The conclusion must be the real question, and not one merely like it.*

4. *Every part must be understood or expressed clearly and accurately,* so that there is no room for doubt as to what the exact things are. For, if this condition is not complied with, violations of the others cannot be detected.

The strength of a chain is not greater than that of its weakest part; and if one material link fails, the whole is worthless.

In order to test the validity of a syllogism, we must discover the three parts. This is sometimes a difficult matter: for, not only are those parts frequently separated from each other by wide intervals, both in spoken and in written discourse, but we have often to gather the detached fragments of a proposition from various quarters, as it is nowhere distinctly stated.

In many cases, the rapidity of thought leads us to overlook, in analysing, some of the syllogisms of the chain; and hence, when we endeavor to trace the steps by which we arrived at the conclusion, flaws appear in the chain, while, in reality, it was continuous and conclusive. We must, therefore, beware of inferring that the process by which we arrived at the conclusion was certainly faulty, because we have failed to make out all the links of the chain. In many cases, a little careful examination will make the defective links obvious. Yet we ought never to adopt conclusions as certainly true until we obtain the evidence of consciousness, at every step, although it is frequently difficult, especially for persons unaccustomed to analyse thought, to retrace all the mental processes.

Reasoning often starts with propositions that are acknowledged to be true by the party to whom it is addressed: but, to render a conclusion absolutely established, we must commence with truths of immediate consciousness, or, in cases of hypothetical reasoning, with the original suppositions. For discernments and suppositions necessarily form the ultimate foundations of all reasoning. We cannot reason from infinity; and we must begin with what is self-evident, or what, without being so, is known by direct consciousness, or with something that we assume or suppose to be true, although it may be only a probability, or purely hypothetical.

We may examine the validity of a chain of reasoning by tracing it from the primary premise to the conclusion: or we may begin with the latter, and trace it back to the former. In both cases, we must always examine what the proposition under immediate consideration is, and whether it is necessarily connected with those adjacent. If an obscurity or ambiguity occur in any part, the whole chain should be held invalid, until the difficulty has been cleared up: for, otherwise, we cannot ascertain whether the necessary connection exists, and sophistry may occupy the place of sound reasoning. Owing to the defects of language and to loose or inaccurate thinking, it is sometimes difficult to discover the exact import of a material proposition; but this must always be done, before we can ascertain whether the reasoning is valid.

If we begin with the primary premise, we are first to consider whether it requires proof. If it does, the whole chain is baseless: if not, we then examine what the first inference is, and whether it is necessarily implied in the premise. If so, we then examine what the next inference is, and so on, till we come to the final conclusion. If this be not the real question, the whole chain is worthless, since it is beside the subject.

If we begin with the conclusion, we first consider what the real question is, and then ascertain whether this is the very inference drawn from the premise immediately preceding. If so, we then find out this premise, and consider whether it necessarily implies the conclusion. If so, we search for the premise whence the former was inferred, and so on, till we come to the primary premise.

In cases of importance, it is proper to try the validity of a chain of reasoning both ways. In determining whether a premise really implies the inference drawn from it, we need not consider the connective in a more general form than the particular connection demands: but it is desirable to understand its universal nature, as we can thus, in many instances, more clearly discern the connection, and apply the principle more readily and safely in other cases.

Wherever we find an absence of any of the four conditions of conclusive reasoning, we need not proceed farther: for one essential defect invalidates the whole. A chain of reasoning resembles one of iron, employed to move a weight, which must be sound in itself, attached to a moving power of sufficient force, and also fastened to the proper weight, and which fails of effecting the object if there be a defect in any of these respects. The case where the moving power is deficient, or the chain not sufficiently secured to it, corresponds to that of an unsound or inadmissible premise: that of a weak or broken chain corresponds to a flaw or obscurity in the reasoning; and the case of the chain being fastened to the wrong weight answers to that where the conclusion proved is not the proposition in question.

An *argument* is, what is employed to prove a conclusion. It may consist of a single syllogism: but it is more frequently made up of several collateral chains of reasoning, or a combination of such chains, often blended with matters which are assumed as known or true—all converging towards the conclusion. Some of these chains may be either essential parts of the argument, or independent of the rest, and only employed to corroborate or strengthen the conclusion deduced from them.

Several primary premises are often employed to prove a more general proposition, which embraces them all, and is tantamount to the whole of them taken together. When this identity has been ascertained, we infer the

Sec. 3.] Processes and Criterions. 69

truth of the general proposition upon the principle already stated, that the character of a proposition is identical with that of others tantamount to itself. Several of the new propositions, thus established, may be employed, in the same way, as premises to prove a still more comprehensive conclusion, and so on, the whole process resembling the confluence of streams, where rivulets flow together to form brooks, and several of these unite to form rivers, while the independent parts may be compared to streamlets that flow directly into the sea.

In such cases, we first observe the various things already proved or known, which we desire to embrace in the more general proposition, then search for a suitable expression, which we compare with them, in order to ascertain whether it comprises them all and no more. When that has been done, we infer that the comprehending proposition is true, and employ it accordingly.

Such a process may be termed *combination*. Although it merely embraces in a more general proposition particular or narrower ones previously known, established, or assumed, yet it is of much importance, as it greatly aids Memory and Reason in establishing comprehensive conclusions. Without such aid, the Attention and the Memory would be so confused that the bearings of the previous cognitions could not be discovered, and we should consequently be unable to connect them with the conclusion.

Every distinct part of an argument is to be tested in the manner already pointed out, as if it formed the whole. But we should carefully observe whether the things combined are tantamount to the comprehending proposition to which they are assumed to be equivalent: for, in many instances, they are so only with certain restrictions or modifications, which are apt to be overlooked, or they contain less than is assumed.

The nature of a complex argument may be illustrated by the following figure.

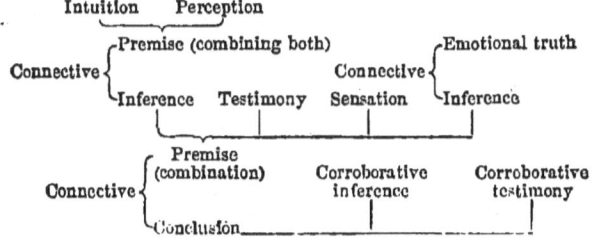

The fallaciousness of an argument sometimes appears from its proving something which we know to be untrue, equally with the conclusion which it is employed to establish. Arguments of this kind are said to "prove too much," and are evidently invalid: but we must beware of assuming that they are so, merely because this is alleged; for opponents sometimes make such allegations when the argument is, in reality, irrefragable. An argument which is conclusive in itself, cannot possibly prove too much: otherwise the same proposition would be both true and false.

All attempts to impugn the faculty of reasoning are fallacious: for they necessarily assume its faithfulness, while they profess to prove the contrary. We cannot proceed a step to impeach it, without first assuming its entire reliability; and if we can trust its conclusions in one case, we are evidently bound to receive them in all other cases equally unobjectionable. A difficulty sometimes occurs from its leading to apparently contradictory conclusions: but, as it is self-evident that contradictions cannot both be true, there must be some fallacy in one or other of the processes, which a careful examination will always detect. Consequently the discrepancy is only apparent; and such difficulties only prove that we are liable to reason erroneously or inconclusively, a truth of which we have frequent proof. Yet it is possible to test reasoning, so that we shall certainly know whether it contains any fallacy.

Although language and other signs of thought are frequently of the utmost use in reasoning, yet we cannot reason closely and conclusively, in all the most difficult subjects of investigation, without discarding all such signs, and directly considering the things signified: otherwise we may possibly be reasoning merely about words, and continue ignorant of what they profess to denote, a thing which has very frequently occurred. Words are merely signs of thought: and unless we discern a necessary connection between the things signified, independently of their signs, our conclusions may possibly hold true only of the latter. For words are frequently of dark or doubtful import, or wholly unintelligible to the party addressed; and the difficulty can be removed only by determining the nature and necessary relations of the things signified, which cannot frequently be done without considering them, wholly apart from language.

Whence once we have clearly understood the exact nature and extent of the things denoted by language, it enables us to substitute signs or symbols for the things signified; and thus we can arrive at conclusions otherwise unattainable, while, in other cases, the process of reasoning is greatly facilitated and abridged.

Memory aids reasoning by enabling us to substitute similitudes for their prototypes, and thus to reason about things absent as if they were present, while language enables us to discard even similitudes, for the time being, and substitute in their place mere visible or audible signs or symbols of them. Such aids are generally requisite, in order either to establish or retain general or recondite truths.

Comprehensions form the starting point of all knowledge, since they are requisite to rouse our intellects into action, and, at the same time, they furnish the fundamental elements of all contingent knowledge. Abstraction is also requisite in all reasoning regarding matters which present any difficulty. Hence the faculties of Apprehension, Memory, Abstraction and Reasoning are designated by the common term *Judgement*, which differs from *Intellect* in excluding Conception. This faculty is comparatively feeble in the greater number of mankind, and much more rarely used in discovering truth.

The conclusions drawn from premises whose character has been investigated by the aid of the Judgement, are termed *judgements*.(9)

CHAPTER III.

OF THE PRIMARY MEANS OF ACQUIRING CONTINGENT KNOWLEDGE.

§ 1. REALITY OF APPREHENSIONS, AND MEANS OF AVOIDING THE PRIMARY ERRORS WHICH THEY OCCASION.—Origin of Errors attributed to the Senses.—How Apprehensions may be distinguished from Ideas.—Specters.—Apprehensions necessarily real.—Distinguishable from their causes.—Inferences from them often erroneous.—How these may be tested, and errors avoided.

ALTHOUGH the senses are frequently occasions of error, yet, strictly speaking, they never deceive; and all the errors attributed to them arise either from confounding apprehensions and ideas, or from drawing unwarrantable inferences from the former.

To avoid errors of the first class, it is only requisite to attend to the peculiarities of ideas and apprehensions. The former are readily distinguished from the latter, wherever the apprehending organs are sound, not only by their shadowy and fleeting nature, but also by their being generally under the control of the Will. If we think of a well-known tree, which is at the moment invisible, the likeness of it which we discern is much fainter than if we actually viewed it; and it vanishes altogether, in a very short time, unless a conscious effort of the Will detains it; and if we so will, it vanishes at any instant, whereas the apprehension is not only much stronger and more distinct, but it cannot be willed away, while we behold the tree.

In some cases of very forcible apprehensions, or a diseased condition of the organs of sense, ideas acquire unusual vividness and permanence, and are, therefore, peculiarly liable to be mistaken for their prototypes. This remark applies particularly to objects of sight and hearing, which are, in such cases, termed *specters*, or *spectral illusions*. Yet, even here, the shadowy character of ideas is still discernible. Error arises chiefly from the attention being so concentrated on the idea that the difference between it and the apprehension is overlooked, although some palpable difference always exists. Thus, even after looking at the Sun, when the specter is still seen, in spite of all efforts to will it away, the difference is so palpable that no person need confound it with seeing the real disc of the Sun.(10)

The presence of something closely resembling an object, is apt to produce the same illusion as disease, especially where the organs are placed in unfavorable circumstances, and the mind is affected with any strong emotion. A dark log seen in a wood at night, may lead a timid person to think that it is a robber, lying in wait for his victim, the excitement probably causing an unusually vivid idea of some grim countenance to arise, and complete what the real object lacks. But, in all such cases, we have only to view the object calmly and attentively, in order to avoid erroneous conclusions. Wherever disease affects the apprehensions, there are discoverable indications of its presence, to put us on our guard; and we have only to attend carefully to all that is actually discerned, in order to avoid error.

In some cases, disease causes objects to affect us differently from what they do in health. But such changes do not in the least affect the reality of the apprehensions: they only warn us to be cautious in assuming that objects are what they appear to a diseased organ.

To mistake an apprehension for an idea, is a very rare occurrence, because the characteristics of the former are so palpable that a very slight degree of attention suffices to identify it. Hence such a mistake hardly ever occurs, except where the thing apprehended is so strange or unexpected as to astound us, and consequently withdraw the attention from the actual apprehension.

When we have observed the characteristic marks of an apprehension, any mistake regarding its reality is impossible. When we feel heat or cold, for example, or perceive certain colors, the only question is, whether they are not mere ideas, as the reality of the comprehension admits of no doubt. What causes it, is quite another matter, which should not be confounded with its reality. In order to test this, we have only to ascertain whether it possesses those peculiarities which distinguish it from ideas. If it does, its reality is certain. If a man should seriously offer to prove to us that we felt cold, saw certain colors, or heard certain sounds, when we actually did so, we should justly consider him deranged; and such an offer would be still more ridiculous if we did not so apprehend. It is self-evident that apprehensions cannot exist without being real, that they exist only because they are apprehended, that they are precisely as they are apprehended, and that, unless they were apprehended, they could not possibly exist. Their being apprehended, therefore, necessarily implies their reality, just as they are apprehended.

Apprehensions should not be confounded with their causes, which are widely different things, but with which they are very liable to be confounded. Apprehensions are purely mental phenomena, while their causes are external things totally different from the mind, and sometimes at a great distance. Two persons may be very differently affected by the same objects; yet this does not render the apprehension of each a whit the less real. When we see and smell a rose, the colors we perceive exist only because we perceive them, just as much as the odor we feel exists because we smell it. The flower has nothing either in it or on it like the colors we see, any more than it contains something like the odor which we feel, but only something widely different, which causes these apprehensions.

In regard to the inferences deducible from apprehensions, we are liable to err, as in the case of other inferences; and this is the real source of most of the errors commonly attributed to the senses. When a person touches a bullet, with the points of his fingers crossed, he thinks there are two; yet he does not actually perceive two. His apprehensions, however, partly resemble what he generally perceives when there are two; and hence he hastily draws an erroneous inference. So, when we first see the Sun in the east, and gradually more to the west, we are apt to assume erroneously that we see it moving westward, whereas a little consideration will show that we see no such thing. All that we observe is, a change in the relative position of the sun and the direction of sight, while we hold the same apparent position. Now there is such a change: but whether it is owing to a motion of the latter or of the former, or of both, the senses say not. Nor do we see the road running away, when we look out behind a vehicle in which we are traveling. We can see things only as they apparently are at the present moment; and, therefore, we cannot possibly see a body moving. We think we do so, only because we confound our inferences with our perceptions.

In all such cases as the preceding, the error has generally been attributed to the senses, whereas they are errors of reasoning, and the senses indicate nothing but what is strictly true. To guard against such errors, therefore, all we require to do, is, to test the validity of the reasoning, as in other cases. When this has been done with the requisite degree of care, the certainty of the inference is established, beyond the possibility of a rational doubt.

The validity of inferences from apprehensions can be tested, not only by a strict examination of the reasoning processes, but also by comparing our conclusions with those of other persons, or those of one sense with the evidence of another. Thus, if I see the likeness of a deceased person, I may feel with my hand, or a cane, whether there is really any such being where he appears; or I may inquire of others whether they see anything there; and if I suspect that I hear a sound, without any impression being made on the ear, I may look whether there is any sounding body within hearing, or ask others whether they heard such a sound. We can also, in many instances, indirectly determine the truth, without any such appeals as the preceding. Thus, if I suspect that I see single objects double, I have only to look at my right hand, or some other object which I know to be single, and observe whether I see it double or single.

§ 2. PRIMARY MENTAL PROCESSES BY WHICH CONTINGENT KNOWLEDGE MAY BE ACQUIRED.—Our own Existence implied in our Apprehensions.—These often independent of our Volitions.—Necessary inference.—Changes caused by our Volitions.—How we distinguish Ourselves from other beings — learn the condition of our organs, through one sense—and move them at pleasure.—How we know the existence of other Substantial Beings.—Why mankind attribute the phenomena of Apprehension to their true causes, notwithstanding certain errors.—What these are.—Extrinsic and Intrinsic Properties.—Principal kinds of each, and how learned.—Particular means of learning Intrinsic Properties, from simple Observation.—Causes of the Contingent Properties of Inanimate Substances, and of Living Beings.—Connection of Apparent and Real Similarity.—Means of enlarging our personal Experience.—Acquisition of Language.

My apprehensions, of which I am immediately conscious, necessarily imply the existence of the substantial, living and thinking self; for I know intuitively that thought cannot exist in a nonentity, and that a being which thinks, must be substantial, living and thinking. When I view the sky or the fields, for example, I am immediately conscious of certain apprehensions which, I know intuitively, cannot be discerned by a nonentity, and must exist in a substantial being. This being must be capable of discerning the phenomena, or, in other words, he must be a living being; else he could not discern them. A dead rock or piece of wood cannot discern anything, and much less can a nonentity, or mere vacuity.

Some apprehensions are pleasant, and others painful. The former excite a desire that they should continue; and the latter excite aversion, or a desire that they should cease. Yet the actual result is often otherwise: the former cease, and the latter continue, in spite of my volitions to the contrary. If a person passes me with a basket of fragrant flowers, the agreeable odor ceases when they are removed, however much I may will that it should continue; and when a wasp stings my hand, the pain does not cease for some time, although I strongly will that it should. In many cases, changes occur while I have no desire or volition either way, as when I view a flowing stream, while I am quite indifferent whether it flow or not. Thus I learn that many of the changes which occur around me, are wholly independent of my wishes or volitions.

Every change which I experience, must originate either from my own volitions, or from spontaneous motions of my parts, or from one or more other beings; and, as no being destitute of thought can originate motion, my parts will not move spontaneously, unless they are the seat of distinct thoughts. In that case they would cease to be a part of myself, since a being possessing distinct thoughts and power of motion, must evidently be a distinct being, and form no part of myself. All those changes, therefore, which I experience, independently of my volitions, are produced by one or more beings distinct from myself.

On the other hand, many changes are produced by my own volitions, including all my words and voluntary actions. This appears conclusively from their uniformly following and conforming to all my volitions. For the phenomena exhibit so much uniformity and regularity, in an endless variety of circumstances, that casual or chance agencies are wholly excluded, and no other being would so obsequiously anticipate all my wishes, and expend such an immense amount of skill and labor in deluding me, to no purpose, as he could have no possible motive for the deception. Moreover, the supposed being would possess contradictory attributes. He would be benevolent, since he often operated to procure me enjoyment—and he would be malevolent, since he often deluded me, and led me into severe and lasting pain, by gratifying my wishes. The being could not design to improve me by discipline, as I should be only a passive recipient of whatever he chose to bestow or inflict.

Those changes cannot be caused, without any external reality, by different beings, some benevolent and some malicious: for either the stronger would exclude the weaker from all control, or, if they were equally strong, the one under whose power I fell, would retain his supremacy. An indifferent being would evidently not interfere with me at all.(11)

I learn the limits of my own person by observing that its parts are all firmly connected, and that they uniformly and immediately obey my volitions, without any apprehensible command or request, which no other object does. If I will to move my foot, under the belief that it exists, and is connected with me, I notice a change in the position of the colors, and feel some new sensations, as if I had actually moved it: but if I will similarly regarding any distinct object around me, there is no such change.

So, if I will to move from the chair on which I sit to another, my limbs all move in harmony to my new position, while my former seat remains precisely as it was. Those objects are evidently parts of myself which are inseparably connected with me, and immediately obey my volitions; and all other objects are not parts of myself, as they are unconnected with me, and do not move uniformly and directly in accordance with my volitions, save when they are attached to me by artificial means, or in immediate connection with some of my parts.

I learn the position and other peculiarities of my organs, at any instant, through a particular sense, by first noting the perceptions and sensations which I experience when some other sense informs me of those peculiarities. Thus, I know the position of my right hand, at any instant, independently of sight or touch, by first marking the apprehensions discerned when I either see its position or feel it with my left hand; and these apprehensions afterwards inform me of its condition, without any aid from other senses, as I justly infer that they are the same as when my apprehensions regarding it were precisely similar.

When I know the condition of an organ of motion, a little experience enables me to move it at pleasure, as I know the very thing to be willed, in order to effect the required motion.

The various changes I experience, which do not originate with myself, must result either from one being or from several. The former supposition is known to be absurd, just as I know that my actions are caused by my volitions. Hence those phenomena must be owing to several beings; and they may produce them either by communicating various motions to me, by means of intervening substances, or by coming into direct contact with me. They can affect me only in one or other of these ways: for it is self-evident that they can cause no change in me without actual contact or a medium. I cannot evidently be immediately conscious of the presence of any external object, unless it is directly in contact with my living self. The immediate causes of those phenomena may be very different from the ultimate, or even the remoter, causes.

The various phenomena presented by the substances around me, imply that some of them possess thought, like myself, and others do not. For the former exhibit changes similar to those which precede and follow my own volitions, while the latter are either uniformly inert, and never act or move save when they are affected by some other substance, or they do so only in one particular way, thus indicating that they are always passive. The supposition of the former's motions being caused by substances distinct from themselves, involves the absurdity already pointed out; and it implies the further absurdity that these beings employ immense pains and skill to delude me into the erroneous belief that there are other beings like myself, without any motive for their doing so.

The only admissible inference, therefore, is, that I am surrounded by a great variety of real beings, some animate and some not, the former of which feel, will and act as I do myself. Each class presents numerous kinds, which differ widely; yet the essential distinction between the two is generally well marked throughout. There is one species of the former to which I evidently belong, as it is precisely like myself, while all the rest differ, some more and some less.

On applying the tests furnished by my various senses, I am only confirmed in those conclusions. If a friend is speaking to me, and walking near me, with a fragrant flower in his hand, I hear his voice and the sound of his footsteps, see his person and movements, and smell the odor of the flower, while he answers my questions or remarks; and if I lay my hand on his head, I feel it, as soon as I see the colors come in contact. At the same time the variations in the tones of his voice, his footsteps and motions, and the odor of the flower, correspond to those in his appearance, as if he were at different distances and in different positions. Sometimes the apprehensions are multiplied by the presence of many persons at the same time.

These phenomena admit of no rational explanation except that they are caused by those things to which they are usually attributed: for the few other possible explanations involve gross absurdities. The phenomena exhibit a degree of regularity and uniformity which wholly excludes the supposition of their being casual or accidental, and that of another being designedly producing them, labors under the difficulties already pointed out. Hence the only obvious supposition regarding the causes of apprehensions, is the true one. It not only accounts for them, with perfect precision, but every other supposition involves impossibilities.

Mankind generally attribute the phenomena of apprehension to their true causes, since these alone are obvious; but this is done, from early infancy, with such ease and rapidity that they overlook the process of inference, and take that to be an immediate discernment which is, in reality, an inference. This error causes difficulties when the subject is attempted to be investigated, because those inferences are sometimes false, which the real phenomena of consciousness never are. Men also fall into the further error of overlooking the other possible explanations that may be given of the phenomena; but this is of no consequence, since a careful analysis shows that these all involve absurdities.

The contingent properties of substances, or those which are not self-evident, are of two kinds, *extrinsic* and *intrinsic*. The former consist of those which are known only by their causing in us certain apprehensions essentially different from anything inherent in the substances, and apparently dependent solely on the form and arrangement of their molecules or atoms. The principal of these are, color, and those which produce the apprehensions of smell, taste and sound. These are made known to us directly, through the proper organs. The color of an object is learned by the eye—the smell, by the nose—the taste, by the mouth—and the sounds which they give forth, by the ear.

The *intrinsic* properties of substances consist of those which are inherent in them, and are not dependent merely on their molecular structure and discovered solely by certain apprehensions produced in us in consequence of that structure. The principal of these are, particular form or shape, size, position, weight or gravity, inertia, mechanical texture, and the various qualities dependent on it, temperature, electric and chemical properties, life, and the various properties connected with it.

We must distinguish the particular and actual qualities of a substance from the general properties of the same kind known by Intuition. We know intuitively that every substance must have some form, size and position: but the actual form, size and position of a substance are contingencies which we must learn from experience, and which cannot be ascertained by Intuition. This class of properties is mostly inferred from phenomena, by various processes, of which the following are the principal, belonging to the subject of this section.

The eye perceives nothing but various expanded colors, which frequently change their apparent forms and positions; and we learn the actual forms, distances and positions of the colored substances by drawing inferences regarding the causes of the apprehensions which they produce.

If we feel with our fingers any object, such as a book, table or chair, and view it in different positions, we find that the outline of the colors presented to the eye corresponds exactly to the outline of the real form, as determined by the touch. The time which the fingers take to move over its different parts corresponds to their apparent size, as exhibited to the eye. The apparent form is such as would arise from the rays of light passing in straight lines from every part of the object to the eye; and as this, in all ordinary circumstances, happens uniformly, we learn that these rays move in straight lines.

We also notice that the shades of color vary according to the particular form of that part of the object which is in sight. We readily distinguish a ball from a flat disc by observing its darker hue towards the edges, while the latter exhibits no such difference; and a little experience renders us familiar with the peculiarities and causes of the variations in the shades of color. Thus we learn to determine form, with the utmost rapidity, from the visible outline and shading alone, without any application of touch.

Painted imitations may sometimes deceive us; and the hues of distant objects are so indistinct that we are apt to draw erroneous inferences regarding their forms. But cases of this kind do not affect the accuracy of our conclusions regarding ordinary objects, within moderate distances; and even there, we have generally some reliable means of ascertaining the true form, without the aid of touch. A painting may often be distinguished from a solid by its not possessing the vividness of nature, and its not changing as we vary the position of the eye. We also frequently know that the circumstances are such as to exclude the supposition of any painting being visible. So the different appearances of the dark spots on the surface of the Sun, as it revolves on its axis, show us, notwithstanding its great distance, that it is a solid body.

We can form an estimate of the size of a body which is quite close to us, by comparing it with that of our hand or foot: and when it is within a moderate distance, we form a judgement from its apparent size, the degree of distinctness in its color and outlines, the number and magnitude of intervening objects, and comparing it with a body near it whose dimensions are known. A little experience shows us the modes in which the appearance of an object varies with its position. But such methods furnish only approximations in any case; and where the body is very remote, as the Sun or Moon, they wholly fail. The exact dimensions of objects can generally be ascertained only by actual measurements and calculations.

The distance of an object cannot possibly be apprehended directly, since space is invisible; and it is estimated in the same way that we judge of its size. Indeed the two properties are so related that a knowledge of one assists us in determining the other. We either form an estimate of its distance from its appearance, and then judge of its size, on the assumption that it is at such a distance, or, if we know its actual size, we can form an estimate of its distance from its appearance.

Another criterion, which may be frequently applied to ascertain the distance of an object, is, the nature of the sounds which proceed from it: for this varies with the distance, which can consequently be determined approximately from noting the character of the sound, as it strikes the ear. The differences in the sounds are learned by noting their character when the distance is known by some other means.

The direction of a visible object may be ascertained by observing its bearing, compared with the line whence the direction is reckoned. Sometimes the direction of an invisible object is determined by observing the quality of sounds proceeding from it, as these affect the ear differently according to the directions in which the sonorous undulations are moving; and we learn the various modifications by observing the character of the sound where the direction is known by sight, touch, or any other means. The form of the ear is such that sonorous undulations affect it variously, according to the directions in which they move.

The direction of one distant object from another may be roughly estimated by determining the distance of each, and then observing the angle which they form with each other, measured from the eye of the observer.

In a void we cannot distinguish one direction from another; and hence the direction of an object can be determined only by observing its position in relation to three or more fixed points. While, therefore, this position continues apparently the same, a change of direction is imperceptible, as when we sit in the cabin of a vessel which changes its direction, we are not sensible of the change, because everything we see around us preserves the same relative position. Where there is an evident change of position, it is frequently difficult to ascertain which object has moved. When I look over the side of a ship, and see the water apparently moving astern, I cannot directly say whether the vessel is moving against the flood, or whether she is at anchor, with the tide or current flowing past. I can ascertain the real fact only by looking at the shore or some fixed object.

The same remark applies to two railway trains which cross each other: and the apparent motions of the heavenly bodies furnish another instance, which differs from these, however, in there being no fixed object to remove the difficulty. We are apt to think that the bodies move, and that the Earth is at rest, because they are apparently much smaller than the latter, and every object around us preserves the same relative position. Our persons appear to stand in the same vertical direction throughout the day; and if we turn towards the pole-star, our two hands seem to point always in the same directions.

When we know the distance and direction of an object, we know its position in space, in relation to ourselves: but we cannot determine absolute position, because we cannot distinguish one part of space from another, and all objects around us may possibly be in motion, and yet constantly preserve the same relative position.

The weight of a substance may frequently be known approximately by observing its momentum or moving force, to which it is always proportional, for a certain velocity; and, in all cases, it may be estimated by observing the effects which the substance has produced by its motions.

The *inertia* of a substance is its tendency to continue in its present state of rest or motion, and its requiring the application of force to produce any change in that state. The general property is learned by simple observation; and accurate measurements show that it is exactly proportional to the weight of the body. Hence the amount of the former is always known from that of the latter.

The mechanical texture of a substance is frequently learned from simple observation. Thus we see that ice is solid, water fluid, and steam gaseous, and that iron is tough and rigid, glass hard and brittle, and moist clay, soft and plastic.

The temperature of several substances may frequently be loosely determined from our sensations, or observing their heating or cooling effects on other substances, and sometimes from their very appearance. Thus, we know that ice is generally colder, and steam warmer than liquid water.

Some of the chemical and electric properties of substances may be learned by observation; but most of them are discovered only by means of experiments.

The peculiar characteristics of living beings may be learned by observing their modes of acting. Life is distinguished by the power of originating or stopping motion, independently of external application or mere inertia; and the differences between the various classes of animals are learned from the modes in which they act, or are affected by the various circumstances in which they are found. Thus, some exhibit cunning, and others, simplicity; some are gentle, and others, fierce; some are strong, and others, weak, and so forth.

We know the nature of our own comprehensions by direct consciousness; and we infer that other beings of our own species discern as we do, when they are placed in the same circumstances with us, and exhibit the same appearances which we present when so situated. We reason upon the principle that the results are the same, where the determining conditions are the same.

The causes of the contingent properties of inanimate substances must be owing to the form and arrangement of their atoms, and their being variously affected by other substances. But we can seldom trace a particular property to any of these causes. We cannot show why gold reflects only the yellow, and grass the green rays, or why nitric acid corrodes iron and silver, while it does not affect gold or platinum.

In order to trace the causes of those properties, we should require to know the atomical constitution of matter: and this we can never do; for however much microscopes may magnify, one of greater power might show that to be porous which formerly appeared to be solid. For the same reason, we cannot trace any connection between the different properties of substances, although such a connection may possibly exist. We cannot determine the tenacity of a metal from its color, nor its fusibility from its specific gravity.

With regard to the properties of living beings, we are, if possible, still more unable to trace them to their causes, farther than we can do by pure Intuition. The causes of the origin and ultimate conditions of life, seem to baffle all human efforts to trace them. We readily learn that certain conditions are requisite to life, and that death ensues when they are violated: but why this is so, nobody can tell. A small quantity of an apparently harmless substance causes speedy death, while a much larger quantity of another substance, apparently much more injurious, fails to do so. All that we can here effect, is, to establish the existence of certain intermediate conditions or causes; and even this is done chiefly by indirect means.

With regard to the question whether substances apparently quite similar are so in reality, we justly argue that they must generally be so; otherwise some of their discoverable attributes would differ. Diamond and quartz are frequently similar in touch and appearance; but chemical processes soon show that they are totally different in composition: and, by passing polarized light through them, we may discover some differences in the texture even of two diamonds. Nor can we certainly say that there may not be other differences which we cannot detect; yet there must be a general similarity in the structure of all diamonds; else they could not possess so many common properties as they do. The same remark applies to all similar cases: for where all the effects are alike, the causes must be alike; otherwise there would be effects without any adequate causes.

Our personal comprehensions necessarily furnish all the primary elements of our contingent knowledge: but we avail ourselves of the observations, reasoning, and experience of others, by means of signs of thought. These are chiefly spoken and written language, by means of which the knowledge of one person may be communicated to all his contemporaries, and transmitted to the most distant times.

We first learn spoken or oral language by a close observation of the usages of those around us. The child learns the names of the visible objects around him, by hearing them repeatedly applied, where he knows the object designated; and he learns the names of qualities, by hearing them expressed by certain terms, where the things meant are obvious. The significations of verbs are acquired by observing the words applied to denote what is present to the senses, or to give an order which is immediately executed: and the significations of the less abstract words belonging to the other parts of speech, are acquired in the same way. These attainments amply suffice for extending his knowledge of the subject, by means of information derived from others, or marking their usages either in spoken or in written discourse.

A knowledge of the vernacular enables us to learn the experience of those around us, and to avail ourselves of their comprehensions as if they were our own, while a knowledge of writing places within our reach all the most important facts known to mankind.

§ 3. PRIMARY EXTERNAL PROCESSES BY WHICH CONTINGENT KNOWLEDGE MAY BE ACQUIRED.—Simple Observation frequently insufficient.—Standard of Weights and Measures.—Method of repeating and taking a Mean.—Requisite to render it satisfactory.—On what assumption based.—Its Advantages.—Method of Approximation.—Method of Extension.—Sometimes combined with Repetition.—Means of measuring very small spaces.—Things which cannot be accurately measured.—Aids of Sight and Hearing.—Various means of testing results.—Experiments.—Of two kinds.—General objects of logical Experiments.—One often subservient to several objects.—Where Experiments are generally requisite.—Relation of Experiments to Comprehensions.—Use of visible Symbols.—Curves.—Application of Symbols.—Tangible Writing.

The methods discussed in the preceding section do not, in many cases, furnish a sufficiently precise and extensive knowledge of the subject; and, therefore, we require the aid of several external processes. In observing quantity, for example, we can generally form only a rough estimate of its amount from simple apprehension; and when we wish to ascertain the exact amount, we must have recourse to numeration, measurements, and calculations.

If a man sell a flock of sheep, at so much a head, he can possibly tell, at a glance, that there are more than one and less than two hundred: but neither he nor the buyer can tell the exact number, by this means. In order to do this, the sheep must be counted. This process must be adopted whenever we desire to ascertain the exact number of single things contained in an aggregate of individual objects, exceeding a few: and, in effecting it, direct and continuous numbering may often be abridged by means of the processes of Arithmetic. Thus, the population of a town is ascertained by first counting that of the various subdivisions, and then adding together these items, the sum of which is the total amount. So we may ascertain, with sufficient accuracy, the number of pores in the whole surface of the body, by counting the number within a square inch, in different parts, till we ascertain the average, and then multiplying this by the number of square inches in the whole area of the skin.

The rules of Subtraction and Division are equally serviceable, when we require to ascertain differences and quotients or aliquot parts: and, by combinations of the simple processes, all ordinary numerical problems can be readily solved, from facts learned by direct numeration or counting.

Sec. 3.] Primary External Processes. 87

In order to measure quantity accurately, some uniform standard is requisite: and we are furnished with it by means of the unvarying time which the Earth takes to perform a revolution on its axis. This enables us to determine the mean length of a solar day, which is the more immediate standard of time. We have no direct perception of the flow of time; and hence we do not know what time is occupied by a revery or a dream. We can, indeed, form some estimate of the lapse of time by noting the number of objects of which we have thought during the interval, or the amount of work done, or the distances traveled by us, or bodily sensations, which indicate a particular period of the day or year, and so forth. But such methods do not possess the accuracy required for many purposes. In order to this, we employ time-keepers, which are regulated by the apparent diurnal motions of the fixed stars, corresponding to the real diurnal revolution of the Earth.

By taking a pendulum which swings so many times in a mean solar day, under specified circumstances of position, heat and atmospheric pressure, we are supplied with a standard of length, which serves equally for measuring surfaces and solids. Then, by taking a certain solid measure of pure water, of a given temperature, and under a given atmospheric pressure, we have a standard of weights.

The ordinary modes of measuring and weighing are sufficiently accurate for common purposes. But, in many scientific processes, the instruments require to be constructed with the utmost accuracy; and, after using them with all practicable exactness and care, some expedients are still employed to eliminate errors.

Accuracy is frequently obtained by *repeating* a certain measurement, with all possible exactness, and then taking the *mean* of the whole. Thus the diameter of the Earth has been determined from various accurate and independent measurements, no two of which gave precisely the same results, although they differed by less than one tenth of a mile; and, by taking the mean of all those measurements, a result is obtained more reliable than that deduced from any single measurement. So the parallax of the Sun has been similarly determined, to a great degree of accuracy, from different observations of the transits of Venus over its disc.

To render this method quite satisfactory, one measurement must be as reliable as another; for, if some were made carelessly or with inferior instruments, they should evidently be excluded from those employed in determining the mean. The process is based on an assumption which experience shows to hold true, and which the circumstances of such cases warrant us in assuming, namely, that, in a great variety of measurements, all performed with equal care, with equally good instruments, and without any peculiar difficulties one way more than another, errors in one direction are very nearly compensated by those in the contrary direction, some falling just as much short as others go beyond the truth.

This method may obviate errors due to defects of instruments, as well as those arising from inaccuracies of observation; and it sometimes enables the observer to eliminate errors due to defects in his instrument even without any aid from a second. Thus, if we carefully measure the angular distance between two stars, on different parts of a graduated circular arc, and take a mean, we obviate any small error arising from inaccuracy of graduation. So, if we weigh a body in the different scales of the same balance, we can determine its actual weight, although one arm of the balance should be longer than the other. Again, if we first balance the thing to be weighed with sand, and then replace it with weights, we determine its exact weight, independently of all the defects of the balance, provided only that it is easily moved by a very small weight; for, as the circumstances are the same in both cases, the sand must balance equal weights.

Approximation is another method of aiding apprehension. The quantity sought is first found approximately from observation; then, by means of this result, we find another quantity, which differs less from perfect accuracy, and so on, to any required degree of exactness. Instances of this method are furnished by the ordinary modes of finding the successive figures of dividends and roots.

Another method of aiding apprehension is, to *extend* our observations, so as to include many similar cases, or such as are separated by wide intervals of time or space. Thus, if we wish to know whether granite is of igneous origin, we examine the whole series of similar rocks, and notice a gradual change, through the serpentine and trap, till we come to the modern lavas, which are directly known to be of igneous origin, whence we conclude that granite had a similar origin. So, in determining the exact length of the year, if we have two observations made at an interval of a thousand years, the errors of observation will be so distributed that the result will vary from the average length of the year, during the interval, only by the thousandth part of their sum, whereas, if the two observations were made at an interval of one year, the result would vary from the truth by the whole amount of those errors.

Sometimes the method of extension may be combined with that of repetition, so as to secure the advantages of both. Thus, in measuring angular spaces with the reflecting circle, the angle is repeatedly measured in such a manner that the error arising from defects of graduation is constant, while the final measurement is the result of all, and secured against errors of observation as in the method of simple repetition; and as that of graduation is equally distributed through all the measurements, it may be made, by means of the extension, as small as the observer pleases.

Very small spaces are generally measured by means of such contrivances as a vernier and a micrometer. But the same purpose is sometimes effected by particular artifices. Thus, the diameter of a very slender thread or wire may be determined by laying ply beside ply, till they exactly cover some small known space, as the sixteenth part of an inch; and if we find it takes thirty of them to do so, we know that their diameter is the four hundred and eightieth part of an inch.

Many things hardly admit of any greater accuracy of measurement than simple observation affords, such as the intensity of a color, roughness and smoothness. With respect to all feelings, whether sensations or emotions, they evidently admit of no measurement. But, in all such cases, Comprehension enables us to determine the greater from the less within narrow limits; and this is all that we require, in such cases, for practical purposes. Thus, although we cannot ascertain that one green hue is twice or thrice as deep as another, yet we can distinguish the various shades, with great accuracy. So, we know that our sensation of pain is much stronger in the case of a severe burn, than in that of a slight abrasion of the skin.

Sight and hearing, the two most prolific sources of apprehensional knowledge, are directly assisted by instruments. The speaking trumpet, by concentrating the aerial undulations, and the hearing or ear trumpet, by collecting them, enables us to hear distinctly sounds otherwise inaudible: and the simple device of changing the direction of the rays of light, by means of some refracting or reflecting medium, enables us, on the one hand, to discover around us innumerable wonders otherwise invisible, and, on the other, to explore the regions of immensity, and countless systems of worlds unseen by the naked eye. The microscope is employed, not only as a means of discovering things otherwise indiscoverable, but also to measure small spaces with accuracy; and this application of it forms a marked period in the history of Astronomy.

The accuracy of results may often be tested by comparing them with those obtained by different processes for effecting the same end, or by observing whether they lead to known truth or falsehood. Thus, astronomical calculations of eclipses and other celestial phenomena, may be compared with subsequent observations, and arithmetical calculations are verified by reversing the operations, and observing whether we arrive at correct results. So the accuracy of a whole trigonometrical survey may be verified by comparing the calculated with the measured length of the last line, which is, therefore, termed "the base of verification."

Another process, which is not only a means of testing results, but frequently a most important means of acquiring a knowledge of primary facts, is, *experiment*, which consists in operating with things, or placing them in peculiar positions, that we may mark the result, and thus illustrate a proposition, solve a difficulty, or discover some unknown truth, or some new means of effecting a known end. Experiments are either *didactic* or *logical*. The former consist of those which are performed for the purpose of illustrating or demonstrating known truth to learners: the latter comprise such as are made for the purposes of discovery or invention; and it is with these alone that we are concerned at present.

The immediate objects of most logical experiments are, to determine the amount of a certain thing, or one or more of its intrinsic properties, or what causes produce known effects, or what effects are produced by known agencies. Many experiments subserve two or more objects. Thus, experiments on the composition of water determine the component elements, the quantity of each, and the circumstances under which they combine and separate.

Experiments are generally requisite wherever properties, agencies or operations are hidden beyond the reach of direct observation or measurement, and are discoverable only by testing or trying them, with the aid of all that we previously knew of the subject.

The nature of the phenomena of Comprehension can be known only by direct observation. Thus, nothing but the actual apprehending can enable us to know the nature of our apprehensions when we smell a flower or hear a sound; and the ultimate processes of apprehension are an inscrutable mystery. But, in determining the causes of phenomena, experiments are frequently of much use. Thus, if I doubt whether the table before me actually exists, I may attempt to strike it, and observe whether I experience new apprehensions when the colors of my hand and the table come in apparent contact. So we may sometimes learn how certain things affect the mind, by exposing it to their influence; and, by excluding some particular agency, we can occasionally ascertain how much is due to its influence in other cases.

In many investigations, such as the processes of Mathematics and Physics, little progress can be made without the aid of visible symbols, owing to the great difficulty of otherwise remembering the various steps of a process. Thus, ordinary geometrical propositions or dynamical theorems cannot be satisfactorily investigated without the aid of figures or symbols denoting quantity. So the laws of many variable quantities can be neither discovered nor effectually remembered, without expressing the several values either in tables, or by curve lines, whose distances from a point or a straight line vary as the quantity. Such are, the variations of atmospheric temperature and pressure, of the magnetic needle, tide-waters, the expectation of life at different ages, and the progress of population in a community. Thus, if we express by a continuous curve the height of the thermometer, at every hour of the day, we can form a correct estimate of its diurnal variations; and a similar curve, representing its daily average height, furnishes the same advantage in regard to its annual variations. So a knowledge and remembrance of the leading events of History, are much facilitated by synchronistic tables.

In all cases of this kind, we can dismiss from our minds, for the time being, the things denoted by the symbols, and concentrate our attention on the latter, while we can return to the former whenever we require to do so. We remember the general import of the symbols, during the investigation, so as to use them aright; and we can afterwards recollect the particular things which they denote. Thus, in an algebraic investigation, we can remember which letters denote the known and which the unknown quantities: and, after the investigation is concluded, we can easily return to the particular quantities which every symbol denotes.

Such devices as the preceding are frequently employed to determine many points which might seem to be matters of direct observation. Thus, Kepler discovered that the planets revolve in ellipses, with the Sun in one of the foci, by representing their distances from it, in various parts of their orbits, on paper, drawing a curve line through all the points of observation, and then determining the nature of the curve, and the exact position of the Sun within it.

Curves are extensively employed in those sciences which treat of variable quantities. They first assist the observer, not only in discovering the laws of variation, but also in eliminating errors of observation: for, as abrupt transitions seldom occur, in such quantities, a mere inspection of the figure will often enable him to detect errors, by the want of symmetry and regularity in the curve. Thus, if one point of a planet's orbit be found a little without, and the part immediately adjacent a little within an ellipse, the apparent discrepancies might be safely assumed to arise from errors of measurement, since no such deflections from its former course can be attributed to the motions of the planet; and hence it might be concluded that the true path is an ellipse. After the laws have been discovered, those devices facilitate both the remembrance of them and an understanding of them by others.

Some representations are as perceptible to the touch as to sight; and these are ingeniously applied to communicate knowledge to the blind. Thus, by means of raised, instead of colored, letters, these unfortunate persons are furnished with books which they can read by running their fingers over the letters, instead of seeing them; and practice enables them to do this with surprising facility.

CHAPTER IV.

OF THE PRIMARY MEANS OF RETAINING KNOWLEDGE.

§ 1. RELIABILITY OF MEMORY, AND MEANS OF AVOIDING THE PRIMARY ERRORS WHICH IT TENDS TO PRODUCE.—Use and phenomena of Memory.—Their possible sources.—How the true one is established.—Indirect proofs of the reliability of Memory.—Common errors.—Nature and requisites of Recognition.—Cases in which these generally exist, and in which they fail.—Phantasms.—Why Imaginations are sometimes mistaken for Ideas.—How this error may be avoided.—Consequence of other Similitudes recurring like Ideas, and of the reality of all Similitudes.

OUR immediate knowledge of all contingent truths is confined to the present moment. Thus, we can neither see nor hear past or future events, which are made known to us only indirectly, through their being connected with the present, by means of Memory, Reason, Abstraction and Conception; and the aid of the first of these faculties is generally requisite, in all such cases, to enable us to pass beyond the present.

When objects of thought formerly apprehended, are again presented to our observation, the ideas of them arise spontaneously before the mind, generally accompanied by those of other objects apprehended at the same time. These ideas sometimes completely correspond to the present reality, and sometimes there are slight differences: but there is generally a close and marked resemblance. In apprehending objects for the first time, no such ideas present themselves. Many such ideas, again, occur spontaneously, according to certain laws, whereas, in order to form conceptions of things never apprehended, we find that a conscious, if not a laborious, effort is requisite. These facts admit of no other rational explanation except that we previously apprehended the prototypes of the spontaneous ideas, but not the other class of objects, and that we are still the same persons.

The spontaneous ideas must evidently arise from some

fortuitous peculiarity of the organization, or from some conscious being designedly producing them directly either in the same person who formerly apprehended the prototypes or in a different person, or from our having actually apprehended the prototypes of the various ideas, and our being so constituted that ideas of objects once apprehended recur spontaneously, in consequence of being related to some other thought. This last supposition accounts for all the phenomena; and each of the other two involves an absurdity.

The corresponding ideas are so numerous, and, in many cases, consist of so many different parts, that a fortuitous production is absurd, since it would be a change without any adequate cause. This becomes very evident from the fact that a long succession of ideas often arises before the mind while we actually apprehend their prototypes, and in exactly the same order. Thus, if we view a well-known landscape, the ideas of the various objects spontaneously arise, as the eye beholds the successive parts, until we see the whole at a glance, when the mental likeness becomes equally complete. The supposition is rendered yet more absurd by the fact that the apprehending of an object often calls up, not only a single idea of it, but also the similitudes of all our previous discernments regarding it. Thus, the sight of a well-known scene recalls the many views we formerly had of it, all of which present themselves to the mind simultaneously, or in very rapid succession.

The second supposition, also, involves absurdities. For the being which produced the ideas in us, would be benevolent, since he often caused us joy, and also malevolent, since he often caused pain. He must also be desirous that we should know, since he labored so much to produce elements of knowledge; and, at the same time, he must be desirous to mislead us, since we actually infer, as the most obvious explanation, that we apprehended originals of those ideas. These objections apply still more forcibly to the supposition that he produced, in the minds of one person, ideas corresponding to the apprehensions of another.

That ideas correspond to their prototypes, is proved directly by experience; for we often find that the apprehensions were actually such as the ideas indicate. Thus, I have the idea of a certain writing, which I made in a book yesterday; I turn to the book, and there I find it, exactly as the idea indicates. Again, I have the idea of a book laid in a certain place; I turn towards the place, and there I see it. So the ideas of numerous instances, in which Memory was found faithful, often arise before our minds, and confirm its faithfulness.

As the true explanation of the phenomena of Memory is the only obvious one, and we are habitually accustomed, from our earliest years, to draw the necessary inferences, with the proverbial rapidity of thought, we are apt to think that we are immediately conscious of the reality of things remembered: but it is self-evident that we cannot be conscious of a past, any more than of a future, contingency. Mankind act here precisely as in the case of apprehensions: they draw legitimate inferences, but mistake them for immediate discernments, and overlook other possible, though really absurd, suppositions or explanations.

We *recognize* an object when we find, on comparison, that the idea of it exactly resembles it, or very nearly so. If the object is not subject to change, there must be complete resemblance: but if it be subject to gradual changes, like most organic beings, we consider whether the difference between the idea and the apprehension is not such as time may have produced, since we apprehended the object. Thus, if we have not seen a boy for three years, we make allowances for his change of stature and general appearance.

When the idea is not very clear or complete, we are apt to commit mistakes. Thus, we frequently take a person not well known or long absent, for one who closely resembles him. In order to recognize an object with certainty, it must possess some peculiarity which distinguishes it from all similar objects; and we guard against the error of mistaking one for another by noting carefully those peculiarities, so that they may be remembered, and afterwards observing whether the object in question possesses them. Thus, a person otherwise greatly changed in appearance, is often recognized by some scar or mark, which distinguishes him from all others. In living beings, distinctive peculiarities are generally found without difficulty, every one having something in form, color, voice, gait, or aspect, by which it can be readily identified. This is also the case, to a great extent, even in the vegetable creation, and in most inorganic natural objects. There are no two trees or valleys in the world which cannot be readily distinguished from each other. But works of art frequently resemble each other so closely that we cannot, with certainty, distinguish them. In such cases, however, mistakes are generally, though not always, of little consequence.

Similitudes of conceptions recur like those of ideas; and thus we know what were our former conceptions. But as conceptions are composed of similitudes or their modifications, their phantasms are nearly as vivid as the original elements; and hence error is apt to arise from mistaking them for ideas. This is particularly apt to occur where the phantasms have long been considered attentively, so that they acquire the vividness of ideas, for which consequently they are sometimes mistaken, as where a man gives a fictitious account of his own personal adventures, with an evident belief in their reality.

To avoid such errors, we have only to recall the ideas of the circumstances under which we first conceived the prototypes. Conceptions are always produced by conscious efforts of the Will, which distinguish them from apprehensions or ideas. When our remembrance of the prototype is so faint that we do not certainly know whether it is the similitude of a conception or an apprehension, we cannot determine simply by Remembrance whether we originally apprehended or merely conceived; and we must have recourse to some external means, in order to determine the truth of the case.

The similitudes of all other thoughts follow the same laws of recurrence as those of ideas; and hence Remembrance enables us to know all our former thoughts.

We are as conscious of the reality of similitudes as we are of that of their originals; and, therefore, we can reason from the latter with as much confidence as we do from the former, while error must arise solely from drawing fallacious inferences.

§ 2. PRIMARY PROCESSES BY WHICH KNOWLEDGE IS RETAINED.—Means of knowing past Contingencies.—How we know the Time and Place of apprehending.—Forgetting.—Different simultaneous ideas of objects.—Recollecting.—Various kinds of External Signs.—Principle of their operation.

Past contingencies are known by means of things present which are signs or indications of them, the things to be remembered being so connected with the signs that a knowledge of the latter leads to a knowledge of the former. These signs are either *Internal* or *External*.

Internal signs consist chiefly of similitudes, the general nature and operation of which have already appeared.

The particular time and place of apprehending are determined by means of the ideas of objects apprehended simultaneously: and if these do not appear, we know only that we apprehended the object, and cannot say when or where. The faculty of remembrance, being wholly dependent on similitudes, cannot act where the latter cease to arise, in which case we are said to *forget* the apprehension.

Where things have been repeatedly apprehended, the several ideas of them which appear simultaneously, sometimes differ. Thus, when we see a person whom we have seen in health and sickness, the ideas of his different aspects frequently appear together. But the ideas of the other objects apprehended on the different occasions recall the various circumstances, and thus rather strengthen remembrance and corroborate its testimony, than produce confusion or difficulty.

Although ideas arise spontaneously, they are always suggested by some other object of thought, which is so related to them that thinking of the latter leads us to think of the former. This peculiarity enables us to recall ideas indirectly when we have lost the power of doing so directly. Thus, we may have forgotten where we saw a certain person, so that we cannot directly determine the place: but we may know it was on such a day, and, by recalling its transactions, the idea of the person may be brought up, with that of the place where we saw him. In such cases, we are said to *recollect* our apprehensions.

External signs consist chiefly of direct likenesses of the things to be remembered, symbolic representations, either of the things or of speech, and phonetic signs of words.

Direct likenesses consist of sculptures and casts, which are formed precisely like their originals, and of drawings, paintings, engravings, or photographs of the things to be remembered, which only represent, on a smooth surface, their appearance in certain positions.

E

Symbolic representations of *objects* represent them by means of some analogy or relation which they bear to the thing represented, as where a science is symbolized by a female figure, or a curve is employed to point out the different values of a variable quantity, or a great event or character is commemorated by a monument, or periodic acts and ceremonies.

Symbolic representations of *words* represent them by their having some real or fancied analogy to the thing denoted, or their being arbitrarily chosen for that purpose, such as—&, ?, !, 1, 2, 3, +, —, √.

Phonetic signs consist of characters which represent, not the objects of thought, but the simplest elements of speech, as a, b, c, &c. As those elements are by no means numerous, a few characters suffice to represent the whole of spoken language.(12)

All external signs operate on the principle that *the perception of the sign reminds us of the thing signified.* The two things are so connected that when we perceive the one, Memory calls up the other.

CHAPTER V.

OF GENERALIZATION.

§ 1. NATURE OF GENERALIZATION.—Definition of Generalization.— Conceptions always particular.—All science dependent on Generalization.—Distinction between the formation of Conceptions and the use of General Terms.—What the latter denote.—How their meaning is learned.

Generalization is, discovering, by means of particular cases, general truths, or propositions which hold true of a whole class, such as—" the lion is carnivorous"—" the Roman emperors possessed absolute power"—" men are mortal," and—" fish live in the water."

Not only is every real object in nature individual, but our conceptions also are equally particular. Not only is there no general tree, river, house, or bird, in the world, but we cannot even conceive such things: we cannot form a notion of a tree that has no form, size or color, nor of one that has several forms, sizes and colors. Such conceptions are evidently impossible: and when we conceive of a thing as having particular attributes, the conception is as particular as any apprehension. We cannot conceive either of a substance destitute of attributes, or of one that possesses incompatible attributes.

Other things denoted by common names are no more general than substantial beings. Thus, there is no general red, blue, hardness, death, justice, fraud or geometry; and we cannot form a conception of any such thing. We cannot form a conception of a red color unaccompanied by any particular substance that is red, or of death apart from any particular scene of death, or of justice apart from any particular act of justice, and so forth. Hence it appears that, without the aid of Abstraction and Generalization, our knowledge of nature would be confined to individuals, and science could not exist.

We must not confound the formation of conceptions with the use of general terms, or words that apply equally to every one of a class of objects. Such terms are often used without our having any immediate comprehension of what they denote: but this does not, in the least, prove that there are general conceptions. When the word "mountain," for example, is mentioned in discourse, we may possibly think of some particular well-known mountain, or of several mountains in quick succession, or think of no mountain at all. The last supposition frequently holds true, where something besides the thing meant by the word is forced on our attention, as in the expression—" the word *mountain* is of the singular number." Here the attention is apt to be wholly occupied with the words; and a similar remark applies to those cases where we do not require to refer to the meaning of certain signs, during an operation, after fixing them at the beginning of a process of reasoning, as generally happens in Algebra.

A *general term* is simply a word which is equally applicable to a certain attribute, action, or relation, wherever it exists, or to every individual of a class. The meaning of such words is learned either from observation or definitions. After noticing a few cases or specimens, we learn the nature of the thing signified; and formal definitions often answer the same purpose.

§ 2. PRINCIPAL PROCESSES OF GENERALIZATION.—(1) Abstracting regarding some common observed attribute.—Naming Classes.—Empiricisms and Inductions.—Requisites to the latter.—Comparison.—(2) Generalization from identity of agencies or conditions.—

100　　　　　　Generalization.　　　　　[Chap. V.

Requisites, in such cases.—On what this process is based.—(3) Reasoning from the attributes which an individual possesses in common with a class.—(4) Proving that all individuals of a class have certain attributes in common.—Attributes embraced in Definitions.—Uniformity resulting from uniformity of the Determining Agencies.—Consequences of the Character of the Deity.—Exceptions.—Limits of physical Inductions.—Common Error.—How the Uniformity and Stability of Nature is logically established.—Means of distinguishing Specific from Individual Peculiarities.—Attributes common to a Species.—Means of detecting Anomalies.—Principles by which Inductions are established.—Why Intuitions do not require generalization.—What constitutes an Induction.

The following enumeration includes the most common and important of the processes of generalization:

· 1. We examine several objects, and compare them, either by direct simultaneous inspection or by the aid of Memory, and notice, by means of the faculty of Abstraction, the attributes which they possess in common: and then we express these in a general proposition, by the process of combination. In surveying animals, for example, and abstracting as to their external forms, we observe that some, though differing widely in other respects, all agree in having four feet, as oxen, sheep, dogs and cats; some have four hands, as monkeys and baboons; some have two feet and two feathered wings, as hens and hawks; and man alone has two hands and two feet.

We may now express the possession of the common attribute by a general proposition, including all the particular ones, and give every one of those classes a name, equally applicable to every individual belonging to it, and distinguishing it from all others. Thus, we may call the first class *quadruped* or *four-footed*—the second, *quadruman* or *four-handed*—the third, *feathered*—and the last, *biman* or *two-handed*. Any other intelligible words would suffice for the purpose of generalization, such as *simia*, instead of *quadruman*, *bird* or *avis*, instead of *feathered*, and *man* or *homo*, instead of *biman*. There is a manifest advantage, however, in having a term which denotes the essential peculiarity of the class, although this cannot frequently be obtained without inventing a new word.

In the same way we generalize regarding actions, modes of action and relations. We see, for instance, a man, a dog, and a horse, all moving swiftly with a peculiar step, and we denote this kind of motion by the term *running*. The process of naming usually follows the discovery of the general attribute: but it is no essential part of it, and properly belongs to what is termed classification, of which we shall treat hereafter.

By this method we generalize regarding all that we have actually comprehended, which constitutes *empirical* generalizations, or *empiricisms:* but to pass beyond these, to *scientific* generalizations, or *inductions*, requires other methods.

An *empiricism* is, a generalization which includes only cases actually experienced. An *induction* is, a generalization which is proved to extend to the whole class to which it relates. "All the horses that I ever saw, were four-footed," is an empirical, and "all horses are four-footed," is an inductive, proposition. To establish an induction, it is requisite, not only that all the things observed should harmonize with it, excluding only such as are mere exceptions to the general rule, but that we have some satisfactory reason for extending the empiricism to the whole class, in which case alone it becomes a scientific generalization.

In comparing external objects, some or all may be absent: for the ideas of them furnished by Memory, if sufficiently distinct and accurate, answer the same purpose as the original perceptions. In *comparing*, we only observe two or more things simultaneously, and examine their agreements or differences, by the aid of Abstraction: and hence the faculty of Comparison is only a combination of Comprehension and Abstraction.

2. As the same conditions or agents, operating in the same circumstances, must always produce the same results, we generalize regarding these by ascertaining what they are in one instance. Thus, if we find, by experiment, that a certain degree of heat has melted iron, we know the inductive proposition that such a degree of heat melts iron.

In investigating the facts of the particular instance, continued observations or repeated experiments are sometimes requisite, in order to ascertain the real conditions or results: for, if we had only one example, a doubt might arise whether a result was not dependent, at least in part, on other conditions; and, therefore, it is necessary to continue our researches, till we have ascertained the real conditions or results. But where there is no room for such doubts, one observed case is sufficient.

3. By taking a particular object, and reasoning from those attributes which it possesses in common with all others of the same class, we arrive at conclusions which hold equally true of the whole class, on the principle, already stated, that conclusions established independently of certain peculiarities, are not affected by any change in these peculiarities. Thus, to prove that "the square of the hypothenuse of every right-angled triangle is equal to the squares of the two sides containing the right angle," we take any right-angled triangle, and prove that its sides have this property, independently of its particular form, size, color, or position: and then, as the reasoning is independent of these peculiarities, we know intuitively that the conclusion holds equally true of all triangles which agree in being right-angled, however much they may differ in other respects. So a multiplication table is constructed by first counting certain numbers of particular objects; and as the amounts must always be the same, as long as the numbers are the same, the results hold true universally.

Inductions regarding nature are established in the same way. Thus it is shown that "whales are mammals, and not true fish," by examining a single whale, and finding that it breathes air, has a double heart, with warm blood, and suckles its young, which are the essential characteristics of mammals. As all whales have these peculiarities, the particular species examined, its size, age, and so forth, are matters that do not, in the least, affect the truth of the conclusion.

By this means we make an individual represent the whole class to which it belongs; and thus, without actually examining more than a few individuals, we obtain a knowledge of properties common to the whole class.

4. In the preceding case, it was assumed that every individual of a class has certain attributes in common; and we now inquire how this is known.

With respect to such attributes as are embraced in the definition of the class, they must evidently possess these; otherwise they would not belong to it. Every right-angled triangle must have a right angle; and every perfect quadruped must have four feet; otherwise it would not be a quadruped. But nature is independent of our definitions; and, therefore, we must look beyond these, in order to generalize satisfactorily regarding other attributes.

We first observe several of a class, till we are satisfied that the properties which they possess in common, are not owing to individual peculiarities or malformations; and we then infer the generality of these properties from the nature of the agencies which operate to produce them.

We know, from the phenomena which everywhere present themselves to our view, that the same agencies operate throughout the visible creation. Hence it follows, from the principles of causation, that the same regularity and uniformity prevail in things not observed, which have been found in those actually examined: otherwise effects would occur without adequate causes, and different effects would result from the same agencies, operating in the same circumstances. Thus, men now possess essentially the same physical constitution that belonged to the dead, because they have the same origin; and they are surrounded and influenced by the same agencies: hence we justly infer that the living will all die, like their forefathers. As the circumstances and agencies are the same, the results must be the same; and thus we arrive at the induction that "all men are mortal." A better observance of the laws of health will prolong the average duration of life: but it cannot essentially alter our physical constitution, or wholly neutralize the action of the various causes which operate to destroy it, so that all the differences in the circumstances and conduct of individuals, only affects the time when they will die.

This process is applicable only to the ordinary course of events, with which it assumes that no extraneous or peculiar agency interferes. But the Deity may occasionally, for special and important reasons, adopt a peculiar and extraordinary mode of proceeding, contrary to the usual course, or allow certain extraneous or unusual agencies to interfere with the ordinary results: and the occurrence of monsters shows that the latter supposition is true. Hence it is only in reasoning from intuitions, hypothetical assumptions or abstract definitions, that our inductions are rigidly and universally true. Yet, as the Most High is eternal, immutable, omnipotent, omniscient and supremely benevolent, he can never act from caprice, ignorance, weakness, malice, or any new resolve. Hence exceptions to the general laws of nature must be of very rare occurrence, so that they detract nothing from the practical or scientific value of inductions. The unconformable cases are only very rare exceptions to a general rule; and the circumstances in which even these occur, frequently enable us to know that a particular case is no exception.

The divine character informs us that the same uniformity and harmony which we now behold, must have prevailed since the present order of things began, and will continue till its termination, which appears, by various proofs, to be still very distant. But our inductions regarding nature do not extend to what may have been, before the present order of things began, or to what will be, after this planet has ceased to revolve in its accustomed orbit: and even within these limits, there are few physical inductions which have no exception. Most substances have weight; but light, heat and electricity have none: ponderable substances generally tend to move towards each other; but excited electrics, the similar poles of two magnets, and the parts of all compressed elastic bodies, including gases, are repelled from each other: and fluids generally contract, as they cool; but water near the freezing point is an exception.

Where a phenomenon is the result of a constant and uniform agency, its generality is proved by showing the existence and nature of the cause. When we trace the changes of day and night, and the succession of the seasons, to the two motions of the Earth, we discover that those phenomena are as constant as the course of nature; and the uniformity and stability of this is shown in the manner just indicated. Thus are established such general propositions as—"day follows night"—"summer follows winter"—&c., which will hold true till the Sun ceases to shine, and the Earth to move, as they now do.

The constancy and uniformity of nature has been frequently assumed from its having been observed by a very narrow experience, without any clear or thorough knowledge of its real foundations: and hence has resulted the error of the individual making his own experience and views a standard for determining the laws of nature. These can never be logically established without a reference to the character of the Author and Ruler of nature. This is ascertained from particular facts or empirical generalizations; and, when once ascertained, it is applicable to establish, not only the general uniformity and stability of nature, but innumerable inductions regarding particular phenomena. Although not always expressly mentioned, it is always referred to in such cases: otherwise certainty would be unattainable; and we could establish only probabilities. An atheist cannot logically establish any physical or mental science, without assuming propositions which directly contradict the essential peculiarities of his creed.

In distinguishing specific or generic from individual or accidental peculiarities, we are aided by knowing that certain properties are essential to the perfect structure or well-being of a class, while others are not, even although they may have been found common to every one of a class hitherto known. Hence the character of the Deity proves that the class must possess the former, while many of its species or individuals may lack the latter. A traveler of ordinary veracity would be entitled to credit, if he should relate that he had discovered, in some hitherto unexplored region, a new species of crows which were white, or a breed of sheep as small as rabbits: for color and size do not affect the well-being of a species, whose individuals often differ widely in these respects. But, if he should further relate that he had discovered a species of bird which spent the winter in a dormant state, under water, or a kind of sheep which lived exclusively on flesh, we should have good reason to conclude that he was either mistaken or desired to mislead. For such animals could not endure, without a perpetual miracle; and, therefore, we might justly infer that the Creator never formed such beings.

Among the properties common to animal species, may be reckoned the structure of the nervous system, bones, muscles, blood vessels, organs of sense, and digestive apparatus. In vegetables the bark, the wood, and the form and arrangement of the branches, leaves, flowers and seeds, are always alike in a species. In inorganic bodies, the individual attributes are generally common to the species, except form and magnitude, although there are some striking exceptions. One piece of pure iron or lime is distinguished from another chiefly by its form and size. In crystals, even the form is uniform, in the various species.

A moderate acquaintance with the general structure of organic beings, and the harmony of their parts, in a normal state, enables us to detect anomalous productions, with little difficulty. A naturalist who had previously known nothing regarding the horse, would readily see that one with only three legs wanted a limb, while one with five had a superfluous organ. He has abundant proofs that the Ruler of nature is able to render every being harmonious and perfect in its organization; and, therefore, when he finds some manifest deviation from that harmony and perfection, he justly infers the being is of monstrous or defective formation, and by no means a normal representative of its species, while, on the other hand, a perfect specimen is known by its exhibiting no such deviations.

All inductions are established by means of the principles regarding determining conditions and agencies stated in Chapter II.; and all the processes consist solely of ordinary reasoning. They are only inferences more comprehensive than the premises. We infer that a proposition known to hold true in a few individual cases, is equally applicable to the whole class to which these belong. We may reason from probable, as well as from certain, premises; yet this does not affect the nature of the process, which is as uniform as that of sight or hearing, whether the subject be Mathematics, Physics or Metaphysics. The only difference is, that the individual is a perfect representative of a class, which are all alike, in one case, but not always so, in others. If we had a horse, which was as like every other horse as one right-angled triangle is like another, we could demonstrate the properties of every horse as conclusively as it is proved that the squares of the sides containing the right-angle are together equal to the square of the side opposite.

Intuitions are known to be true by direct consciousness, even when they are stated in a universal form; and, therefore, they require no process of generalization, while it is by their means alone that we ascend from the particular facts of comprehension to a knowledge of general contingent truths.

It is not the number of observed cases which constitutes an induction, but the proof that the observed attribute is general, which requires that there be no unconformable cases, except such as are anomalous or irregular. Although new-year's day should be found to have been stormy, for ten years in succession, this would not establish it as an induction that it is always so, whereas the chemical decomposition of a substance, in a single instance, may establish its composition as an inductive truth. In the former case, there is only a fortuitous coincidence, while, in the latter, the previously discovered laws of chemical composition exclude such a supposition. It is even possible that all the observed cases may be contrary to the induction, as where a person who has never seen a normal individual, is shown a number of monstrous or diseased specimens collected by a friend. So one who never saw any horned animal except Shetland sheep, might imagine that three or four horns is the usual, whereas it is the exceptional, number.(13)

§ 3. EXTENSION AND USES OF GENERALIZATION.—Superior and Subordinate Generalizations.—Mode of establishing the latter, and why they are usually discovered first.—Use of Empiricisms and of Inductions.—Advantages of extending Generalization.—Empiricisms distinguishable from Laws of Nature.—Evils arising from confounding them.

In the progress of science, it is often found that generalizations which were formerly considered ultimate and independent, are only particular cases of a more general law, or common effects of the same agency. In such cases, the more general law is termed the *superior* or *higher*, and the more particular law or effect, the *subordinate* or *special* law. The law that "terrestrial bodies tend towards the center of the Earth," and that "the planets tend towards the center of the Sun," are only particular cases of the law of gravitation.

The processes by which we ascend from the subordinate to the higher laws, are similar to those by which we establish the former. But, as the higher include many particular cases, a wider range of observation is generally requisite, and the connecting points are not so easily ascertained. Hence subordinate laws are usually the first discovered. In some cases, however, this order is reversed. The doctrine of gravitation was established before anything was known regarding many of its subordinate laws. As the general includes the particular, the establishing of the superior implicitly establishes all its subordinate laws; but these are not explicitly established, till they are specifically known.

Empirical generalization is necessary, in order to our remembering and reasoning about any considerable number of the countless multitude of objects which fall under our notice: for, if we attempt to view every individual as an isolated whole, and do not attend to those points in which it resembles others, we shall be confounded, amid such a chaotic maze, and our reasoning will be confined, in a great measure, to individuals. The difficulty is surmounted by forming empiricisms, which enable us to investigate an individual as a proper representative of a class, and to comprise many particulars in general propositions.

Empiricism is also, in many instances, a necessary preliminary to induction, which is requisite in order to our performing many of the duties of life, since experience teaches us nothing regarding the future, or that wide and important part of nature to which our experience has never extended. Hence we must go beyond empiricisms, in order to know what will result from certain circumstances, or what we shall find, immediately beyond what we now comprehend.

The advantages of extending our generalizations, so as to include as many particulars as possible, are much greater than we might expect. Not only is the Memory assisted, by being freed from the necessity of retaining a great variety of special generalizations, but the path to knowledge is rendered shorter and more secure, while the results are frequently more abundant. For we have only to deduce all the logical consequences of the general principle, in order to establish them as truths, instead of performing many independent investigations, and frequently relying on doubtful observations or testimonies. Thus, the laws of motion and gravitation enable the investigator of nature to establish a great number of important laws in the physical sciences, many of which would otherwise require treble the labor, or could not be established at all.

Another great advantage of extensive generalization is, that it enables us to take a wider view of the relations which things bear to each other, without which a knowledge of special laws is sometimes of comparatively little avail. For, as these often result from a more general law, the power which we derive from a knowledge of the latter, is as much superior to what we could obtain from a knowledge of special laws, as the power of permanently stopping a fountain is greater than that of drying for a time one of several rills, while their common source continues to flow. If we could discover the cause of some deadly epidemic, and an effectual means of removing it, this would be of much more consequence than the discovery of a medicine which would cure only some forms of the disease.

The extension of induction is of much importance, in many instances, by exhibiting a greater variety of means for effecting the same end. By obtaining, for example, a more general view of the causes which produce a particular effect, we are frequently furnished with a variety of agents for accomplishing the same end, so that when one is not within reach, we can employ the other.

We should carefully distinguish empiricisms from laws of nature, which are always the expression of an induction, because much evil has arisen from confounding the two kinds of generalization. The mere fact of our having found a certain uniformity prevailing within the range of our observation, does not warrant us in assuming that this uniformity holds throughout nature, unless this can be established by some satisfactory proof. Many lives have been lost, owing to medical empiricisms being mistaken for inductions, and physicians consequently adopting an improper course of treatment, because it had been found beneficial in other cases, apparently similar but really different.

CHAPTER VI.

OF HYPOTHESES.

§ 1. NATURE AND USES OF HYPOTHESES.—Definition of Hypotheses.—Their general value.—Why their importance frequently overlooked.—How they promote close and careful investigation.—Indispensable, in many cases.—Their importance as guides.—How they aid Memory and Classification.—Such uses independent of their correctness.—Use of Hypotheses partially incorrect.—Refutation of objections.—Application to things inconceivable.—Use of Hypotheses in common life.—Abuse of Hypotheses.

AN *hypothesis* is, a supposition, made on a subject of which we have some knowledge, regarding something still unknown, which we desire to find out.

Wherever the character of the proposition in question cannot be ascertained by direct observation, experiment, or reasoning, we must form an hypothesis, and then proceed to test its truth. The importance of this course is evinced by the history of the physical sciences, many of whose truths could have been discovered by no other means.

The confirmation frequently given to established hypotheses by subsequent discoveries, is apt to make us overlook the means by which they were first proved, because all other suppositions now appear so baseless that we forget some of them once appeared more plausible than the true. When this has been ascertained, the discoverer often confines himself to the proofs in its favor, without informing us of the means by which he first made the discovery.

Hypotheses indirectly aid the progress of discovery by their leading to close discussions and extensive observations and experiments: for their authors and upholders exert themselves to ascertain facts which may confirm their own views, and to detect misstatements or fallacies on the part of their rivals, while the latter follow a similar but adverse course.

In many inquiries, hypotheses are both unavoidable and indispensable, because the apparent differ widely from the real facts, while the latter can be ascertained by no direct means. When we view the heavenly bodies, for example, we must form some supposition regarding their real motions; and we can determine these from their apparent motions, only by making several suppositions, and ascertaining which is the true one. In all cases of this kind, it is impossible to think of the subject at all, without forming an hypothesis, until some particular hypothesis has been established.

In some cases, hypotheses, although not indispensable, are of great importance, because they enable us to ascertain what course is likely to succeed, and what not. Hence much time and labor are saved, which would otherwise be wasted to no purpose. For, by considering the various admissible suppositions, and their relative degrees of probability, we shall ascertain the truth much more readily than if we proceeded without any guidance.

Hypotheses also aid Memory, and assist us in classifying a great variety of phenomena, according to some

SEC. 1.]   USES OF HYPOTHESES.   111

logical principle, and ascertaining their relations to each other. The observable phenomena of many subjects are so numerous, and apparently so unconnected, that we can remember and classify the facts and their manifest relations to each other, only by forming some suppositions regarding their causes and connections. Thus, the classification of the various divisions of organic nature, the remembrance of the peculiarities of each, and a knowledge of their relations, are greatly facilitated by supposing some archetype or general model, which was variously modified, in order to produce the different classes, and adapt them to specific modes of living and acting: for we can thus most easily understand and remember the characteristics of every kind, and what relations it bears to others; after which a proper classification becomes comparatively easy.

Such uses of hypotheses are quite independent of their correctness. Thus, in the case just mentioned, the supposition that the Creator took some form as a model, and variously modified it, to produce the different classes, is not only destitute of proof, but most probably false. Man requires such helps, to produce a variety of similar forms; but we have no reason to believe that the Deity needs any such aid. Yet it is nevertheless true that the various classes are formed *as if* such a model had been employed; and, therefore, the supposition is as useful for the purposes just mentioned as if it were actually true.

The Ptolemaic system of Astronomy is another instance of this kind. Although fundamentally erroneous, it served to introduce some degree of system and order into the complex phenomena of the heavenly bodies; and some of its suppositions are so convenient, for several purposes, that they are still retained in various operations, although they are well known to be false.

If an hypothesis fundamentally erroneous may be useful, much more may one which is only partially untrue, and correct in the main, such as the views of Copernicus regarding the motions of the planets. He was right in supposing that the Sun was comparatively at rest, in the center of the system, with the planets revolving round it; but he erred in supposing that they moved in epicycles. This hypothesis accounted for most of the phenomena then known; and its errors were such as could not fail to be detected by further observations, made with better instruments than those of that age. An hypothesis fundamentally correct generally indicates the course to be adopted in testing its details; and hence the principal difficulty is usually surmounted as soon as such an hypothesis has been established, although it may be erroneous in several minor details.

Hypotheses also enable us to refute objections urged against a proposition of whose truth we have satisfactory evidence. We meet the objection by showing what *may* be, even when we cannot prove that such is actually the case.

By means of language, hypotheses further enable us to transcend the powers of Conception, and thus to discuss things inconceivable or impossible: for we can suppose, for the purposes of argument or examination, whatever can be expressed in words. Thus, although we cannot conceive of an endless series of events, or an eternal being, we can reason about them, as long as our reasonings do not require us to form any conception of them. Much of the utility of hypotheses arises from this power of applying them to any proposition: for, in many instances, it is only by investigating and testing all the possible kinds of suppositions regarding the subject under consideration that we can determine which is the true one.

As the term hypothesis is most frequently applied to suppositions on scientific subjects, we are apt to overlook their great importance in common life. Yet this is not the less real. In all those cases where we have to choose between several practicable courses, we may suppose them successively adopted, and trace the several advantages and disadvantages attending each, after which we can compare these, so as to adopt the most eligible. Incalculable evils have arisen from either totally neglecting this process, or performing it very hurriedly and inaccurately. We are often so anxious to act, and so impatient of delay, that we overlook the serious disadvantages of the course we adopt, and the much greater advantages of another. Hypotheses of this kind are often employed without the individual being aware of it, because they are made, and conclusions drawn from them, with such extreme rapidity that these are taken for immediate discernments.

Hypotheses are liable to the great abuse of adopting

SEC. 2.]   TESTING HYPOTHESES.   113

them as established truths, while they are only plausible suppositions. Before adopting or admitting any hypothesis as a truth, it should first be conclusively established by proof.

§ 2. METHODS OF TESTING HYPOTHESES.—Tests of mathematical and some other Hypotheses.—Four characteristics of correct Hypotheses regarding Phenomena.—When a phenomenal Hypothesis is established.—Theory.—Principal means of testing phenomenal Hypotheses.—Effects of increased knowledge.—Anticipation of future discoveries.—How erroneous Hypotheses are sometimes exploded.—Influence of discoveries in other departments.—Superior kind of Hypotheses.—Cautions.

In many cases, hypotheses are proved by simply finding that they are necessarily implied in something already known to be true; or they are disproved by our discovering that they involve errors or absurdities. It is by such means that all mathematical hypotheses are tested; and the same processes are often applicable in other cases also. Thus, the hypothesis that nature abhors a vacuum, is disproved by the fact that a vacuum occurs wherever there is no force in operation to fill it.

To establish an hypothesis regarding phenomena as a truth, it must possess the four following characteristics.

1. The phenomena or appearances which it is employed to prove or explain, must be real or actual, and not merely supposed or assumed. 2. It must not be inconsistent with any cognition. 3. It must account for all the phenomena in question. 4. It must be the only hypothesis which possesses the second and third characteristics.

If we overlook the first of these conditions, we attempt to account for what may have no existence; and we may only strengthen error by withdrawing attention from the nature of our assumptions. The second condition is requisite to render the hypothesis correct, so far as it goes. The third is necessary to render it complete: for an hypothesis which accounts only for some of a group of connected phenomena, must be at least defective. The fourth condition is necessary in order to exclude other hypotheses: for, if one accounts for all the facts as well as another, and contains nothing inconsistent with known truth, it is as well entitled as any to be received as correct.

When a phenomenal hypothesis has been proved to

possess those four characteristics, it ceases to be an hypothesis, and becomes a cognition: and if it consists of a fundamental principle, or series of such principles, it is termed a *theory*, which implies something put forward as a truth, and considered such by the propounder, although it may not, in reality, be satisfactorily established.

The principal means of testing phenomenal hypothesis are, extended observation and experiment. An hypothesis may possess the three first of the above characteristics when it is formed: but the progress of discovery may refute it, by disclosing facts incompatible with it, as happened with the Ptolemaic system of Astronomy.

While an erroneous hypothesis is generally detected by more extensive researches, a correct theory receives confirmation from new discoveries. Thus, the aberration of light, which directly proves the motion of the Earth in its orbit, was discovered by Bradley, in the eighteenth century, nearly two hundred years after the death of Copernicus: and, of the lunar irregularities now deduced from the law of gravitation, more than three fourths were unknown to Newton, who established that doctrine.

A correct hypothesis not only accounts for new facts, after they have been discovered, but it frequently enables us to conjecture them beforehand. Thus, the theoretical astronomer has often made discoveries in his study, which were afterwards verified in the observatory. An erroneous hypothesis, on the other hand, is liable to be exploded by a single observation or experiment. Thus, the hypothesis that dew emanates directly from the ground, is exploded by seeing it copiously deposited on a piece of iron.

An hypothesis is frequently confirmed or refuted by discoveries made in other departments than that to which it refers. Thus, the establishment of the undulatory theory of light explodes the material theory of heat; and the progressive motion of light is proved by observations on the eclipses of Jupiter's satellites.

In determining which of several admissible hypotheses is the true one, we are often assisted by a knowledge of the nature of the various agencies which may possibly be concerned in producing the phenomena under consideration, without further recourse to observation. Thus, the force of gravitation accounts for the ascent of mercury in the tube of the barometer, and of water in the suction pipe of a pump, so that we may at once discard the hypothesis of nature's abhorrence of a vacuum, or of a power of suction in the tube.

Of two hypotheses, otherwise equally probable, that which traces phenomena to a real agency, is entitled to a preference over one which attributes them to some unknown cause. For the Ruler of nature has never been found employing two different agencies when one would accomplish all the results equally well. But in those cases where unknown agencies may possibly interfere with the results deducible from a certain hypothesis, these ought to be investigated: for otherwise we know not how far such agencies may interfere; and the hypothesis is not legitimately established till it has stood this test. Thus, the actual path of a projectile through the air, is very different from what is deduced from an hypothesis that overlooks the atmospheric resistance. A cannon ball, fired off at a certain angle of elevation, will go less than the twentieth part of its range, if it moved in a vacuum.

Before applying hypotheses, we should ascertain that we have a clear and accurate knowledge of the facts involved: for, if our views of these are confused or erroneous, the use of hypotheses will be apt to mislead us, and confirm us in erroneous opinions.

We should also beware of adopting any hypothesis, however plausible, as a truth, until we obtain decisive proof; and we should discard it whenever we have obtained satisfactory evidence that it is erroneous.

PART II.

OF THE PRINCIPLES AND METHODS OF
INVESTIGATION.

CHAPTER VII.

OF INVESTIGATION IN GENERAL.

§ 1. OF DISPOSITIONS AFFECTING INVESTIGATION.—Common Error.
—Advantage of Equanimity.—Two Extremes.—Their common Origin.—Proper Medium.—Diversity of Opinions.—On what our Attainments chiefly depend.—Requisites to successful Investigation.
—How the proper Disposition is to be secured.

THE investigation, even of the most important subjects, is frequently conducted in such a way as to be worse than useless: for it satisfies the individual that he now knows enough on these topics, and thus stifles further inquiry, while it substitutes positive unconscious error in the place of conscious ignorance. In order to act more warily, we must see the necessity for distrusting the correctness of opinions adopted without carefully examining the foundations on which they rest.

If the mind is strongly affected by some emotion, while we are examining a subject, it will both distract the attention, and prevent us from taking a sufficiently close and extensive view of the subject, so that some important points will be considered inattentively, and others will be wholly overlooked. We should, therefore, abstain from investigation, till we are disposed to consider it attentively, and without any passionate emotion: otherwise we shall only fortify ourselves in error.

Many have assumed that truth is very easily discovered, and that men fall into error owing to some defect or depravity from which *they* considered themselves entirely free. This state of mind leads to adopting, without any proper examination of their real nature, the first plausible opinions that force themselves on the attention, if not opposed to the party's prejudices, retaining them with unreasonable pertinacity, and contemning every contrary view.

Many have fallen into the opposite extreme. They have seen the great diversity of opinions that has prevailed, even among the more intelligent portions of mankind: and hence they concluded that truth must be very hard to discover, or that it cannot matter much whether we discover it or not. This class delight in caviling objections, and in oppugning all the foundations of knowledge: they draw futile distinctions, while they overlook important differences, and oppose to one set of opinions the objections urged by the disciples of another, without ever carefully examining whether they are valid or worthless.

Those two extremes both originate from an erroneous opinion regarding the human faculties, and an unwillingness to undergo the labor of proper investigation, combined with various other prejudices and prepossessions. Hence it frequently happens that the same individual is very sceptical regarding disagreeable truths, and equally credulous and dogmatic regarding agreeable errors.

The proper disposition avoids each of those extremes: it is ready to investigate aright the arguments in favor of any opinion which really deserves or requires investigation, and to adopt it, if these be found conclusive, while, on the other hand, it is prepared to reject everything which is refuted by irrefragable proof.

The great diversity of opinions which has prevailed among mankind, shows that truth is not to be discovered without proper investigation: yet all the most important truths are placed within the reach of ordinary abilities; and persons of greater talents fail to discover them, only when they never search for them aright. The young inquirer must not imagine that certainty is unattainable, even where he finds conflicting opinions held by distinguished men. Investigators whose principal object was gain, applause, victory, or the gratification of some darling prejudice, could not reasonably be expected to discover truth; and we cannot rightly attach any weight to the opinions of the many who have either adopted the views of others, without ever bestowing on the subject an independent examination, or who have formed opinions of their own, without having ever investigated it with care and attention.

Our attainments depend much more on the use we make of our faculties than on their natural force; and truth is reserved for those who discard prejudices, and seek it with proper care and diligence, without which great abilities only generate self-conceit, indolence and error. If it were more easily attainable, it would be less valued, and sought more negligently and indolently, whence error would abound more than it does. Although the path of the faithful inquirer may at first appear gloomy, yet the dark clouds raised by ignorance and misconception rapidly vanish, as he advances, and celestial radiance then cheers and guides his course.

In order to successful investigation, the following things are requisite. 1. A correct estimate of the value of truth, and of the sacrifices which must be made for its attainment. 2. A mind prepared to examine properly every important subject, and adopt whatever conclusions truth dictates. 3. A correct estimate of our own capabilities. We must neither think that we can secure truth without a careful, continued and impartial examination, nor that it is placed beyond the reach of ordinary abilities perseveringly and judiciously applied. 4. A patient and active disposition, that we may neither adopt a conclusion till we have sufficiently examined the subject, nor hesitate to investigate carefully and diligently all subjects having an important bearing on our welfare.

The disposition proper for investigation is to be attained by acquainting ourselves with the value of knowledge, and the evils of ignorance and error, the sources of error, and the requisites to successful investigation. For this will produce an earnest desire to secure truth and avoid error.

§ 2. OF HABITS AFFECTING INVESTIGATION. — General influence of Habits.—(1) Attention and Inattention.—(2) Methodical and desultory Application.—Advantages of a general Plan.—Caution.—(3) Wide and narrow views of the subject. — Evils of superficial and partial views.—(4) Properly investigating Proofs, and jumping at Conclusions.—(5) Activity and Perseverance.—Evils of Indolence and Indifference.—Important Laws.—(6) Resisting Prejudices.—Results of being led by our Wishes.—(7) Temperance and Self-Control.—Connection of Habits.—Evils of Self Indulgence.—How good Habits are to be formed.

Habits sway our conduct so extensively that, in order to successful investigation, we must form such as are favorable to that result, and avoid those of a contrary kind. We shall, therefore, consider the former, and notice those which are opposed to their influence.

1. We have already seen that all erroneous opinions are traceable to inattention, as their immediate source:

F

and it also requires close attention, in many cases, in order to discover the properties of things, or their relations to each other. Hence the habit of *close attention to the subject under consideration* is necessary to guard us against adopting errors, and enable us to go beyond the simple elements of knowledge. It is opposed to the habit of *considering a subject carelessly, or while our attention is, at the same time, fixed on something else*, a course which inevitably leads to serious errors. We should, therefore, beware of tormenting ourselves about small matters and imaginary evils, and guard against the practice of skimming along the subject that we are considering, without devoting to any part more attention than it excites spontaneously: otherwise we cannot, without great difficulty, concentrate our attention, even for an hour, on any subject.

2. Several subjects, of the highest importance, cannot be rightly understood without a continued degree of attention. Truth is not generally discovered at a glance, but only by a long examination, and considering the subject again and again. Now if we rove from one subject to another, without devoting sufficient time and attention to each, our thoughts become more or less confused, so that, when we return to the subject whence we deviated, we hardly know where we left off, and find that we must begin again at the starting-point. Hence our progress will be very slow and unsatisfactory. Not unfrequently the attention will be so occupied with the subjects previously considered, that careful examination is impracticable, various things widely different are confounded with each other, and many errors inevitably result.

In order to avoid these evils, we must acquire a habit of *considering, at the outset, the scope and object of our investigations*, and of *continued and methodical application*. This is opposed to the habit of *flying from one subject to another, and taking up anything which strikes the fancy of the moment, without any definite aim or object*, which leads to much time being spent with no other result than acquiring many erroneous opinions, and a habit of relinquishing an investigation as soon as it has lost the charm of novelty. Thus we come to wander from one subject to another, without learning more of any than suffices to make us conceited, and satisfied with the mere husks of knowledge, while our views may be, in many respects, very erroneous, and we may be ignorant of the more important parts of the various subjects over which we have been flying.

We should form a general plan, which devotes a reasonable time uninterruptedly to the various subjects of our investigation, from which we should not deviate without some urgent and justifiable motive, nor farther or longer than the circumstances warrant. We are generally surrounded by temptations to turn aside from investigation to things which are immediately more attractive, especially to persons who do not look beyond the present moment, nor consider the tendency of desultory habits, and occupying the attention with matters of no real consequence. Yielding to enticements of this kind has been a fruitful cause of unsuccessfulness, among those who have desired and attempted to succeed in the pursuit of truth.

We may investigate more than one subject to advantage, in the course of the day. But, in order to this, we must devote a considerable time to each, and not run desultorily from one to another. Increasing knowledge may suggest improvements in our plan, as we advance: but we should beware of proceeding without plan or object, which we shall inevitably do, unless we consider the precise drift and purpose of our investigations. If we fail to do this, we shall often begin an investigation, and never finish it, whence we readily slide into the habit of dipping into every kind of subject, without ever acquiring a solid knowledge of any.

We should, therefore, arrange our thoughts and materials in the most regular manner practicable, and not take leave of the subject until we have done our best to master it. Otherwise we shall fall into confusion of thought, and abandon a subject before we have begun to understand it. Our investigations may be frequently interrupted by things beyond our control: but this need not prevent us from returning to the subject when the cause of the interruption ceases; and the habit of doing so is requisite in order to understand any subject thoroughly.

3. The habit of *viewing the subject in its various bearings, and ascertaining its real character*, is necessary to prevent us from overlooking points of great importance

or attaching undue weight to those which we have considered, to enable us to see everything in its true light, and to know the value and application of our attainments.

This habit is opposed to that of *forming conclusions before we have rightly examined the whole subject, or considering only one of several important bearings, and overlooking the rest.* This leads to the numerous evils incident to superficial or partial views, which frequently mislead as effectually as positive error, an omission being as fatal as a false addition. It also leads us to overlook the recondite and future bearings of a question, which are frequently very different from the obvious and present aspects, and of incomparably more importance. This error is one of the most common and pernicious of all those to which mankind are liable. What is under our immediate view is often assumed to be a fair representation of the whole, when it is the very reverse.

This habit also produces the error of attaching an undue value to the least important parts of a complex subject, while we overlook or undervalue the most important, and fail to discover the true relations of the various parts, so that the shadow is mistaken for the substance, and a most inaccurate and false estimate is formed of the whole subject. Moreover it leads to the great evil of learning important truths, without ever perceiving their practical applications.

4. The habit of *properly sifting evidence or arguments, and adopting no opinion as a cognition without conclusive proof,* is requisite, on the one hand, to prevent us from accepting a proposition as true till we know that it is so, however much we may be prepossessed in its favor, and, on the other hand, from rejecting the proofs offered in its favor, without due examination, however disagreeable it may be to our wishes. This habit will prevent us from stopping short till we have ascertained the foundations and character of the subject, which may be very different from what some allege or believe. It is indispensable to the attainment of real knowledge, and the avoiding of error.

This habit is opposed to that of *jumping at conclusions, before we have ascertained the real nature of the proofs by which they are supported, and assuming propositions as true without any good grounds.* This leads us to adopt the erroneous opinions of others, or the dictates of our own wishes, as established truths. It is very easy to assume that to be true which we strongly desire to be so, or which is believed by those around us, while it is frequently a difficult and laborious task to ascertain where truth really lies.

5. There are so many things to draw away our attention from the pursuit of knowledge, and to retard and mislead us in our course, that, unless we acquire the habit of *working actively and perseveringly, amid difficulties, discouragements and disappointments*, we may rest assured we shall not be successful. It is frequently much more agreeable, either to spend our spare time in pastimes and amusements, or to seek after knowledge in an indolent and slothful manner, than to apply ourselves actively and perseveringly to its attainment. Hence many have acquired the habit of either using no active exertion in acquiring knowledge, but taking everything on trust, that does not offend their prejudices, and rejecting everything that does, or of giving up the pursuit as soon as they get into difficulties, which they speedily do, as a matter of course: and hence much that has passed for knowledge, is little more than a web of errors and misconceptions.

It has frequently been supposed that persons of common abilities need not give themselves any great trouble about the acquisition of truth: for, if men of great talents failed, could they hope to succeed? Undoubtedly they could. If, as has often happened, men of great talents are the slaves of various bad habits, they will fail in acquiring truth, when persons of much smaller abilities will succeed, provided they avoid such habits, and make the most of the powers which they possess. Difficulties are encountered even by the most gifted, who are generally obliged to toil long and arduously, in order to effect great results. If we do our best to acquire knowledge, we shall never be disappointed: for although it is a law of nature that *great good is never secured without great labor*, it is equally a law of nature that *persevering care and diligence will overcome every obstacle to the attainment of the most important knowledge.*

It requires neither uncommon abilities nor wealth, in order to master the most important truths within the circle of human attainments. Bad habits are a much greater bar to progress than want of abilities or money;

and, of these, habits of indolence and indifference are by no means the least powerful. Gold may purchase material wealth: but it neither secures truth nor excludes error. These objects require persevering and careful toil from the rich as well as from the poor; and experience has shown that the disadvantages of the former are greater than those of the latter.

6. Not the least important of good habits is, that of *investigating candidly, without yielding to the influence of prejudices,* which are among the most fertile of all the sources of ignorance and error. They frequently prevent investigation altogether, and still more frequently render it abortive, by leading to the adoption of erroneous opinions.

This habit is opposed to that of *allowing our wishes to turn away the attention from certain parts of the subject, and to view the rest hurriedly and negligently,* a course which inevitably leads to error. Our wishes do not, in the least, change the nature of truth and falsehood; and, in order to discriminate these from each other, we must view a subject as it is, and receive or reject proofs according to their real character. We must receive as true what is sustained by conclusive proof, however unacceptable, and reject whatever is destitute of such proof, however agreeable.

7. Habits of *temperance and self-control,* or resisting the solicitations of desires and appetites which ought not to be gratified, are essential to the successful pursuit of truth. Dissipated and luxurious habits form an insurmountable obstacle to great progress in the attainment of knowledge, while they tend strongly to root out all good habits, and introduce, in their place, those which are incompatible even with the retention, and much more with the acquisition, of knowledge. The investigator of important subjects must, in order to be successful, not only abstain from such things as are pernicious in any degree, but also guard against going beyond the bounds of moderation in what is proper and allowable. It is not enough to avoid the extreme of a debauch: he ought to shun the excesses of gluttony, exciting and frivolous pastimes, and the intoxicating paroxysms and reactive prostrations of ungoverned passions.

We are incessantly tempted to forsake the pursuit of knowledge altogether, or to investigate hurriedly, and without due care and attention, for the sake of present gratification, or escaping from present pain or uneasiness: and, unless we resolutely refrain from indulging such desires, the successful pursuit of truth becomes hopeless.

This habit is not only of great importance in itself, but it is indispensable to the existence of other good habits, which will be speedily undermined, if we habitually yield to the desires of the present moment. The attention will be distracted; the subject will be investigated carelessly and immethodically; we shall be misled by prejudices; our inquiries will be conducted without vigor or perseverance; and consequently error will usurp the place of truth.

Those habits are opposed to that of *self-indulgence, and sacrificing important future good for the sake of a fleeting present gratification, or exemption from transient pain or toil,* which is the most fertile of all the sources of error, and the great barrier to the removal of ignorance. Its votaries dislike study, because it is less pleasant to their dark minds and groveling tastes than gross enjoyments or frivolous amusements; and when the consideration of an important subject is forced on their attention, they gladly avail themselves of any fallacy that first presents itself, to quiet their consciences for not examining it aright. They generally adopt those opinions which are most agreeable to their prejudices, without ever carefully examining whether they are true or false; and if by chance they come to see that any of these is untenable, they usually supply its place with some similar error.

In order to form good habits, we must obtain the disposition proper for successful investigation, then consider the subject till we understand the nature and operation of those habits, and finally practice them, and shun the contrary habits. When once fairly adopted, this course becomes easy, as good habits are much more pleasant, in the long run, than bad, and that which has become habitual is easily done, from the fact that it is habitual.

§ 3. OF THINGS WHICH REQUIRE NO PROOF.—Why Intuitions require no Proof, nor Apprehensions.—Reality of Conceptions and Similitudes.—General expression of what needs no Proof.—Truths which admit of no logical Proof.—Sole pre-requisites to their admission.

As the proof of a proposition is, that which shows that

it is true, either absolutely or conditionally, intuitions require no proof, because they are self-evident; and therefore no proof could add to their certainty, while we could attempt to prove them only by assuming something tantamount to themselves.

With respect to apprehensions, we have already seen that they are matters of immediate consciousness, the reality of which admits of no rational doubt; and proof is as superfluous as in the case of intuitions.

Ideas are equally known by immediate discernment. When the idea of a well known person or place, for example, is present to the mind, we know certainly that we discern such an idea, and that, if we did not, it could not exist, although its origin, or whether it ever had a prototype, is another question.

A similar remark applies to conceptions. When we form a conception of a whale or a sphinx, for example, we know that we actually discern the phenomenon. Whether the likeness is correct, or whether such an animal really exists, are questions which do not in the least affect the reality of the conception.

All our comprehensions, including apprehensions, conceptions, and similitudes, are equally and certainly real; for unless they were so, they evidently could not exist. They are all precisely what they appear to be, without any possibility of error, which affects only the inferences.

Hence it appears that truths of Intuition and Comprehension require no proof, which may be more briefly expressed thus: *discernments require no proof.*

Moreover all attempts to establish these truths are necessarily fallacious, since they must assume the very propositions which are intended or professed to be proved. Thus we cannot proceed a step to prove the reliability of Consciousness without assuming this as true, and all our processes necessarily involve that assumption.(14)

Before admitting a proposition as a discernment, all we require to know is, that it really belongs to this class of truths. We are indeed liable to take for self-evident even things which are self-evidently impossible; and we sometimes assume fallacious inferences to be immediate comprehensions. Yet all that is necessary to distinguish comprehensions from all other propositions, is, an attentive and close consideration. That which is really self-evident, will clearly appear to be such when well understood and carefully considered; and the same process will readily distinguish comprehensions from inferences.

§ 4. OF THINGS WHICH MAY GENERALLY BE ADMITTED AS PROVED.—How mediate Knowledge is established.—Twofold division of Proofs.—Evidence.—Signs.—Testimony.—Deduction.—Primary foundations of sound Reasoning.—Why frequently unnecessary to trace reasoning to these.—Mode of proceeding regarding External Objects.—(1) Principle regarding the causes of Apprehensions.—Chief source of Error.—How avoided.—Indications of Disease.—Circumstances in which the organs are liable to mislead.—Imitations of Nature.—Real source of many Mistakes attributed to the Senses.—General nature of Apprehensional Errors.—(2) Principle regarding Remembrance.—Chief source of Errors attributed to Memory.—How they may be avoided.—(3) Principle regarding Testimony.—Origin of Falsehoods.—When we may assume testimony to be correct.—What things require formal proof.

As all knowledge is either mediate or immediate, that which does not belong to the latter must fall under the former; and cognitions of that class are established by ascertaining that they are necessarily connected with our immediate knowledge, that which shows this connection being termed the *proof*.

All proof is either *evidence* or *deduction*, or a combination of both.

Evidence is, any contingent fact which goes to establish the proposition in question. It consists of *signs* and *testimony*.

A *sign* is, something observed or comprehended by ourselves, which proves the thing in question by being its uniform antecedent, concomitant, or consequent, or by being directly and manifestly incompatible with its falsity. Signs are either *internal*, consisting of the phenomena of Memory and other internal comprehensions, or *external*, which consist of indications discernible by Apprehension. These comprise every apprehension which is known, either by experience or Intuition, to imply directly the past, present or future existence of something different from itself, so that most of our apprehensions belong to this class, and many of them are signs of various things. A sign indicates the past, the present, or the future, according as it is a consequent, concomitant or antecedent of the thing signified or indicated.

Testimony is, a statement made by another regarding what he professes to have comprehended or to think. It may either state the very thing in question from the witness's own knowledge or some sign of it which he professes to have comprehended, or it may only declare what he has heard from other witnesses.

Deduction consists of reasoning, which professes to show that the proposition in question is necessarily implied in something which either requires no proof or is already known, by some means, to be true, or is merely assumed to be so; and it is tested by the criterions of reasoning formerly mentioned.

If sound reasoning be traced to its foundations, we shall always arrive at intuitions, comprehensions or suppositions, as the primary premises, because these are the only possible sources of mediate knowledge. But such a thorough investigation is generally unnecessary: for when certain connections have once been established, they may be taken as proved, in all subsequent inquiries. We assume what has been already proved, and proceed thence, as a new starting point.

When we have once clearly ascertained the connection between certain apprehensions and their external causes, we generalize precisely as in other cases, and thenceforth assume that the same phenomena have the same or similar causes. So, when we have once ascertained that the phenomena of Memory must represent real prototypes, we assume this for the future. When we see certain complements of colors, and hear certain sounds, for example, we assume that these are caused by such and such substances, without or beyond ourselves: and when certain groups of ideas spontaneously arise, we assume that we apprehended their originals, without waiting to discuss the various possible sources of those ideas.

We shall now mention the cases in which such assumptions may be safely made, and the errors against which we have to guard, supposing the requisite degree of attention has been bestowed upon the immediate comprehensions.

1. *Apprehensions may be assumed to be caused by agencies similar to those which were found to have caused such phenomena formerly.* If, for instance, we see those colors and hear those precise tones, which we formerly apprehended, as Remembrance shows, when a certain friend addressed us, we may assume that the same friend

is now present and addressing us. Here the principal source of error is, assuming that the same phenomena must have resulted from agencies precisely similar, if not the same, whereas these may be, in some respects, different. We are apt to attribute a phenomenon to the very same agencies which produced similar phenomena in our previous experience, overlooking the possibility of their being caused, in some cases, by agencies in several respects different.

To guard against this error, we must ascertain that a phenomenon can be rationally attributed only to one cause or set of agencies. Such is the uniformity of nature, that deceptions of this kind form only rare exceptions to a general rule. The principal sources of illusion are, a diseased state of the apprehending organs, and intentional imitations, which exhibit phenomena that are usually caused by a different combination of circumstances. When our organs are healthy, the various beings that surround us affect them with great uniformity. Thus we readily distinguish the countenance of one friend from that of another; and we are in no danger of mistaking the smell of a rose for that of an apple, although both belong to the class of fragrant odors.

Disease of our organs is generally indicated by comprehensible signs; and it is only when the morbid affection is violent or extensive that we are apt to be misled at all, while, in such cases, the indications of disease are generally palpable. Thus, a slight inflammation of the eyes does not prevent us from accurately distinguishing colors, and ordinary affections of the ear only increase or diminish its usual degree of sensibility. When disease has been violent and long continued, all mistakes are obviated by the organ's ceasing to perform its functions. The blind make no mistakes regarding colors, nor the totally deaf, regarding sounds. It is, therefore, only in particular cases of disease that mistakes are liable to be committed; and, in these cases, we are put on our guard by palpable indications of the affection.

With respect to imitations of nature, we generally know in what circumstances they are to be expected, and they are seldom so complete as to escape detection, on a close inspection. A juggler uses sleight of hand; but this is out of the question in our ordinary intercourse with friends. So, a distant figure, seen at the end of a long passage, may possibly be a statue; but if we notice a similar appearance in a lonely uninhabited place, we justly infer that it is a real person. Again, a skilfully executed painting may be mistaken for its original, especially when seen from a distance or in an obscure light; but, on a close inspection, the resemblance is hardly ever complete; and even when viewed from some distance, its unchanging appearance, as it is surveyed from various points, distinguishes it from a solid object. A ventriloquist may imitate the pitch of the voice, to some extent, so as to deceive us regarding the distance; but as he cannot affect the directions in which the undulations strike the ear, without turning his voice into a mere echo, the origin of his utterances may always be detected.

Intentional imitations must proceed from some motive; and as they generally require time and skill, they are seldom attempted, without some motive stronger than a momentary whim. An extensive acquaintance with the laws of nature, both mental and physical, would obviate every mistake of any consequence, likely to arise from this source.

Many of the mistakes which are attributed to illusions of the senses, arise solely from inattentive observation, without apprehension being in the least at fault. Thus, we sometimes mistake one person's voice or countenance for another's, which it resembles, although a little close observation would show that there are peculiarities by which they can be easily distinguished.

2. *The clear presentations of Memory may be assumed to represent real originals, and prove our personal identity.* The processes by which we ascertain the connection between ideas and their prototypes, incidentally proves our personal identity, so far back as Memory extends. For it consists in continuity and similarity of thoughts, the corporal changes which incessantly occur having no influence on it; and it would be as little affected by similar mental changes, even if they existed, which they probably do not. Our identity, and the reliability of Memory, having been once ascertained, they are afterwards rightly assumed, without any formal repetition of the processes by which they were first discovered.

The principal sources of error connected with Remembrance are the following. (1) Overlooking the peculiarities which distinguish ideas from phantasms, whence they are confounded, and we suppose that we apprehended what we only conceived, (2) Overlooking the peculiarities which distinguish one similitude or apprehension from another. As the power of Memory is much stronger than that of Conception, mistakes of the former sort are very rare. They occur chiefly where a lively conception is formed of a particular thing, and its phantasm is frequently considered, while the ideas of the things apprehended when it was formed are forgotten. Mistakes of the second kind are very common. We often mistake one thing for another, for instance, because we overlook the peculiarities which distinguish them, and which close observation would readily detect.

In all cases where errors are attributed to Memory, it will be found that they are, in reality, attributed to unsound reasoning, or to hasty and inaccurate observation. When a thing is quite forgotten, Memory cannot mislead us regarding it, any more than if we had never known it; and when the idea is faint, or unaccompanied by those of the other things apprehended at the same time, yet it is real, and its character suggests caution in drawing inferences.

In our own case, a very moderate degree of attention will show us to which class a certain similitude belongs. If it be clear, and accompanied by equally clear similitudes of the other things comprehended simultaneously, we justly infer that all were so comprehended. When the case is otherwise, we should suspend a judgment till we can obtain some extraneous aid, such as that of notes or memorandums made by ourselves, or some evidence proceeding from others. By this means we may avoid errors proceeding from our own Memory.

In the case of other persons, the only difference is, that they may misrepresent the actual nature of their similitudes. This point may possibly be determined from cross-questioning, from their general character, from the special relations which their views and interests bear to the testimony, from comparing their statements with those of other persons, or from facts ascertained by any other means.

3. *Things testified by credible witnesses may be assumed as true, so far as their actual experience extends.* To invent a falsehood requires an effort, as it never arises spontaneously, the similitudes of things actually compre-

hended being always presented, except where the Will interferes with this result; and the utterance of falsehood is always followed by thoughts more or less painful, even where it escapes detection. Hence no person will relate a falsehood, without some particular motive; otherwise it would be an effect without any adequate cause: and, therefore, where there is either no motive or none that could operate, there will be no wilful falsehood. Erroneous statements must arise, either from the relator's unwillingness or his inability to state what he knows to be true, or from his having been misled; and hence, when there is no room to doubt the veracity or the ability, memory and opportunities of a witness, we may safely assume that his testimony is correct, as to what he actually comprehended. The inferences deducible from that, is quite another matter.

A proposition which does not belong to any of the classes now considered, requires formal proof. If it is neither a matter of immediate consciousness, nor one of the simple and ordinary inferences from the phenomena of Apprehension or Memory, nor related on the testimony of a credible witness, it has no title to be classed with cognitions, till it has been brought within the bounds of one or other of these classes, or of hypothetical truths.

§ 5. GENERAL MODES OF DETERMINING THE VALIDITY OF PROOFS.—Direct and Indirect Proofs.—Characteristics of each.—(1) Means of testing Signs.—When conclusive.—(2) First requisite regarding Testimony.—Frequent error and difficulty.—What must be ascertained.—Second requisite.—(3) Requisite in the case of Deductions.—Hypothetical cases.—Reasoning from Probabilities.—Requisite in case of Absolute Conclusions.—Frequent Error.—(4) Arguments.—Proper course in cases of inadmissible premises, and of modifying or distorting the truth.—Primary premises requisite in investigations regarding Substances.—Certain Sciences not independent of experience.—Cases where the investigation of an argument is needless.—Caution.—New premises.—Temporary cessation of Attention.—General and partial Abstracts.—Testing final Conclusions.—Things to be constantly guarded against.

Proof is either *direct* or *indirect*. The former establishes the proposition in question by direct means, and the latter, by showing that it cannot be false, or by establishing something inconsistent with any other supposition. Thus, when it is shown that a man is guilty of a crime, by the testimony of a witness who saw him commit it, the proof is direct; and when it is proved that another is innocent, by showing that he was at a great distance when the crime was perpetrated, the proof is indirect. Both kinds of proof are often combined in an argument, the one corroborating the other. Thus, it may be argued that a proof is worthless, by demonstrating its inconclusiveness from direct analysis, and then by showing that it equally proves something which is otherwise known to be untrue.

Either kind of proof is conclusive, when it conforms to the proper criterions: but the direct is usually preferable, because it is more concise, more easily tested and less liable to error. In some instances, however, indirect proof is more satisfactory, because it is less easily forged, and more liable to detection when it is so concocted.

1. In determining the character of signs, we must ascertain whether they are not compatible with the falsity of the proposition which they are supposed to prove: for they frequently establish only a slight probability. Thus, certain symptoms are often assumed as proofs that a person is affected with a particular disease, when in fact these symptoms accompany several other diseases. It is only where the signs are clear and unequivocal, that they are entitled to be received as conclusive proofs. If the mercury stands unusually high in the barometer, it is a sure sign that the atmospheric pressure is unusually great: but if we assume this proves it will be fair weather, we go farther than the sign warrants. Those cases where a phenomenon can be shown to prove a conclusion only by means of a chain of reasoning, come under the head of deduction.

2. In cases of testimony, we should first ascertain whether it is conclusive, even admitting it to be true: for it is often assumed to prove a proposition when it only proves something like it, and essentially different. In many instances, it is rendered suspicious by our inability to determine whether the statement is not a doubtful inference from what was actually apprehended, as when a man says he was cured of a certain disease by using such a medicine, when, in fact, it is doubtful whether he ever had the disease, and still more doubtful whether the medicine had any beneficial effect. Unless we know the phenomena actually comprehended, the testimony is always inconclusive, except in those cases where there is no ground to suspect erroneous inferences.

If we find that the testimony conclusively proves the point in question, supposing it to be true, we must next determine whether it actually is so; and, in order to this, we must ascertain whether it possesses all the requisites of reliable testimony. If it does, it proves the point in question: otherwise it does not.

3. In order to render deductions satisfactory, it must be ascertained that they possess the four characteristics of valid reasoning, formerly stated.

In hypothetical cases, the primary premises are sometimes mere suppositions; and the only thing to be ascertained regarding them is, whether they are expressed so clearly and precisely that there is no danger of mistaking one for another. If not, they are inadmissible: for it would be impossible to determine whether the subsequent reasoning were valid, since we could not ascertain to which of several things it applied.

Sometimes the primary premises are only probabilities; and the conclusion consequently partakes of the character of its bases. We reason from the premises as if they were cognitions, and then attribute to the conclusions the same degree of probability which, we believe, pertains to its foundations. We should mark the degree of probability belonging to the premises, and limit the conclusion accordingly. The reasoning, in other respects, is tested precisely as in cases where the primary premises are certain.

In all cases where the conclusions require to be proved absolutely, and not merely hypothetically, or as probabilities, the primary premises, besides being clear and precise, must also be true. This is to be ascertained by observing whether they belong to that class of propositions which need not be proved, or whether satisfactory proof has been already obtained, either by conclusive testimony, or by deducing them from the former class of truths. If the premises are objectionable in this respect, they are inadmissible, since otherwise we should assume as true what is possibly false, and thus build on an unstable foundation.

It frequently happens that the premises are true in one respect, but not in the sense in which they are assumed. It is true, in one sense, for instance, that "human testimony is fallacious," meaning "some human testimony:" but it is very false in the sense that "no human testimony is conclusive," which is the sense sometimes attributed to it. This kind of error is greatly fostered by the use of metaphorical or figurative language: for as it is often ambiguous, it admits of different interpretations; and those who employ it have frequently no clear conception of what they mean, and slide unconsciously from one signification to another.

4. In arguments, various assumptions are generally made, and different conclusions are deduced from them, in the first instance, but all going to establish the final conclusion. Some of these premises may be admissible, and others not. If any of the latter be an essential part of the argument, this vitiates the whole: but, in some cases, the undue assumption is immaterial. To which of these classes a premise belongs, can frequently be determined only after some progress has been made in the examination of the subject. Here we ought to proceed until we can ascertain whether the undue assumption is material, in which case the whole argument is worthless, or whether it be not immaterial, in which case we must beware of rejecting the whole argument as worthless, which would be like inferring that the main channel of a great river must be dry, because we have found some of its tributary rivulets dried up.

Sometimes the assumptions are partially correct, and only exaggerations or modifications of the truth. Here we adopt the same course as in the preceding case, and observe whether the subsequent reasoning is supported by the real facts, when stript of the distortions to which they have been subjected. If not, the argument is futile.

In all legitimate investigations regarding substances, the primary premises must consist, at least in part, of comprehensions or inferences from them: for intuitions alone teach us nothing regarding the existence of such things, and hypotheses here serve only to aid us in deducing necessary consequences. Hence all attempts to construct physical or mental sciences, independently of experience, must necessarily fail: for the reasoning will either be inconclusive, or occupied with puerilities and questions beyond the reach of human investigation.

In some cases, the argument is known to be invalid from the conclusion being either self-evidently false or quite incompatible with some known truth. Wherever this is, in reality, the case, it is unnecessary to test the argument, as it must be fallacious. But we should be cautious in making such admissions, since incompatibilities of this kind are sometimes alleged or supposed, where they do not exist, the proposition said to be incompatible being either false or not, in reality, inconsistent with the one in question.

Wherever new premises are assumed, in the course of an argument, we should test their character like that of the first assumptions.

The attention should be closely fixed on the subject, while we are examining an argument: for if this flags only for a little time, we may then encounter and fail to detect, a fallacy which vitiates the whole. This is particularly apt to happen in examining long arguments; and it is a good rule, in such cases, to stop whenever our attention cannot be prevented from wandering except by a strong effort of the Will, and not resume the investigation till we are able to do so with unflagging attention.

In difficult cases, much aid may be derived from writing an abstract of the whole argument, omitting all details of facts, and inserting only the essential parts, which never occupy a large space, even in the longest arguments. This will enable us to take a general view of the argument, and attend particularly to the more important parts. We are apt to think that we thoroughly understand an argument, merely from reading or hearing it, while reducing the substance of it to writing may show us that we do not rightly understand even its essential parts.

Where the difficulty lies in a particular part of an argument, this alone may be written as fully as its nature requires. Millions have been misled by fallacies which cannot fail to be readily detected by any person who simply writes them down and attentively considers them. But, in doing this, we should attend to every essential part: for the strength of a fallacy frequently lies in its assuming something as true which is never expressed, and which a careful consideration will show to be false.

If we find that a conclusion is logically established, we should then ascertain whether this is the proposition in question. In order to this, we must compare the one with the other, and ascertain whether they are virtually, if not completely, identical. If so, the argument is sound: otherwise it is worthless, because it is beside the subject.

Throughout every investigation of proof, we must beware of being misled by vague or indistinct thoughts, or by obscure, ambiguous, or unknown expressions, both of which are frequent sources of fallacy.

CHAPTER VIII.

OF STUDY.

§ 1. NATURE AND USES OF STUDY.—Knowledge Original and Secondary.—The latter acquired by Study.—Why first considered.—Its Importance.—General objects of Study.—Various other advantages.

KNOWLEDGE is either *original* or *secondary*. The former consists of that which is acquired without any direct assistance from others, or in addition to the previous attainments of mankind, by means of what is termed *original investigation:* the latter consists of that which has been previously acquired by others, and is communicated to us by means of some known signs, the investigation of which, for the purpose of learning what they profess to convey, is termed *study*. This claims our first attention, because it is of easier acquisition, and we are seldom in a proper position to enter upon original investigations of much consequence, until we know what others have already accomplished. It is only after this has been acquired, that any person can reasonably hope to make important additions to the stock of human attainments.

The knowledge acquired by study is by far the most important of our attainments, subsequent to those of early childhood, since the original acquisitions, even of the most gifted and favorably circumstanced, form but a very small fraction of the whole body of human knowledge. This holds true of necessary, as well as of contingent, truth, as to the latter of which we cannot evidently go beyond the bounds of our personal experience without the aid of others; and although we might, by possibility, learn much of the former by our own unaided efforts, yet, considering the great labor and difficulty of establishing many of the conclusions, we cannot expect to master much of the subject without extraneous aid.

The general objects of study are, to discipline the faculties, form proper habits of investigation, and communicate to the learner the previous attainments of mankind. But it frequently effects several other purposes. Obscurities and ambiguities are removed; fallacies are detected; that which was locked up in an unknown language is set forth in the vernacular; the fragments of knowledge formerly disconnected and hidden from the view even of intelligent men, are collected and made to illustrate each other, so that we obtain a more extensive and accurate knowledge of the subject than was hitherto possessed by any, and thus we see various parts of it in a new light, and ascertain what is defective or redundant. By these means, study not only renders previous attainments available, but it also prepares the way for new discoveries and inventions, so that it frequently leads to success in original investigations.

§ 2. SUBJECTS, MODES AND GENERAL RULES OF STUDY.—Selection of Subjects.—Order of Study.—Proper course regarding certain Subjects.—Advantages of previously considering the Objects of Study.—Preliminary Studies.—Extent of Study.—Three modes of Study.—Advantages of Written Communications, of Conversation, and of Lectures.—Disadvantages of the last.—Proper course, on controverted Subjects.—Rules regarding Prejudices.—Meaning of Terms.—Evils of disregarding them.—Requisite in order to be instructed by Language, and understanding it aright.—Advantage of a good knowledge of the Language.—Mode of dealing with Difficulties.—Evil of deviations, and of cramming.—Proper course. —Common mistake of beginners.—General Rule for surmounting Difficulties.—Advantages of careful and thorough Study, and evils of the opposite course.—Testing Statements.—Requisites before receiving an Argument as Sound.—When another instructor should be sought.—Sources of information which should be used.—What subjects should be studied simultaneously.—Recreation.—Laws of Health.—Final reviewing.

In selecting subjects for study, we should first attend to those which ought, on account of their general importance, to be understood by all, whatever be our peculiar pursuits or tastes, and then to those that relate to our particular vocation. Both of these are frequently studied, to some extent, under teachers, in early life. But this by no means dispenses with independent study, in order to render our knowledge of them accurate and extensive. We may afterwards attend to those studies and investigations to which we are led by our individual tastes and circumstances. These ought not to be taken up sooner, lest the former should not receive due attention, and our attainments be most extensive on subjects of comparatively little importance.

Sec. 2.] Subjects and Modes of Study. 141

The only modification requisite, in investigating a branch already studied under a teacher, is, that we may hurry over those parts which we already sufficiently know or understand. But a little self-examination, after really mastering the subject, will frequently show us that these form a much smaller part of the whole than we had supposed, and that, while our views of some parts were confused and erroneous, we were totally ignorant of others.

Before commencing the study of a subject, we should consider the objects which are to be accomplished by mastering it. If we cannot find any good object, we should give up all thoughts of studying it: for if we do, the only results will probably be, that we shall spend time to no purpose, and form a habit of studying in a careless manner, and roving from one subject to another, without acquiring more of any of them than a slight smattering, which only fosters a groundless conceit of varied knowledge.

If, on the other hand, important objects will be secured by properly studying a subject, a view of them will increase our diligence and attention, and a review of them will revive these, when they begin to flag. We shall thus secure the greatest degree of attention to the most important subjects, while a contrary course is apt to produce effects precisely the reverse, since that which is least important, frequently strikes the fancy of an ignorant person more readily and powerfully than the most weighty subject.

A preliminary requisite, in many cases, is, a competent knowledge of those branches on which the one in question is based, and which it assumes and applies as already known. Thus, we cannot master Dynamical Astronomy without a knowledge of Dynamics; nor can we acquire a knowledge of the latter till we have studied Mathematics. We should not enter upon a subject till we know that it will reward our labors, and we have obtained a sufficient acquaintance with the other branches requisite to successful study: and when we have taken it up, we should not lay it aside till we have mastered as much of it as our circumstances permit or require.

Communications from others may be *written* or *oral*, and the latter may be *colloquial* or *formal*.

Written communications possess the great advantages

of enabling us to consider any part as often and as deliberately as we please, to compare the various statements of different persons, to trace carefully every part of a chain of reasoning, and to make a correct abstract of the whole, a thing which ought to be done, in all cases of difficulty and importance. Another great advantage of writing is, that it enables us to test every statement, without any risk of mistaking it. This is frequently impracticable in oral communications, on account of our inability to remember the very words employed.

Conversation is more lively, and allows us to propose difficulties, to have ambiguities or obscurities removed directly, and to make inquiries which will give us a fuller and more correct view of the subject. This is on the supposition that the instructor understands it thoroughly. If he does not, questioning will detect his deficiencies more readily than formal lecturing, which possibly may only echo another's expressions, that were misunderstood by the speaker.

Public addresses enable a person to convey his thoughts to many simultaneously, while they unite, in some degree, the liveliness of conversation with the precision and regularity of written discourse. But they lack the means of careful examination and review supplied by the latter, and the advantages of questioning the instructor afforded by the former, so that it is impossible to master any science or any difficult subject by this means. One part is apt to withdraw our attention from another, which may be equally important, while the strong sympathetic emotions frequently produced by the presence of a multitude, tend in the same direction. Hence we wholly lose much of what is said; some parts are forgotten, and others are misunderstood. Consequently lectures are better adapted to amuse, or rouse emotions, than to convey solid instruction.

On controverted subjects, it is desirable to examine some of the best authorities on each side: for, in such cases, men are apt to conceal or misrepresent, so that we cannot obtain a correct view of the subject without examining both sides. Even when we find, as frequently happens, that one party is right, in the main, and the other as much in the wrong, an examination of the objections and arguments of the latter will give us a clearer and more extensive view of the whole subject than we could obtain by studying the productions of one party exclusively. Truth always gains by a close and candid examination: for the more it is investigated, the more certain will it appear, while error appears more baseless and absurd, the more closely it is examined. Hence truth can never suffer from proper investigation, which generally explodes error.

We must constantly guard against the influence of prejudices, which tends to make us overlook several things altogether and to consider the rest without due care and attention. A little reflection will generally show us the nature of the bias against which we have to guard: and when the statements made agree with our wishes, we should be particularly cautious in receiving them as true, without the most conclusive proof. On the other hand, we must beware of rejecting as unproved a statement which conflicts with our prejudices.

We are apt to take arguments which chime in with our cherished views or wishes as conclusive, when possibly they may be quite worthless, while we are equally disposed to reject, as fallacious, arguments which militate against our desires. Hence we should never receive the former as satisfactory, till we have tested them most rigidly, while we should never reject the latter as inconclusive, till we have proof that they are so, beyond the possibility of a rational doubt. We can generally ascertain whether we have done so, by submitting the proofs to intelligent persons whose prejudices run counter to our own, and fairly weighing their objections, or by comparing them with the arguments of those who hold opposite views of the subject.

The precise significations of obscure, ambiguous or unknown terms should be ascertained as they occur; for otherwise our labors will only fill our minds with misconception and error. We may have often heard or read many words and expressions of this class, without ever knowing their real signification; and few things have tended more to produce and perpetuate error, than the free employment of such terms by writers, and their ready acceptance by readers, without any precise definitions of the senses in which they were employed or understood. By this means we are liable to think that we understand a subject, when our actual knowledge of it is nearly confined to a string of unintelligible or vague phrases, to which we do not attach any precise and accurate signification, and we have hardly ever looked beyond the mere words.

We should avoid the common error of assuming that a knowledge of names and definitions is tantamount to a knowledge of the things named or defined. In order to understand the exact import of terms relating to phenomena, we must first know, by our own experience, the nature of their primary elements, and then determine what combination of these is denoted by the particular term in question; and we must always consider the thing signified, apart from its name : otherwise, we cannot rightly understand what it is, and our supposed knowledge of the subject will be little more than a compound of ignorance and positive error.

We cannot acquire real knowledge by means of terms whose import we do not know: and, therefore, although it frequently requires time and labor in order to ascertain the exact and proper meaning of words, it ought always to be done; else the study will probably mislead and injure, instead of enlightening and improving. In effecting this, it is often necessary to consider the things denoted, apart from language, and then compare them with the usages and formal definitions of those who best know and most accurately employ the terms by which they are expressed.

The student should possess a good general knowledge of the language employed by the person whose expressions he is studying: otherwise he will be very apt to misunderstand them. He should not presume that he possesses a competent knowledge of the language, because it is his vernacular, of which we learn only the smaller and easier part without the aid of regular study. We should, therefore, be properly prepared on this point, before commencing the study of any subject of importance, in order that we may know what is said, and its real import, which cannot be discovered till we first know precisely what it is.

When the student meets with statements and reasonings which he cannot readily understand, he should consider them deliberately, and compare one part with another. If he should not then see through them, the difficulty should be reduced to writing: for this concentrates the attention, so that it sometimes vanishes by the time it is written down. If it still continue, reference may be made to some other authority; and if this does not suffice, it should be noted, and reconsidered at some future time. A more extensive acquaintance with the subject, or a better frame of mind, may render it very intelligible. Difficulties which resist our efforts to master them, in the course of formal study, frequently disappear when we casually think of them calmly, while we are not particularly engaged otherwise. The mental vision resembles the physical, in our sometimes seeing an object better by means of an indirect glance than by a direct gaze.

The learner is always liable to be drawn away from his proper subject to collateral, but distinct, matters which it suggests; and he must beware of being thus led wholly off his path, or spending so much time on devious questions as unduly to retard his progress towards his proper object. Otherwise he will be apt to run from one subject to another, and resemble a traveler who incessantly deviates to the right and left, so that he never reaches his journey's end.

We should not study longer than we can do so with steady attention; and every part should be considered deliberately, without caring whether this requires more or less time. When a person hurries impatiently forward, anxious only to finish the subject as soon as possible, and merely crams his Memory with what he reads or hears, without observing its character or bearings, or even comprehending it aright, he will generally accomplish nothing of any value, and only acquire bad habits of study. What is apparently acquired, is partly misunderstood and blended with erroneous views; things widely different are confounded with each other; and the little that is really mastered, will be soon forgotten, while the hurried and irksome process is apt to give the learner a dislike for the subject, and to unfit him for careful and thorough investigation.

We should study the successive parts deliberately and attentively, ascertain the exact import of the expressions, master the arguments, observe the connection between one part and another, and ascertain their general bearings and relations. We shall thus avoid the habit of careless and inattentive study, and at the same time obtain a good understanding of the subject of our labors,

G

and remember permanently, at least its more important parts.

When a subject is studied with the requisite degree of care and attention, the learner's progress is apt to be slow at first, especially in studying science; and he is inclined to think that it forms an endless task. But a better acquaintance with it will dispel this error; and he will find that perseverance, care and diligence will enable him to surmount, within a reasonable time, every difficulty which he encounters. In order to this, however, he must begin by concentrating his attention on one short and simple point, and mastering it before proceeding to the next. The art of surmounting difficulties of study lies in this, *attentively considering the simple elements separately.* However complex a subject may be, it always consists of very simple elements, just as the largest book is made up of a few letters, and the most complex machine consists of a few simple elementary structures.

A careful and thorough method of study from the outset, is the quickest, easiest and most pleasant means of acquiring a good knowledge of the particular subject, while it aids us in forming proper habits of investigation, whereas a contrary course not only prevents us from ever acquiring a good knowledge of that subject, but also, by fostering habits of hasty and careless investigation, tends to keep us ignorant on other subjects. It leads us to adopt the errors of others, and add many of our own.

Those who skim along the surface, or run from one study to another, without rightly finishing any, can acquire a good knowledge of none of them, while the bad habits they form, become inveterate. In order to master the subject, train the intellect aright, and form proper habits, the student must proceed deliberately, and avoid hurry and superficial work. These lead to time being lost in acquiring narrow and erroneous views, which must be unlearned before he can acquire an accurate knowledge of the subject; and this now becomes more disagreeable than at the first, because it has lost the charms of novelty.

The statements and reasonings should be tested by the proper criterions, as they occur. Thus, when assumptions are made, on the authority of Consciousness, we should observe whether they are really discernments; where statements are made on testimony, we should try whether it is satisfactory; where a conclusion is inferred from certain premises, or held to have been already legitimately established, we should examine whether this is actually the case; and where a conclusion is logically proved, we should ascertain whether it is the real point in question, or the one which ought to have been proved.

Our inability to detect a fallacy should never lead us to receive an argument as sound, unless every essential part is found, upon proper examination, to accord with the criterions of truth. We are not warranted in assuming that it is sound, simply because we do not see that it is otherwise; and the proper course is, to assume that everything is fallacious, for the truth or correctness of which we have not the evidence of Consciousness, at every step, and in every part: otherwise we shall frequently adopt gross error as demonstrated truth. For the most pernicious errors can be sustained by plausible arguments, which sound and look well, and whose real character cannot be detected without a close and attentive examination. Some attempt to avail themselves safely of others' labors, by simply adopting their conclusions, without any proper examination of their arguments: but this course is always liable to mislead; and it has very frequently led to the adoption of the most pernicious errors.

The best accessible sources of information should always be used; and the student should never rest satisfied with a doubtful authority, when he can consult one which is unquestionable. Thus, he should not take second-hand accounts of an author's doctrines, when he can consult his own writings; a prevaricator should not be listened to, where a person of strict veracity testifies on the point; and he should not pay any regard to hearsay or flying rumors, when he may obtain the statements of a credible eye-witness. So, he should not employ a teacher who expresses himself vaguely, obscurely or unintelligibly, reasons fallaciously, or arranges his materials confusedly, when he may have recourse to another who is free from such defects, and in no important respect inferior.

We should never study at a time more than one subject requiring close and deep thinking, since otherwise we cannot devote the requisite degree of attention to each, because they have a tendency to draw the attention to themselves, even when we are engaged with other matters. But such studies should be intermingled with those which entertain the mind, without seriously impairing its energies.

Study should be diversified, at moderate intervals, with pursuits requiring active, but not toilsome muscular exercise, and forming an agreeable relaxation for the mind: otherwise what is learned is apt to be mastered imperfectly, and speedily forgotten.

The student will be quickly incapacitated for successful efforts, unless he attends to the laws of health, and avoids deleterious agencies and practices: for disease not only injures the senses, and the power of accurate observation or continued application, but it also distracts the attention and impairs the Memory, so that it generally renders successful study impracticable.

Before finally discontinuing the study of a subject, the whole should be reviewed mentally, and reference made to a teacher or a book only when we are at a loss. The foundations, scope and connections of the various parts should be considered, until we clearly understand and remember them. We should also ascertain what properly belongs to the subject, mark the parts of which our knowledge is defective, and rightly employ the best means within our reach for supplying the deficiency. Without some such course as this, it is generally impossible for most persons to possess a good and extensive knowledge of the subject, or to see the amount of their real attainments. We are often apt to think that we know a subject well, when this process will show us the contrary, and convince us that our knowledge of it is much less extensive and accurate than we supposed.

§ 3. SELECTION AND STUDY OF BOOKS.—Great Importance of Writings.—No privileged road to valuable Knowledge.—Means of discovering good and detecting bad Books.—Characteristics of the best works for Beginners.—Two evil Practices.

In every study of any extent, we require the aid of written or printed materials, in order to understand the statements aright, and impress them on the Memory, and also to follow the chains of argument, and test their validity. To be convinced of this, we need only consider the extreme difficulty of remembering the substance of a discourse, after one hearing, or following an ordinary mathematical demonstration, without the assistance of visible words and symbols.

There is no privileged easy road to valuable knowledge of any kind. It is not enough to sit down and passively read or listen to the words of a teacher: we must actively attend to what is said, test it properly, and commit more or less of it to Memory. The best teacher can only assist the learner: the subject must be mastered, and the chief difficulties surmounted, by his own individual efforts. Hence the necessity of having recourse to books, in order to obtain a correct knowledge of any subject.

On most subjects of importance, numerous works have been published, of very different degrees of merit. Some are of such a character that time spent in studying them would be thrown away; and even where several are good, one may be much better than any of the rest. Hence care should be taken to procure the best: and, for this purpose, the advice of judicious persons well acquainted with the subject and its literature, is of great service. But we must beware of regarding the opinion of a distinguished man, on a subject with which he is not well acquainted. A great chemist may be a bad guide in Mathematics; and one well skilled in Elementary Geometry may know very little of Mathematical Analysis.

It is equally requisite to ascertain whether the person to whom we apply is not under the influence of strong personal, party or national prejudices, which may render his opinion unreliable. The opinions of anonymous and unknown critics cannot be safely followed, because they may really know little of the subject, or be influenced by strong prejudices or sinister motives.

In the absence of any reliable advice from others, we may frequently detect a worthless book by certain characteristic marks. Sometimes the very title shows that the writer knows little of the subject, as where it promises a refutation of the Copernican system of Astronomy, or a disclosure of the means of living as long as we please, or of constructing machines that can be moved by weights without ever being wound up. Occasionally the author evidently commits a fatal error at the outset, such as assuming, in effect, the very thing he professes to prove, or attempting to prove something self-evidently impossible. In other instances, he expresses himself in such a manner that we cannot discover his meaning, on fundamental or material points.

In many cases, the character of a book may be discovered by special tests. Thus, the impartiality of an historian may be tried by observing how he handles a subject with which we happen to be well acquainted, just as we test the accuracy of an Atlas, by looking at the representations of parts with which we are familiar. So, the reasoning powers and ingenuity of an author may be ascertained, by observing how he discusses some point of peculiar difficulty; and, in the same way, we can often determine whether he is credulous or sceptical, indolent or industrious, learned or ignorant, modest or conceited, loose or precise in his statements, and so forth.

On first beginning the study of a subject, we should, if possible, choose a work which avoids the dryness of brief outlines, on the one hand, and tedious details and discussions on the other. The former are generally uninteresting for their meagerness, and frequently substitute the shadow for the substance, while the latter weary a beginner by their prolixity. The best book for a beginner is, one which goes systematically over the whole ground, and refers to no other source for rudimentary instruction, which is fair and accurate in its statements, sound in argument, clear and precise in style, and which exhibits the subject in its most improved form, without omitting anything of great importance which falls within its scope.

Having selected a text-book, we should confine our attention to it, till we have perused and understood it, except where we meet with some difficulty which resists our best efforts to surmount it, or some defect which must be supplied from another source. Flying from one treatise to another distracts the attention, and is apt to produce erroneous views of the subject. But, after properly finishing one good general treatise, we may advantageously consult other works, which contain views or matter not found in the former. Our previous attainments will render it unnecessary to do more than run over these productions, and attend chiefly to what is new to us.

We should avoid the pernicious custom of running from one part of a book to another, and not studying it continuously from the beginning, which is the proper method. What precedes, is generally the foundation of what follows, especially in scientific works Hence we cannot rightly understand the subject, unless we peruse the book in regular order; and even the little that we do learn, is so disconnected that it is mostly forgotten, after a short interval.

CHAPTER IX.

OF ORIGINAL INVESTIGATION.

§ 1. GENERAL CHARACTER, USES, PREREQUISITES, AND METHODS OF ORIGINAL INVESTIGATION.—Why Original Investigation is more difficult than Study.—How it differs from Study.—Why its fields are gradually narrowing.—Advantages of original Researches and new Discoveries.—Superiority of personal Observations.—Why those of others should not be overlooked.—Relation of the preceding to the present subject.—Means of selecting subjects.—Seven Prerequisites, and remarks.—Two methods of proceeding.—Subdivision.—Six Rules regarding the course of proceeding.

ORIGINAL investigation presents greater difficulties than study, because it is more indefinite. While we consider only cognitions, we may easily know the precise thing to be done: but when we enter on the region of the unknown, we are often at a loss which way to direct our course. In study, words and direct representations guide us to the knowledge of the things in question: but here we should begin with the latter, and end with the former, which should not be employed till we know the precise thing which we denote by them; and when that is done, we should write this down, wherever there is any danger of our afterwards forgetting or mistaking it.

While the fields of study are becoming more extensive, as time adds to the truths already known, those of original research are becoming gradually narrower, not only because less remains to be discovered, but because new additions to the stock of knowledge frequently render it more difficult to place ourselves on the vantage ground for making further additions. Yet mankind are far from having reached the boundaries of possible attainments; and if new discoveries increase the preliminary labors of subsequent explorers, new inventions may aid them in a corresponding degree, while the wider generalizations and more compact and regular arrangement of materials, resulting from increased knowledge, frequently facilitate future acquisitions more than this difficulty retards them. When the materials are confusedly arranged, and consist of unconnected fragments, intermixed with many errors and redundancies, as is often the case in the earlier stages of a science, it may require more time to learn the little that is known, than it will take to master the whole, after original investigations have given it a more regular and connected form, and banished the crude speculations, positive errors, and prolix irrelevancies of its rudimentary stages.

Original researches or observations are evidently the only means of making additions to the sum of human knowledge, or discovering something that nobody knew before. They also detect errors, or confirm previous researches, even when they fail to make any such additions. Many instances occur, in the history of science, where things assumed for ages, as established inductions, were found, by subsequent research, to be totally erroneous. Of this the old opinion regarding the influence of the Moon on the weather and vegetation, is a good instance.

In other cases a proposition, true in the main, may be found to have been either too wide or too narrow, or to have involved some degree of positive error. It is also no small advantage that disputed points should be established, by an irresistible mass of evidence, beyond further cavil or contradiction, from any person who has properly examined the subject, even although the matter in question may have been virtually established before.

Original observation also gives us, in many cases, a more lively impression and more exact knowledge of things, than we could obtain from descriptions, as Conception is more feeble than Apprehension, and at the same time more liable to mislead us, to say nothing of the numerous mistakes arising from language. But we should not forget, on the other hand, that a person well acquainted with the whole subject, may give us a more extensive and accurate knowledge of it than we could acquire by our own unaided observation. We should, therefore, never fail to examine the statements of other intelligent men who have examined the same objects.

They will generally furnish important aid towards a good understanding of them.

What was said, in the preceding chapter, regarding the number of subjects that should engage our attention at one time, the propriety of exercise and relaxation, and observing the laws of health, are equally applicable to the subject of the present; and, in the higher departments of original research, a careful attention to them is still more requisite, since the mind is more absorbed, and the physical energies more prostrated than when we are only following in the footsteps of others. Here vigorous thinking is generally requisite to success, since feeble and superficial efforts will accomplish nothing of consequence, though continued for years.

Few persons will care to turn their attention to everything worthy of investigation; and knowledge is best advanced by every one's attending to what most interests him, as this is generally the subject to which his previous pursuits and attainments qualify him to do justice. In selecting subjects, the best course generally is, to take up those which we can investigate with the greatest zeal and perseverance; for unless we delight in the pursuit, our labors will generally prove abortive.

The following are the principal requisites preliminary to original investigation.

1. *The mental and bodily faculties requisite to success in the particular department of investigation.* A blind man cannot extend the science of Optic, nor a deaf man that of Acoustic. So, one who is unable to reason accurately and continuously, should not attempt new discoveries in Mathematics; nor should one who is unable to analyse thought, attempt to extend mental science. We should possess those talents which are indispensable to success, before attempting any original investigation: otherwise we shall labor in vain.

2. *A precise and accurate conception of the very thing of which we are in search.* Unless we determine this, we must work, in a great measure, at random, and spend our time and labor without knowing the exact object of our pursuits, so that we shall generally effect nothing of any consequence.

3. *A subject worthy of our investigation, and within the reach of the human faculties.* Otherwise we shall pursue trifles, or attempt impossibilities, and our efforts

will prove abortive. The difficulty of a subject is no criterion of its value. It is harder to chew the shells than the kernels; yet the latter are by far the more nutritious; and the case is frequently the same with knowledge. That which is obtained with most difficulty, may possibly be the least valuable.

4. *A knowledge of previous attainments regarding the subject of investigation.* Unless we know what has been already accomplished by others, we may toil in repeating what others previously achieved.

5. *An acquaintance with the various branches of knowledge which are related to the subject of examination.* Otherwise we lack a good guide to direct our investigations, since one department always throws light on kindred subjects.

6. *A knowledge of the history of that branch to which the subject under consideration belongs.* This will prevent us from attempting what has already been found to be impracticable, and at the same time furnish various hints regarding the best mode of proceeding.

7. *Any instruments, tools, or apparatus which may be requisite in investigating the subject.* No progress can be made in astronomical observations without telescopes, accurate time-keepers, and instruments for measuring small angles. So, the chemist requires retorts, heat, and reagents. Indeed most of the physical sciences require some apparatus in order to extend their boundaries, without which success cannot reasonably be expected.

In making original investigations, we may proceed by two different methods: 1. We may simply observe and record phenomena, including immediate and obvious inferences, as when we travel through a foreign country, and learn its geography, or discover new species of plants and animals. This method may be termed *direct* investigation. 2. We may commence with discernments, assumptions, or the truths acquired by the preceding method, and proceed to trace latent causes or remote consequences and inferences, forming *indirect* investigation. Such are, the discovery of the real motions of the Earth and the invention of a calculating machine. Here it is frequently requisite to employ long chains of reasoning, and hypotheses, or experiments. This method may be subdivided into—*indirect discovery*, where we ascertain new truths otherwise than by direct discovery—and *in-*

vention, where we contrive a new combination, or devise some new means, for effecting a definite object or known end.

The following are the principal rules applicable to both methods:

1. *Obtain clear and distinct comprehensions*, so that you may know the exact thing comprehended, and distinguish it from any other which it may resemble: else your knowledge of the subject will be vague and dim, or you will mistake one thing for another.

2. *Obtain complete comprehensions;* otherwise you will think that you understand the whole subject, when you see only one side of it; and serious error will result. In order to this, we must look closely to every part of the subject, and not turn away from anything because it may be displeasing.

3. *Distinguish probability from certainty, and draw no inferences but such as necessarily follow from your comprehensions:* for otherwise you will inevitably fall into error.

4. *If you form hypotheses, test them by the proper rules for that purpose*, that an erroneous hypothesis may not be substituted for a demonstrated theory.

5. *In classifying your knowledge, adhere to the principles of classification*, so that you may avoid the confusion which will otherwise ensue.

6. *Write down properly, and with the least possible delay, the results of your labors*, so that you may avoid errors of memory. When some interval has elapsed, previous to writing, it should be noted; and we should employ the most appropriate terms for expressing those results.

2. OF DIRECT DISCOVERY.—Requisites to Success.—Common Errors.—Proper Course.—Important requisite, in certain Investigations.—Sphere of Direct Discovery.

This kind of investigation requires sound senses, careful observation, and an application of the simpler processes of generalization, with sufficient judgement to direct the mind to proper objects. It is also necessary to possess habits of close attention, and of distinguishing phenomena from inferences or conceptions, so that one thing may not be mistaken for another, and what is actually seen confounded with what is only inferred or imagined.

To guard against such errors, we must use the means formerly pointed out, for determining what is really comprehended, and what are the direct legitimate inferences.

In all investigations in the physical and mental sciences, we must beware of the once common practice of assuming that things are as we think they must or ought to be, instead of ascertaining how they actually are, which may differ widely from the former. We are tempted to this course by the comparative ease with which disputed or doubtful points may thus be apparently set at rest. Another common error is, to gather up all that leads to a favorite conclusion, and overlook everything of a contrary tendency, which may possibly be more cogent than the former.

We should observe closely and attentively all that is within our reach which has a bearing on the subject, so as neither to overlook anything of importance, nor distinguish things that are essentially alike, or confound things that are otherwise: we should record clearly and accurately, with the least possible delay, what admits of description and is liable to be forgotten: and we should draw no unwarrantable inferences. In order to avoid misdescription, we must rightly understand the thing to be described, and employ suitable terms for that purpose. Where we have failed to make a record at the time of observation, we should beware of stating more than we distinctly remember, and note the interval between the time of observing and writing: otherwise various errors will probably ensue.

In those investigations which involve quantity as a material element, exact numerations or measurements are of much importance, in enabling us to avoid error. Many false theories, adopted for centuries, would have been at once exploded by the application of this test.

Although unexplored regions or subjects furnish the widest fields for direct discovery, yet there is often ample room for them elsewhere; and much that is unknown and important may often be learned from proper observations, made in places often traversed, but never thoroughly explored, or on subjects often considered, but never thoroughly understood.

§ 3. OF INDIRECT DISCOVERY.—Requisites to Success.—Principal Fields.—Uses of a knowledge of kindred departments.—How proper Subjects are discovered.—Next Step.—Extension of Generaliza-

Sec. 3.] Indirect Discovery. 157

tion.—Analogy.—Proper Course regarding it.—Its Uses.—Abuse of Analogies.—Usual course of Indirect Discovery.—Importance and means of readily discovering the most probable Hypothesis.—Indications of such Hypotheses.—Usual results, where no Probability guides the investigation.—How to be avoided.—Importance of analysing the subject mentally.—What is frequently the greatest Difficulty.—Means of effecting the Analysis.—Mode of proceeding where we cannot employ Premises strictly True.—Failures.—Mode of proceeding where there are several Independent Means of testing the Conclusion.—Rules regarding Experiments.—Probable Consequences of disregarding them.—Requisites before drawing a final Conclusion.

In addition to the prerequisites mentioned in the preceding section, the successful prosecution of this kind of discovery requires that the faculties of Intuition and Conception should be vigorous, and that the investigator should possess great perseverance and the power of continued attention. Activity of conception will produce only wild theories, unless its operations are controlled by close reasoning, and its productions tested by careful observation or experiment. Acuteness of apprehension and a strong memory are advantageous; but they are not indispensable.

The principal fields of indirect research are, the sciences and the arts. The former are so connected that, in order to be placed in a favorable position for extending their boundaries, the investigator should possess a knowledge of the elements of all that have a bearing on his immediate subject, in order that the various things which must be considered, may suggest themselves to his mind at the proper time. This requires that he familiarize himself with them, by repeatedly considering them closely and attentively.

The investigator should also be well acquainted with the various parts of the subject, as it now stands, and able to view them in all their principal relations; for a knowledge of one part is often necessary to a right understanding of another, apparently unconnected with it.

It is only by means of the known that we ascertain the unknown: and, therefore, a knowledge of the way in which former discoveries were made, of the suggestions thrown out by their authors, and of the failures that occurred, generally furnishes indications of the means by which a doubtful or disputed point can be either established or disproved, while it guards us against courses which are likely to prove unavailing.

Proper subjects for investigation are suggested by the deficiencies of some science or art, or the manifest advantages of further knowledge, on some point regarding which little is still known.

The next thing is, to ascertain the limits within which the proofs are likely to be found; and here the subject itself generally indicates their source, as they are usually found in other parts of it, or in kindred branches of knowledge.

Some of the most important additions to knowledge consist of the extension of generalization, or of bringing a certain law under a superior law. In effecting this, we are often aided by *analogy*, which is, a resemblance transcending direct apprehension, such as similarity of functions, origin, tendency, or any other relation. Visible resemblances have seldom a place in these investigations, because they are exhausted in forming the lower generalizations.

In searching either for analogies or closer resemblances, we should first refer to those things which most resemble the one under consideration, as these are most likely to furnish faithful similarities. It is proper afterwards to extend our comparisons, in order to reach the higher generalizations, or remove doubts or difficulties. Thus, some obscure points in Physiology have been cleared up by having recourse to Phytology, or that part of Botany which treats of vegetable structures.

The reality of supposed analogies is to be tested by observations or experiments; and the nature of those required is generally indicated by the analogy: but the proper mode of conducting them often requires great skill and dexterity, which is obtained only by some practice and a good acquaintance with the general subject.

The discovery of mere analogies is sometimes a thing of great importance as an end, since they furnish probabilities, which are all that require to be known, in many cases. But they are more frequently important as a means of extending generalization: for, by discovering such resemblances, we are often led to find that things apparently very different, are still of the same class. The polarity of a magnetic needle has no apparent resemblance to the phenomena of common electricity; yet, when we observe that the like poles attract, and the unlike repel, each other, as in the case of two excited electrics, we might suspect that both phenomena belong to one general class, a conjecture which has been fully established.

Analogies also guide us in testing hypotheses: for as every analogy furnishes a probability of the truth of the hypothesis based on it, the closest and most numerous analogies indicate the most probable hypotheses, and thus aid us in ascertaining the truth.

Analogies are very liable to produce a belief that the resemblances extend farther than they really do, and that resemblance in one thing proves a like resemblance in another, or that things which perform similar functions are themselves similar. They have thus led to various errors: and, therefore, we should note the differences, as well as the points of resemblance; otherwise the analogy will frequently mislead us. In comparing a bird with an ox, we should not infer that the two are of the same class, because they have a skull, a spinal column, and a double heart, or that the bird's wings are used in moving upon the ground, because they are analogous to the ox's forelegs.

The usual course of indirect discovery is, to detect similarities between two or more things, and then test the extent of the actual resemblances by means of reasoning, observation, and experiment. If the knowledge of the mere resemblances be not the ultimate object sought, we form hypotheses; and, after rejecting those which either involve absurdities or are incompatible with known facts, we determine which is the most probable, by means of analogies, and then test it, in the manner already indicated. Ready skill in discovering analogies is generally acquired by means of an accurate and extensive knowledge of the particular, and kindred, subjects, combined with habits of close and careful observation, and a vigorous power of conception.

The most probable hypothesis is generally that which is most consistent with known truths, and which accounts, in the fullest and simplest manner, for all the phenomena. It may possibly be false: yet the best course usually is, to test it, before trying any other.

When there is no probability to guide us, we may still observe and experiment: but our labors will generally prove abortive, or end in something different from the object of our search. A person sometimes stumbles on the right hypothesis at the outset, without ascertaining the most probable: but the case is usually otherwise; and indirect discoveries of any consequence are very rarely made by persons devoid of judgement and perspicacity, while they are never accidental.

In order to success, we must generally analyse the subject mentally, and thus ascertain its various parts. Things are mostly presented to our immediate observation in so complex a form that it is impossible to discover the object of our investigation, without considering the various parts in succession: for nobody can consider several points simultaneously, when every one of them requires close attention from the same faculty.

This analysis is equally useful in detecting analogies, deducing remote inferences, or establishing hypotheses. The greatest difficulty generally lies, not in actually accomplishing the results, but in discovering clearly how they are to be effected, and in performing the initiatory preparations; and, in order to this, the mental analysis is generally indispensable. It enables us to resolve a complex or difficult problem into a number of easy steps, every one of which prepares the way for effecting that which follows; so that we pass from one to another without any great difficulty, verifying the different processes as we proceed.

The analysis is generally effected by concentrating the attention on a particular point, considering it, in its various aspects, till we clearly see its bearings, then proceeding to another point, and so on. The points to be selected for consideration depend on the nature of the subject and the object of the investigation.

In many cases, it is impossible to reach our final object from any premises which are strictly true. We, therefore, make an assumption which is partially inaccurate, then deduce the consequences, and afterwards discover the modifications to be made in our conclusions, in order to render them quite accurate. The Astronomer cannot determine the future place of a planet by starting with the truth regarding the various forces that affect it, as no human intellect can solve so difficult a problem. But he first assumes that it is influenced solely by the Sun; and he afterwards determines what is due to the disturbing influence of the other planets, for which he then makes proper allowances.

Much judgement is often required in discovering the best course to be adopted, which is frequently done only after several other methods have been tried, and failed: and, therefore, such failures are to be expected, and must not discourage the investigator. They have often preceded important discoveries.

Where we are furnished with several independent means of testing the point under consideration, the best course generally is, to trace carefully the result to which we are led by each, irrespective of the others. If all conduct us legitimately to the same result, the conclusion is firmly established. If the results are inconsistent, there must be some fallacy in our proceedings, since one sound conclusion cannot possibly be inconsistent with another; and we should search till we detect the error. If some of the processes merely fail to give any reliable result, while others conclusively establish the point, the negative results are entitled to no weight.

Experiments frequently furnish important aid in this class of discoveries; and the following general rules will assist us in their application:

1. Form a clear and precise notion of the *object of the experiment*, or the purpose which it is designed to accomplish.

2. Remove, as far as practicable, every source of *doubt or uncertainty* regarding the result, so that it may be as unequivocal as possible.

3. Test the *inferences* which the result seems to warrant, by the principles of sound reasoning, so as to obviate fallacious conclusions.

4. Where a *result is equivocal*, or otherwise unsatisfactory, repeat the experiment, with such variations as seem best calculated to remove the difficulty.

5. If the result should still continue unsatisfactory, *adopt*, if practicable, *a different kind of experiment*, which promises to obviate the difficulty.

If these, or similar, rules are not observed, the results will be more likely to rivet error and conceal truth, than to dissipate the former or establish the latter, since prejudice will persuade the investigator to adopt its own dictates, and reject hostile, though necessary, inferences.

Before drawing a final conclusion, we should have satisfactory proof that we know everything requisite to a right view of the subject under investigation: otherwise we shall probably adopt some theory which misrepresents the subject. Most of the objectionable theories ever broached were owing, in a great measure, to ignorance of some important parts of the subject, which were overlooked.

§ 4. OF INVENTION.—Relations of Invention to indirect Discovery.— Use of Analogy.—Inventions of two kinds.—Principal fields of Invention.—Mathematics.'—Physics.—Education.—Social Institutions.—Requisites in Inventions immediately regarding Man.— Why Legislative and Social have not kept pace with Physical Inventions.—External aids of Invention.—How the place of Models may be supplied.

Invention resembles indirect discovery so much that the remarks made in the preceding section are mostly applicable to it. The principal difference is, that we have here to contrive some means of solving a given problem, instead of discovering a proof or refutation of a given proposition, or deducing remote inferences. Hence there is rather less occasion for simple observation or comparison, and we are more dependent on an extensive and familiar acquaintance with the subject of investigation, and the power of readily forming conceptions and testing them by reasoning and experiments.

Analogy is generally less available than in discovery: yet it is often of great use, particularly in the earlier stages; and many important inventions originated in observing the modes in which certain ends are effected by natural means, a source of assistance which cannot easily be exhausted.

Inventions are either *physical* or *mental*. The former consist chiefly of tools or machines, such as saws, planes, files, ships, clocks, and steam-engines. The latter consist of means for enabling us to acquire or retain knowledge, or to perform some process with correctness and dispatch. Such are, the ordinary rules of Arithmetic, and all those arts in which the processes are not obvious or dependent on machinery, and also the modes of influencing emotions and opinions.

The principal fields of invention are, the mathematical and physical sciences, education, social institutions, and the arts.

The objects of mathematical invention are, the solution of problems, or the application of Mathematics to extend the boundaries of Physics. In some cases, modes of effecting the desired results are already known; but they are too tedious, or not sufficiently accurate, or too liable to lead us into error. In such cases, the inventor's labors are requisite, until he has discovered the means of effecting the desired results with the greatest attainable quickness and accuracy.

In the physical sciences, the fields of invention are widest where the application of Mathematics is most extensive. But there is room for improved methods of making observations and experiments in others also; and these are the principal means by which they can be corrected and extended. A wide field for physical invention is furnished by the application of the various natural forces or powers to produce desired results, including those of gravity, expansion and contraction of substances by differences of temperature, chemical action, heat, light, and electricity. The inventor will derive much aid from studying the effects produced by these agencies, under the various circumstances in which they operate.

The principal subjects of invention, in education, are, the best means of disciplining and instructing youth, so that all the faculties may be properly developed, while the pupil, at the same time, acquires the elements of useful knowledge. No system of education is satisfactory, unless it effects both of these objects, which are so related that the same system which best secures the one, is also the most favorable for accomplishing the other.

In social institutions, the principal subjects of invention are, public and private law. The aims of invention, in relation to the former, are the two following: 1. To ascertain that form of government which will be most favorable to the enactment, enforcement and perpetuity of the best private laws. 2. To determine the best laws for regulating the intercourse between different nations and their subjects, so that it may be as safe, free and mutually advantageous as possible.

In all inventions of which man is the immediate subject, the investigator should be guided by a correct and extensive knowledge of human nature in general: otherwise his inventions will possibly be suitable only for imaginary beings. Another important requisite, in all inventions of this kind which are intended for immediate practice, is, a knowledge of the peculiar characteristics and circumstances of the parties for whom they are designed. A system of education or government, for instance, may be the best possible for one community, and decidedly objectionable for another, owing to the wide disparity in their characters and circumstances.

Improvements in social institutions have not generally kept pace with those in physical science, because the laws on which they should be based are not so easily discovered, and the pursuit has generally been conducted against the influence of strong prejudices and external disadvantages. Yet these difficulties do not render progress impossible.

In all inventions, we may be aided by writings, drawings, and experiments; and, in several, chiefly of the physical kind, we derive much assistance from models, or actual representations of the proposed invention. The faculty of Conception is seldom so powerful that it can figure forth and clearly discern, at a single effort, all the parts and relations of a complex invention. Hence the inventor first makes a representation of his present conception; and when he has this clearly before him, improvements and additions are easily planned and effected. A carefully constructed model, after being fairly tried, generally gives us a good view of the defects and merits of the invention, as it now stands, and suggests improvements better than anything else, short of a long practical trial.

In many cases, experiments and the construction of models are impracticable; yet simple observation may answer the same purpose. Thus, if a certain scheme of social organization has occurred in History, in circumstances which gave it a fair trial, and it was found to fail, no model or experiment is requisite.

CHAPTER X.

OF CAUSES AND EFFECTS.

§ 1. SOURCES AND APPLICATIONS OF THE KNOWLEDGE OF CAUSES AND EFFECTS.—Two classes of Causes, and means of ascertaining them.—Experience, when requisite, and when not.—Advantages of knowing the Laws of Nature.—Inadequate Agencies.—Means of determining a Cause indirectly.—Important application of the Laws of Causation. — Advantages of a knowledge of Controllable Causes—and of Causes beyond Control.—Importance of knowing what Effects will be produced by a given Agency.—Tracing the implied Consequences of a known Agency.—Moral bearings of a Knowledge of Causes.—Cause and Effect signs of each other.—Important Distinction, and Practical application.—Caution.

CAUSES may be divided into *necessary* and *contingent*. The former consist of those whose existence is discerned intuitively, without any direct aid from Comprehension: the latter include those whose existence and nature cannot be known without the aid of experience, although, as causes are distinguishable from mere phenomena, they are never discovered without the application of Intuition.

Contingent causes cannot generally be discovered without the aid of observation or experiment, since there are several agencies to which the effects or phenomena may be attributed, and experience alone can inform us which is the actual cause. Thus, if a rock roll down from its former fixed position, on a mountain's side, we know intuitively that there must be some cause for the change: but what the particular cause is, must be learned from the circumstances of the case.

On the other hand, the existence and general nature of necessary causes are known solely by Intuition. Thus, we know that the voluntary acts of a man are caused by motives, which must be objects that he deems desirable. So we know, independently of experience, that the causes of our ordinary apprehensions are, external substances, distinct from ourselves, and that an exquisitely finished time-keeper is the production of skill, and neither of chance nor of a bungling artisan. Wherever the phenomenon is such that Reason can demonstrate it could proceed only from a particular kind of agency, we do not require the aid of experience to teach us that it is the effect of such an agency.

Although experience generally lies at the root of our knowledge of the particular cause of an occurrence, yet we need not have recourse to it, in every case, in order to ascertain the very agency concerned. For, when once we have found out that all the changes which usually take place in nature, are immediately owing to certain agencies, and that these act with great uniformity, we can frequently determine the cause of a phenomenon, without any elaborate investigation. Thus, in the case mentioned above, if we know the effects of freezing water, and that the fall of the rock was preceded by a severe frost, we have at once an explanation of the occurrence; and when we see a rainbow, we know its cause, without any investigation.

A knowledge of the laws of nature frequently enables us, not only to ascertain readily the cause of a phenomenon, but also to exclude possible causes, as being inadequate, when the particular cause is still undetermined. An agency is *inadequate* when the laws of its operation, or its want of skill or power, are incompatible with the supposition of its being the cause of the phenomenon in question. Thus we cannot attribute warm weather to the position of the Moon, because it neither gives out nor produces any heat itself, nor does it influence that imparted by the Sun, in any measurable degree.

A knowledge of the requisites necessary to produce a certain result, enables us to determine the nature of a cause of which we have no direct knowledge. Thus, we judge of the skill of a mechanic solely from a view of his handiwork; and we ascertain the force of volcanic action by witnessing the effects which it has produced. Thus, also, we ascertain the motives and characters of men, when there are no reliable means of doing so directly, from continued observation of their conduct. The nature of the inapprehensible cause is learned from its effects, as the quality of an unseen tree is known by its fruits.

A knowledge of the cause of a phenomenon enables us to modify it at pleasure, wherever we can control the former. Thus, if a certain disease is found to be caused

SEC. 1.]   APPLICATIONS OF KNOWLEDGE.   167

by miasms, arising from a marsh which can be drained, we can stop its ravages whenever we like; and if a public speaker knows what causes will produce certain emotions in his hearers' minds, and can apply them, he may produce these emotions at pleasure.

Where the cause is not directly under our control, a knowledge of its nature may still enable us to counteract its influence. Thus, we may neutralize the effects of miasms which we cannot remove, by avoiding those exposures and excesses which render us peculiarly susceptible of their influence; or we may directly counteract them, by artificial heat or warm clothing.

When the object is, not to avoid, but to secure, effects whose causes are not under our control, a knowledge of these causes aids us in finding that of which we are in search. Thus, a knowledge of the origin of coal directs us to look for valuable mines only in such rocks as were deposited while vegetation was very luxuriant; and it informs us that no coal whatever is to be found in the non-fossiliferous rocks, and that no extensive deposits can be expected either in the oldest fossiliferous rocks or in those of recent formation.

When the cause is beyond even our indirect control, a knowledge of its existence and nature may still enable us to keep beyond the reach of its influence, and thus avoid the pernicious effects. If a person has contracted a disease, by living in a very damp locality, he may possibly remove to a dry atmosphere, and thus get well, whereas, if he knew not the cause of his ailment, or attributed it to some imaginary agency, he would be more likely to be injured than benefited by attempts to effect a cure.

It is equally important to know what effects will be produced by a given agency, acting in peculiar circumstances. A knowledge of the properties of steam and electricity enables us to produce results that would once have appeared incredible; and, by knowing the effects of certain mental agencies, we can influence the views, feelings, and conduct of others, and remove what is bad, or improve what is weak or defective, in our own characters.

A knowledge of the nature of a certain agency often enables us to ascertain the existence of effects, of which we have no other evidence except that they must result from the agency. Thus, many facts in Astronomy were discovered, as necessary effects of the force of gravity, before they were known by observation.

By deducing the implied consequences of a known agency, we can often discover other truths, beyond its mere effects. Thus, the effect of the valves in the blood-vessels is simply to check a retrogressive or backward course of the blood, as it circulates in the arteries and veins; and this is, no doubt, the object for which they were formed: but, by investigating this object, Harvey discovered the circulation of the blood; and other important physiological discoveries have been made in the same way.

A knowledge of causes is as important in its moral bearings as in any other. When we view an admirable piece of mechanism, and understand the skill and dexterity necessary to plan and execute it, we partake, through sympathy, in a greater or less degree, of the emotions of him who produced it: for these feelings are evidently dependent on a knowledge of the causes that produced the results, and proportional to its extent and accuracy. It is for this reason that a good artisan beholds a masterpiece with much greater pleasure than one who supposes that such things can be produced as easily as common tools.

The most important aspect of this subject is, that which regards the works of the Eternal Architect. These are incessantly presented to our view, in vast profusion, and under an immense diversity of forms, from the depths of the Earth (whence have so often issued streams of liquid fire, bearing the elements of future fertility, mixed with iron, gems, and gold) to those remote constellations whose numbers, sizes and distances are so great that the contemplation of the whole amazes and confounds the mind.

Not only does a knowledge of the agencies that operate in nature, enable us to supply many defects in our knowledge of its phenomena, and to form a more correct and extensive notion of the creation than is otherwise possible, but, when we understand the difficulties which have been overcome, in accomplishing the results that we behold, and learn the origin and purposes of the various parts, we obtain an effective knowledge of the Most High, and partake of those emotions which unite us to our Creator, and which alone can satisfy man's immortal mind, when all lower enjoyments shall have ceased.

As cause and effect, or the common effects of a cause, are inseparably connected, wherever there is no interfering or counteracting agency, they are signs of each other; and, therefore, when one is known to operate or exist, the other may be inferred. It is thus that the geologist discovers the causes of phenomena long after they have ceased to operate. So, a physician learns the origin, character, and future course, of a disease, from present symptoms; and a statesman sometimes foresees future events, as the effects of existing agencies. In this way we learn many past and future contingencies, since they are connected with the present as cause and effect, or successive effects of a common cause; and we may frequently learn the present unseen from the seen, since the one either produces the other, or both are contemporaneous results of the same agency.

We must distinguish a knowledge of the effects of certain agencies, or of the causes of particular phenomena, from that of the mode in which the causes operate. We may possess the former, while we are quite ignorant of the latter. Thus, we may know that certain poisons speedily destroy life, without knowing how they do so, just as we know that we move our limbs at pleasure, although the mode in which we do so, is a profound mystery.

The value of the former kind of knowledge is evidently independent of the latter, and it may be of the utmost importance, though we should continue quite ignorant of the mode of operation. This, however, is a legitimate subject of inquiry, though we shall often find that we can make little real progress in the investigation. *Words* have sometimes been freely employed, in such cases, while those who used them had no clear idea of what they meant by them.

§ 2. VARIOUS KINDS OF CAUSES.—Efficient and Conditional Causes.—Immediate, Mediate, and Ultimate Causes.— Only Ultimate Causes.—Common Error.—Remarks on the Laws of Nature, and on Ultimate Causes.

All causes are either *efficient* or *conditional*. The former consist of those forces which actually produce the effects, and are otherwise termed *powers:* the latter are, the circumstances or conditions which are requisite to the production of the effects; and they are also termed *occasioning* causes, or simply *conditions*. In popular language, that is generally termed the cause which is most subject to change, whether efficient or conditional. Thus, if a man slips, when walking on ice, his fall is said to be caused by its slipperiness, although this is only a conditional cause, while the force of gravity is the efficient cause.

Causes, again, are *immediate*, *mediate*, or *ultimate*. The *immediate* cause is that which directly produces the result, without the intervention of any second agency. The *ultimate* cause is, the good or evil which originates the volition that leads to an effect. A *mediate* cause is one that intervenes between the ultimate and the immediate cause, and forms a necessary link in the chain. When a blacksmith hammers a bar of iron, the stroke of the hammer is the immediate cause; the thing which led him to will the hammering is the ultimate cause; and the movements of his arm are an intermediate cause.

The causes of volitions are the only ultimate causes, since eternal spontaneous motions of inanimate beings would be effects without any adequate causes; and, if such beings are at rest, they will evidently move only as they are moved. Hence, although they often communicate, they never originate motion, either in themselves or in other beings.

When a phenomenon has been shown to belong to a known class, it is often thought that its ultimate cause has been unfolded, while, in fact, no explanation has been given of any such cause. Thus, when certain motions of the Moon are traced to the force of gravity, many think that their ultimate cause has been explained, whereas the law of gravitation explains neither the nature nor the origin of the force which produces those motions.(15)

Most of the laws of nature are merely inductions regarding occurrences or phenomena, and give no explanation of any cause whatever. When we say, for instance, that "metals expand, on the application of heat," we do not explain why any of them does so. Even those laws of nature which express causes, seldom explain the ultimate cause. Thus, it is a law of nature that "heat expands gases;" but this does not explain why it does so; for heat is as destitute of thought as the gases.

In several branches of Physics, it is not necessary to investigate ultimate causes; but, even there, it is a serious error to suppose that we know more of the matter than we actually do; and, in many physical researches, the consideration of ultimate causes is of much use in promoting discovery. In the mental sciences, the subject is one of the utmost importance.

§ 3. METHODS OF DETERMINING CAUSES AND EFFECTS.—Divisions of the subject.—Inadequate and absent Agencies.—Effect attributable to only one Cause.—Case of combined Agencies.—Means of determining the influence of each.—Application of the fact that Effects follow their Causes.—Proper course when Cause and Effect appear simultaneously.—Application of a knowledge of the Laws of Nature.—Frequent means of ascertaining the existence or absence of possible Agencies, and the effect or intensity of a known Agency.—Caution.—Proper course regarding known and unknown Agencies.—Seven Principles applicable where preceding methods fail, with Remarks.—Proper course where one Principle gives equivocal Results.—Sole and combined Agencies.—Means of ascertaining the nature and extent of a particular Agent's influence.—Use of Analogy.—Common Mistake.—Use of Experiments. —Effects which cannot be traced to any known Cause.—Errors regarding them.—Mode of tracing a chain of Causes.—Means of determining what Causes would produce a supposed Effect.—Modes of ascertaining the effects of a known Cause.—Use of Experiments in such cases.—Proper course where these are inapplicable.— Means of tracing the Effects of Causes which have ceased to operate.—Independent Effects.—Cases of Reciprocal Action.—General Requisites, and Cautions.

All inquiries regarding causes and effects are necessarily either into the causes of known or supposed effects, or into the effects of known or supposed causes; and each inquiry contains three subdivisions, which may be stated as follows:

A. 1. What causes, now inoperative, have produced known effects. 2. What causes, now operating, produce known effects. 3. What causes would produce certain supposed effects.

B. 1. What effects have been produced by known causes, no longer operative. 2. What effects known operating causes are now producing. 3. What effects would certain supposed causes produce.

Both classes of inquiries are based on the same general principles, so that we need not discuss every subdivision separately. In many cases, also, we do not know beforehand whether the causes are still operative or not, as that can be learned only when we have ascertained what they really are.

The two intuitions that a cause must be adequate to produce the effect, and that it cannot act where it is not, except by a medium, frequently enable us to exclude many possible agencies from the list of admissible causes. Thus, the vapors arising from cooking utensils cannot be the cause of heavy rains; and a man who was in China when a murder was committed in New York cannot have had any direct hand in it. In applying those intuitions, however, we should ascertain that the agency in question is inadequate or absent: for this has frequently been assumed where the case was otherwise.

In many instances, the circumstances are such that an effect can be attributed only to one cause. These are chiefly where we know that no other adequate agency operates in producing the result. Thus, the immediate cause of the change of seasons must be solely the different directions of the Sun's rays, since there is no other agency which has even a tendency to produce those changes. But we should know that no such agency exists: for we are liable to assume this, when we are simply ignorant of the existence of any such cause, while we may erroneously assume something else as the cause, which is a mere antecedent or concomitant. The requisite knowledge is generally obtained by extending our observations, or repeating our experiments, until the supposition of casual or latent agencies is excluded.

A certain agency frequently operates in producing a result, while it is only one of several causes, each of which may be inadequate to produce the effect: and, therefore, before we can legitimately conclude that an effect is produced solely by a single agency, we must know that no other contributes, in any degree, to produce the result.

Where several agencies co-operate, it is often important to ascertain their comparative influence. This is to be done, either by reasoning from intuitive principles, or by observing the force of each, where it acts singly. All we can ascertain, in many cases, is, that one agency is much more potent than another; but this usually answers the purpose, and greater accuracy is of little consequence.

Where the agents are different in their nature, and every one is essential to the production of the effect, we cannot even institute a comparison between their respective influences. Thus, we cannot rightly say that food is more essential than water to sustain life, since each is indispensable. But where the agents are alike in kind, and their power admits of measurement, while the effects of one are only cumulative to those of another, we may possibly ascertain the exact amount due to each. Thus, when several steam-engines are employed to drain a pool, it may be easy to ascertain the exact amount of water removed by each.

Where several similar causes unite in producing a certain result, the total effect is generally a combination of the separate results. If two mechanical forces act in the same direction, the combined force is equal to their sum; and, if they act directly against each other, the result is equal to their difference. But experience is often necessary, in order to determine the actual results: for there are various exceptions to the general rule. White light, for example, is a combination of blue, yellow, and red, each of which is much darker than the compound.

Every effect is necessarily preceded by its immediate or mediate cause; and this often enables us to determine whether a particular agency is concerned in producing a certain result. For, if the agency appears subsequently, or even simultaneously, it cannot be the cause. But when an agent acts continuously, and the effect appears in the same manner, cause and effect appear simultaneously, although every part of the effect succeeds the particular act by which it is produced. Here we must compare the time when the cause began to operate with that when the effect first appeared, and the time when the former ceases to act, with that when the latter ceases to appear.

The peculiar nature of many agents excludes them from the supposition of their being concerned in producing certain results. For experience, and the intuitions of causation, inform us that they either are neutral, or that their results differ materially from the object in question, or that they are even the reverse. We cannot consider the geological formation of a country the cause of the change of seasons, or unstratified rocks as the results of aqueous deposits, or light the cause of darkness.

This method of limiting the possible causes of an effect supposes that we already know the characteristics of the various agencies; and hence it becomes more applicable as knowledge advances. Owing to their general ignorance, rude nations have often assigned agencies as causes of results with which an intelligent person would readily know that they have no connection. Pieces of human mummy, ground to a powder, were long considered excellent medicines for certain disorders, although everything of that kind is as useless for any such purpose as it is disgusting.

An extensive knowledge of the laws of nature enables us, not only to exclude certain agencies from the list of admissible causes, but also to conjecture the actual cause, amid many which are not absolutely inadmissible. Thus, by knowing the properties of water and heat, we can readily ascertain the causes of many geological phenomena. A knowledge of those laws also aids us in determining when an unexplained phenomenon is to be attributed to some agency previously unknown, and in forming an accurate conception of the nature of that agency.

The existence or absence of a possible agency may frequently be ascertained by observing whether or not its constant effect is present, on the principle, formerly stated, that the presence of an effect proves the existence of its cause, and the absence of an effect proves the absence of its cause. Thus, the formation of ice on the waters proves that the weather has been cold, and its absence proves the reverse. In applying this principle, we must beware of assuming, without proof, that an effect can have proceeded only from a particular cause, or that the phenomenon in question is a constant effect. We should also ascertain that there has been no extraneous interference with the agencies or phenomena, either accidentally or from design.

In determining the effect or intensity of an agency, we are guided by the intuitions that *like effects will follow, in the same circumstances,* and that *an effect which depends solely on a particular cause, varies in proportion to the changes in the cause; and changes in the effect must have been preceded by corresponding changes in the cause.* Thus, a knowledge of one element enables us to know the amount of the other. But we must know that no other agency is concerned: otherwise our conclusions may be very erroneous. Heat generally expands liquids; but it contracts water, when near the freezing point; and when it reaches the boiling point, instead of farther expanding, it evaporates in the form of steam.

In searching for the cause of a phenomenon, we should

first ascertain whether it is not produced by some known or familiar agency: for, until this is done, we can evidently have no proof that it results from an unknown or new agent. We should make no assumptions in favor of either class of agencies, but be guided by probabilities and proofs. The mere fact that a certain agency operates, does not prove that it is the cause of a particular phenomenon, with which there is no proof that it is connected, and which may be wholly produced by some other agency.

Where none of the preceding methods furnishes the requisite information, we must have recourse to farther observations or experiments: but sometimes the former are sufficient, and the latter are impracticable: in other cases, these afford the readiest means of solving the problem. In all cases of this kind, important aid may be derived from the following principles, which are only modifications of those of causation, already stated.

1. *The effect must always appear where the agency operates freely, and never appear where it has not previously acted.* Hence, if the supposed cause is found to act freely, without being followed by the effect in question, or if this is found to exist, where the agency has not operated, it cannot be the cause.

2. *The commencement of the free action of the agency must be followed by the appearance of the effect, where all other things continue the same as formerly; and the effect must cease to be directly produced, when the agency ceases to operate.* Hence, where the supposed cause commences to act freely, and the effect in question does not begin to appear, or where this does not cease to be directly produced, when the supposed cause has ceased to operate, the agency is not the cause. The continued motion of a body, after it has once been moved, seems to contradict this principle: but, in reality, it does not; for the motion is only a continuation of the body's preceding state, without any additional effect. When a moving body changes either its direction or its velocity, there is a change, for which there must be some adequate cause: but a continued motion, in one direction and with a uniform velocity, is all the unchanging effect of the agency which first communicated the motion.

3. *Changes in the agency must be followed by corresponding changes in the effect; and changes in the effect*

must have been preceded by corresponding changes in the agency. Hence, if one change without any corresponding change in the other, the agency is not the cause. It must be observed, however, that, in both cases, the corresponding changes are not necessarily similar: and, in fact, although they frequently are so, the case is often otherwise. Thus, a certain degree of heat produces a pleasant sensation, while a great degree produces pain.

4. *The presence of the peculiar effects of a certain agency proves its previous action; and the absence of its uniform results proves the reverse.* Hence, where the former are found, the action of the supposed cause is established, and where the latter are wanting, it is disproved.

5. *The time that elapses between the action of the agent and the appearance of the effect, must conform to the nature of the agent and the thing on which it operates.* Hence, if the intervening time be greater or less, the supposed agency is not the cause. Thus, fire directly applied to gunpowder must cause an explosion immediately or not at all; and a man cannot have died from the effects of a little arsenious acid swallowed seven years previously, while the decease of one who dies instantly after taking a few grains of that substance, must be owing to some other cause.

6. *Voluntary acts must proceed from motives known to the agent, and must harmonize with his character.* A motive wholly unknown to an agent, or one to which he attaches no weight, cannot influence his conduct. Thus, brutes are never influenced by a regard for a future state, and a thoroughly selfish man never makes great sacrifices purely from philanthropic motives.

7. *The motive must be adequate, with reference to the agent's views and belief.* A motive may strongly sway one person which would have little influence with another. Hence, we must ascertain, not only that the agent knew of the supposed motive, and that it harmonizes with his character, but that it is one which may have produced the effect in question, under the circumstances. A miser readily makes great sacrifices for money: but he will not knowingly barter everlasting bliss for it, as this is an impossibility; and hence the absurdity of the stories that certain persons deliberately made a compact with the devil, to be his forever, on condition of his imparting to them certain magic powers. Nor can a miser

be supposed to have been influenced by the desire of gain, when he must have seen that the consequent pecuniary loss would inevitably be much greater.

The preceding principles all assume that the agencies operate freely, or without any counteraction, and that no conflicting or extraneous agencies interfere with the ordinary results: and they are applicable only upon these conditions.

There is an evident necessity for an agent's conforming to one or other of the preceding tests, in order to proving that it is a cause, and not merely an antecedent or concomitant, supposing that there is no other satisfactory proof. Its unequivocally fulfilling one of those conditions, may furnish the requisite proof. But the evidence of one is frequently unsatisfactory, owing to the narrowness of the phenomena, or the interference of extraneous or counteracting agencies, or our investigations having been either too limited or not conducted properly. Here we should apply other tests, and continue our researches, till we find some conclusive proof. We should first try the most probable cause, and, if the result be adverse, try the next most probable, and so on, till we either discover the true cause, or find that it lies wholly beyond our former conceptions. The principle that *similar effects generally spring from similar causes* will frequently suggest the nature of the unknown agency; and further aid may be derived from the proper application of hypotheses.

In order to establish an agent as the sole cause, it must appear that the result is unaffected by changes in the other agents that might possibly be concerned in its production. When it is found that several agents are jointly concerned, we should test the interference of others, as if the combined group were a single agent. We find what agents are joint causes of a result, by testing each of them separately, until we have ascertained all that are so concerned.

In investigating the causes of the ordinary decomposition of organic substances, for example, we first find that a certain amount of heat is requisite, as one cause: for wherever water freezes, decomposition ceases. Again, we find that moisture is requisite: for decomposition never proceeds without the presence of water, either in the liquid or in the gaseous form. Lastly, we find that air is another requisite: for wherever it is removed, as in the exhausted receiver of an air-pump, decomposition does not take place.

Thus we find that ordinary decomposition requires the presence of heat, moisture, and air; and, as it always proceeds where these are present in sufficient quantities, and without any counteracting agency, these are the only causes, although other agents may accelerate or retard their influence. The nature of the process is shown by comparing its products with the substances previously in contact. We thus learn that some of the organic compounds are decomposed, and unite with the gases that compose air and water.

The nature and extent of a particular agent's influence, where it co-operates with others, is ascertained by comparing the result produced where it acts with what appears where it is absent or inoperative. Thus, the influence of atmospheric resistance on falling bodies or projectiles, may be determined by comparing the ordinary phenomena with those produced by experiments performed in a vacuum.

In many instances, the agent always operates, but with very different degrees of intensity. Here its influence may frequently be ascertained by the third of the above principles. Thus, it is impracticable to produce a perfect vacuum; yet the influence of the atmosphere may be ascertained from observing the variations in the bodies' motions, as the density of the resisting medium varies.

In searching for the cause of a known phenomenon, we are often aided by analogy, as like effects generally spring from like causes. But we should guard against the erroneous supposition that causes resemble their effects. There is nothing in a musical instrument or its motions that resembles its sounds, nor is there anything in a rose resembling its odor or color; and so of all our senses. Yet this error has prevailed extensively, owing partly to the common tendency to confound cause and effect, and partly to overlooking the wide difference between direct resemblance and corresponding intensity. The effect generally varies as the intensity of the cause: but this no more proves resemblance than the fact that the more a man spends, the less he has, proves that wealth resembles poverty.

Although experiments cannot aid us directly, in ascending from an effect to its cause, yet they frequently enable us to test the influence of a possible agent, in producing the effect: for we have only to put it into operation, and note the results. Thus, if we are investigating the origin of basaltic columns, we may take some of the rock, melt it, and allow it to cool under pressure. If we now find that a columnar structure results, we have ascertained one cause which may have produced the phenomenon in question.

A single experiment of this kind, however, teaches us only that such an agency was probably the cause, except where there is no other admissible cause: and, therefore, it generally requires repeated experiments, to determine the question. Thus, if we are investigating the origin of mineral veins, we may try if we can produce such phenomena from aqueous deposits. When this experiment gives a negative result, we may try chemical precipitation. If this should fail, we may try electro-galvanic agency: and if this fail, we may try the influence of several agencies combined.

When it is found that an effect is not produced by any known cause, it must be attributed to some new kind of agency, whose nature is to be ascertained by observation or experiment and testing hypotheses. Here we must beware of adopting an agent as the cause without conclusive proof, as we are liable to assign imaginary causes, many instances of which occur in the scholastic philosophy. The descent of heavy bodies, for instance, was attributed to a natural tendency downward, just as the phenomena of gravitation have been more recently attributed to a natural tendency of bodies to move towards each other, whereas no such thing exists, the natural tendency of bodies being to remain as they are, and neither to move nor to stop moving, except as they are made to do so.

In tracing a chain of causes, the easiest course generally is, to ascertain first the immediate cause, and then the succeeding causes successively, in their order. Thus, the real motions of the planets are the immediate cause of the phenomena which they present to our view. Those being determined, the next point was, to ascertain the causes of these motions; and when this was done, the law of gravitation still remained to be accounted for. Supposing it should be traced to undulations of ether, we might still inquire into the cause of these, and thus proceed, till we came to the direct action of God, which is the ultimate efficient cause of all natural phenomena.

In investigating what causes would produce certain supposed effects, we reason from intuitive principles, combined with our experience as to what causes produced precisely similar effects; and the course of proceeding is essentially the same as that just discussed. We argue on the principle that a cause which has produced a certain effect, will produce the same effect again, if operating in the same circumstances.

In ascertaining the effects of known causes, the processes are very similar to the preceding: but they are generally simpler, and the aid of direct observation or experiment is frequently more extensive. For, as the agents, or others quite similar, generally operate within our view, we can observe the results, whereas the causes of many visible effects have long ceased to operate, and therefore observation may be unavailing. Sometimes, again, a single well-conducted experiment is conclusive, as an agent will always produce what it has once produced, in the same circumstances, while the fact that a certain agent produces effects precisely similar to the one in question, does not prove that this did not spring from an agency in some respects widely different.

In many important cases, however, experiments are inadmissible or impracticable. Thus, we can rarely test a political theory, by starting a community organized on its principles; and we cannot, without imminent danger of sacrificing life, try the influence of powerful newly discovered medicines, by administering them experimentally. In all such cases, we must have recourse to the more indirect methods already explained, the rules for testing effects being substantially the same as those by which we ascertain causes.

Where the causes in question have ceased to act, we may observe the effects of perfectly similar causes which still operate; and, if none such exist, we must extend our observation, and apply such of the preceding rules as are applicable to the nature and circumstances of the case.

The fact that *similar causes generally produce similar effects*, often aids us in determining the effect of a cause similar to one whose effects are already known. But the same caution is requisite here as in the case of the converse principle stated above: for there are numerous exceptions. Thus, not only do different degrees of heat sometimes produce very different effects, but those of a freezing cold strongly resemble a burn.

Where an agent produces several distinct effects, independent of each other, every one of them may be studied as if it were the sole effect, without any reference to the others: and this course is frequently requisite, in order to avoid confusion, and to obtain a correct and extensive knowledge of the particular phenomenon under consideration.

In many instances, the various effects are directly connected with each other, and re-act on their causes, as in cases of many moral and political agencies. Here it is requisite to take a wide view of the subject, in order to obtain a just and adequate knowledge of it. We should first acquire an accurate knowledge of all the phenomena, considered simply as such, and then apply the proper principles, in order to discover their mutual relations and connections. We must not infer that the same cause which produces an effect in one case, may not be the result of this effect in another. Penury sometimes produces vice; but vice more frequently produces penury. Wherever we can discover a primary cause, we should first ascertain its original effect, then trace the reciprocal influence exerted on it by this effect, and afterwards inquire how the latter is re-affected by what it has itself produced.

In all inquiries regarding causes and effects, we should ascertain the real character of the thing which we assume as known: else we shall either fall into error or lose our labor. Thus, if we assume that a bad law is good, and then try to trace its effects, we shall undoubtedly arrive at an erroneous conclusion; or if we attempt to discover why nature abhors a vacuum, or why swallows can live under water during winter, we shall certainly labor in vain, because our assumptions are false. Effects which never existed can have no causes; and imaginary causes can produce no effects.

So we should note and bear in mind the particular object of our investigation, and not confound an inquiry into causes with one into effects, or one into the immediate with one into the ultimate cause.

CHAPTER XI.

OF LANGUAGE.

§ 1. ORIGIN AND PROGRESS OF LANGUAGE.—Causes of Language.—Its two primary Sources.—Its further Progress.—Formation of various Parts of Speech.—Modes of enlarging the stock of Words.—Foreign and vernacular Roots.—Origin of different Significations of the same Words.—Apparent and real Derivatives.—Origin of the Diversity of Languages.—Language not of Divine Formation.—Importance of knowing its Origin.

MAN is very superior to all the lower animals in the power of vocal expression, while he surpasses them still more in the extent and vigor of his thinking faculties. These appear to be the causes that have raised, on a very narrow foundation, a system of phonetic expression which, even among the rudest of the human race, incomparably excels anything found among the lower creation. If we compare the highest attainments of a parrot with what a child effects by his own unaided efforts, we shall have a striking proof of man's superiority: and if we further notice the ingenuity with which the dumb communicate their thoughts to each other, without any training, we shall readily understand the manner in which spoken language must have originated, and infer that it is a natural result of our circumstances and faculties.

Man expressed all his strong emotions and desires by peculiar instinctive ejaculations, and also imitated the various sounds that he heard. From these two sources have sprung all spoken language. When we examine even the most copious original language, such as the Ancient Greek, we easily trace its myriads of words to a few hundred roots, which give manifest indications of their origin. Such words as *eat, laugh, moan, groan, hiss, buzz, hum, crash, crush, rush, crow, roar, low, snarl, hurl, gurgle, murmur, purl, coo, cackle, snap, slap, rap, cut, babble, hop, strike, bull, bee, drum, horse, cuckoo*, &c., &c., evidently originated in onomatopoeia, or an imitation of natural sounds. Most actions are accompanied by certain sounds, an imitation of which would form the verbs employed to signify them. So the original names of animals would consist either of an imitation of their own cries or of the instinctive exclamations uttered on first beholding them.

It will not admit of a doubt that man would soon discover his extensive powers of vocal expression; and the rapidity and precision with which he could thus communicate his thoughts could not long escape his attention. This would lead to the constant use of those sounds, by the head of the family, to denote the objects and actions which they were employed to express in the first instance; and the other members of the household would, of course, adopt his expressions. Proper nouns, or names originally applied to individual objects, would be generalized by being applied to all things of the same kind, just as children call every horse or ox by the names of their fathers'.

Language being thus started, it would gradually be extended and improved, in various ways. At first it would be much assisted by gestures and expressions of the countenance: but as it became more copious, these would fall into disuse, except to give it force and vivacity.

The same words would be frequently used to denote both objects and actions; or, in the language of grammarians, they would be employed both as *nouns* and as *verbs*, a practice of which we have still many instances, as *feed, drink, touch, feel, smell, taste, love, hate, hope, fear, earth, air, fire, water*, and *light*. Which was the earlier use is a question of little consequence. Many words were probably used in both significations from the first, while in some cases the verbal sense probably had precedence, and in others the nominal.

Adjectives arose from using participles or the names of actions and objects to denote qualities, a practice still common. Thus, we speak of a "straight" (that is *stretched*) line, a "brick" house, a "sea" bird, "iron" strength, and so forth. In many instances, the form of the word was changed, to correspond with the different significations, the older being generally retained for the quality. Thus a "red" color is a "roe" color, and "ten" men is "toes" men, as we shall readily see by referring to the corresponding German and Saxon terms. In other cases, the original meaning was entirely lost; and it now appears as a noun or a verb only in some kindred language. Thus, we have the origin of "*strong*" in the Latin *stringo*, stretch or strain; and the source of *weak* is found in *vinc* (past *vic*), subdue or overcome. So we find the origin of *green* in the Latin *gramen*, grass.

Interjections are a part of instinctive language, and must, therefore, have been nearly as numerous in the earliest times as they are now.

Adverbs were originally phrases, adjectives or participles, of which they still exhibit many indications. Thus *to-day* is simply *this day*, as the corresponding Latin *hodie* is only *hoc die* (this day). The same remark applies to *prepositions* and *conjunctions*, although several of these were originally pure verbs, as some of them are still, such as *except* and *suppose*.

The advantages of distinguishing the speaker and the person addressed or spoken of, are so great and obvious that the personal *pronouns* must be nearly as old as the first origin of language. Yet they probably sprung from nouns, participles and adjectives, as we may infer from the abridgements which they have undergone, since the period of the earliest written compositions.

In enlarging the stock of words otherwise than by the original processes, arbitrary terms, entirely new, were rarely introduced, since they would sound strange, and furnish no key to their own signification. Instead of this, other methods were adopted. Old words were gradually changed in pronunciation, till several sprung from one; and the different forms which thus arose, were used to designate different modifications of the same thing, each form being appropriated to what it was thought to express with most precision. Thus, a child *squalls*, and a pig *squeals;* an owl *screeches;* a person suddenly frightened *shrieks;* and a woman in great pain *screams*. In some cases, instead of changing the old word directly, it was adopted from another language, in its altered form, with a new but kindred signification. Thus, the winds *blow*, and the waters *flow*. The latter word is directly from the Latin; but it sprung from the same root as the former. So *aur-ist* is only a Græco-Latin form of *ear* doctor.

A more fertile source of additions, at least in the more cultivated languages, was the practice, still common, of uniting two or more, to form one word. Here the new meaning is a modification of the thing denoted by the principal compounding term, which is particularly indicated by the other word. Thus, in the word *bookseller*, the last syllable is the principal root, or that which denotes the object meant. *Er* is an abridgement of the old word *wer*, a man, so that *sell-er* is, a man who sells, and *book-sell-er*, a man who sells books.

The compounding terms, instead of being adopted from the vernacular, were often taken from some foreign language. This frequently rendered the meaning of the compound much more definite, especially in abstract terms, or such as express a great variety of things. For, as the compounding terms were unfamiliar, the actual signification of the compound depended mainly on the definition, and therefore its exact import was not readily mistaken, so that a word formed in this way was equivalent to the invention of an entirely new term. Compounds formed from vernacular words, on the other hand, directly suggest a meaning, independently of any definition, which is often considered unnecessary, in such cases. But the literal signification of the compounding words is frequently different from the true one, especially in technical and scientific terms; and hence will arise obscurity and error, unless we attend to the actual, and not to the etymological sense.

To illustrate this difficulty, we may observe that *Geometry* literally signifies *land-measuring*, *Geography*, a description of the Earth, and *Geology*, a discourse about the Earth, so that, if we look only to the etymologies, we should be quite misled regarding the first, and we could not distinguish the second from the third. So *Astrology*, etymologically considered, is a more proper term for what is called *Astronomy*, as the former literally signifies the science of the heavenly bodies, and the latter, only the science of their laws. *Alchemy* and *Chemistry* are only different forms of the same word; yet their real meanings differ as much as those of the two preceding terms.

Owing to the difficulty and inconvenience of forming new terms, various significations were frequently attached to the primary import of many words, without their undergoing any change. A common instance of this was, employing words which originally meant physical objects to denote impalpable things, to which they were believed to bear some analogy. Thus, the same Hebrew word denotes both *wind* and *spirit;* and the Latin *animus*, mind, and *anima*, soul, are evidently identical with the Greek *anemos*, wind. So *investigate* originally meant to *track*, and *ponder*, to *weigh*. As objects of sense first received names, because they first excited attention, those of all others originated chiefly in this manner, although the primary significations are now found, in many instances, only in other languages.

Words and expressions were frequently employed figuratively, to denote things which bore some real or fancied resemblance or relation to the original significations, and, in many instances, this usage prevailed so extensively that the new meanings became as familiar as the primary, so that the original metaphor disappeared. Thus a "hard-hearted" man is not considered a figurative expression, any more than a hard rock.

Sometimes the new significations entirely superseded the originals; and, not unfrequently, a third was ingrafted on the former, which bore no resemblance to the latter. Thus, the word *virtue* originally signified *manliness*, being taken from *vir*, a man—then, *bravery*, because that is a manly quality—and, finally, *any good quality*, so that we speak of the virtues of drugs and plants. By this means the same word sometimes came to have contrary significations. The Latin *sacer* means both *holy* and *accursed*, the original signification being *set apart* or separated.

When the vocabulary of a language had thus become comparatively copious, figurative expressions became less common; and the general style became more precise and literal, because there was less occasion for metaphors, while the more extensive application of language for didactic purposes rendered perspicuity and precision more desirable.

Words not primitive are usually divided into *compound* and *derivative:* but there is generally no real etymological difference; for most derivatives are only compounds of which the subordinate compounding terms no longer appear separately in the language. These are generally found, however, either in its older forms or in other languages. Thus the affix *er* is the Saxon *wer*, a man, so that *hunter* is, in reality, a compound as much as *huntsman;* and the common prefix *in* or *un*, signifying *not*, occurs in Hebrew as a distinct word, with the same sense.

Sometimes prefixes and affixes are only common terms slightly altered. *Begird* is *by-gird* or gird round; *mispronounce* is *miss-pronounce*; and *manly* is *man-like*.

The real derivatives of a language are, the various new forms of a word, employed to denote several modifications of the same thing: but as it is difficult to trace the derivation, in such cases, grammarians have frequently considered all the forms primitive words.

Different communities pronounced, compounded, and contracted words variously, and employed different new terms to denote the same thing, while each, in many instances, superinduced peculiar new meanings on the primary. Striking differences arose, also, from some races amalgamating modifying words with the principal terms in pronunciation, such as personal pronouns with the verbs of which they were nominatives, and personal pronouns and prepositions with the nouns which they qualified. Hence originated numerous inflections, as in Latin and Greek, while other races, as the Chinese, always kept the words distinct, whence their languages exhibit hardly any inflections. Other differences arose from figurative or poetical expressions superseding, in some languages, the original terms which others retained.

When we further consider the wide diversity in the circumstances of communities, and in the objects with which they were conversant, the great variety of languages which is found throughout the world, appears only a natural consequence of these numerous causes of divergence. If we observe the rapidity with which a copious language has run into several, within the period of authentic history, we can easily understand how quickly a language, yet rude and barren, might run into many, while the art of writing was unknown, and communities were very small. A patriarch, living alone with his family, would communicate to them his own linguistic peculiarities; and these would increase, more or less, with every successive generation, more especially if every one of the sons separated from the common parent, and held little further intercourse with his kindred, as would very frequently happen in early ages when the greatest part of the world was yet uninhabited. It may further be observed that the names of things which vanished from sight would be forgotten, while new terms would be used to denote new objects.

When a language has become copious, assumed a definite form, and been reduced to writing, subsequent changes of the old elements and constructions may be very slow, although it may receive a host of new words. Hence those languages which sprung from a common source, at periods long subsequent to the general dispersion of mankind, frequently resemble each other very closely, while those which separated much earlier often retain such faint traces of their original unity that this is established only by means of the intermediate languages, which serve as connecting links.

We are told, in the Holy Scriptures,* that God brought the animals to Adam, in order that he might name them, and that whatever Adam called every species became its name. Now animals would excite attention before any other object; and when language had once originated by giving them names, there was no occasion for miraculous aid to continue the process thus begun. The meager foundations on which language is built, the numerous and palpable marks of onomatopœia which it still displays, the absence of terms originally denoting any mental or immaterial objects, and the many defects with which it abounds, and which are particularly observable in its most ancient forms, show that it is of human, and not of Divine, origin.

The Almighty would not encourage mental inactivity by conferring miraculously what he gave us faculties for forming in a natural way. Nor is it credible that he would act thus when the gift was soon to be lost irrecoverably, and superseded by a multitude of rude languages, precisely such as mankind would readily have invented, without any miraculous aid.

A correct view of the origin and progress of language is requisite, in order that we may use it aright ourselves, and properly interpret the expressions of others: for errors on this subject produce fallacious rules both of interpretation and of composition. The fiction of the Divine origin, and consequently primitive perfection, of language, repels scrutiny, misleads the grammarian, and tends strongly to produce and foster the common error that language is a correct representation of nature, and

* Genesis, Chapter ii., verses 19 and 20.

that we can discover truth by the study of mere expressions.

§ 2. USES OF LANGUAGE.—Uses and Defects of "natural" Language.—Original object and superiority of Speech.—Its other Advantages.—Use of General Terms.—Requisites to the proper expression of Thought.—First Rule.—Common Violation of it.—Different styles of didactic and emotional compositions.—Comprehension and Extension of Terms.—Second Rule.—Sources of Ambiguity.—Third Rule.—Common Errors.—Means of testing the amount of our Knowledge of a subject, and of avoiding Vagueness and Obscurity of Expression.—Verbal and Real Definitions.—Why the former only belong to Logic.—Three Rules regarding them, with Remarks.—General characteristics of good Definitions.—When new Terms may be rightly employed.—Ordinary Practice.—Foreign and Vernacular Roots.—Evils of introducing new Terms unnecessarily.—Origin of many new Terms.

Gestures, expressions of the countenance, and instinctive exclamations, which have been termed "natural" language, bear the same relation to speech that pictures do to phonetic writing; and hence they are very serviceable in exciting the passions. But they are incapable of expressing many thoughts at all, and others they express only vaguely, while their power of exciting emotion is positively unfavorable to them, as a means of communicating knowledge, since disturbing emotions are fertile sources of error.

The original object of speech was, simply to express or communicate thought: and vocal language possesses over every other means the advantages of much greater precision and rapidity of communication. Even if we could invent manual or other inaudible signs, to denote every thought, there are none that could be communicated so rapidly as vocal signs or words. Nor could man have invented such signs without the previous assistance of words, as otherwise many of the requisite signs could not be agreed on.

It is by means of words alone that the attainments of one person or generation can be satisfactorily communicated to another, and mankind be both benefited by the experience or labors of their predecessors or contemporaries, and enabled to add to their acquisitions.

Speech is also employed, in a written form, to aid us both in our original researches and in retaining our acquisitions. For we cannot generally follow out a long and intricate train of reasoning, without assisting the

Memory by written symbols, which could never have been brought to any degree of perfection without the previous use of words. But, by means of written speech, we can clearly express every step of the most intricate and subtle process; and thus we can easily retrace our steps, whenever we wish either to recall something which we have forgotten, or to test the accuracy of certain propositions, by a deliberate examination of the various processes through which they have been professedly established, so that none of them may be overlooked.

The aid of language is requisite to retain even a fragment of our former acquisitions, in such a manner that we are safe from error: for we are liable to forget our former conclusions, as well as the processes by which we arrived at them, so that we readily mistake them for others which they resemble, and thus confound what we established with what we only heard, read, or imagined.

Language greatly aids reasoning and remembrance, by means of comprehensive and general terms. Even such simple expressions as *hope* and *fear* imply things which the Memory could not easily retain, without the aid of language, and much more will this remark apply to such terms as *duty*, *law*, and *government*, which distinguish the things denoted by them from others with which they would be confounded, but for the aid of expressions showing the very things meant. Such terms are also requisite in the higher kinds of generalization, that we may be able to remember and reason regarding preceding results: and they also enable us to reason upon assumptions regarding things of which we can have no idea, such as indefinitely small quantities, or even things which cannot exist, as the square roots of negative quantities.

In order to express our thoughts aright, we must possess a knowledge both of the subject and of the language we employ: and when these prerequisites have been acquired, the observance of the three following rules will secure proper expression.

1. *We should say precisely what we mean.* It is not sufficient to avoid expressions totally wide of our meaning: we must beware of saying either more or less than we intend. Terms may be either too general, and thus include something which ought to be omitted; or they may be too particular, and consequently omit something which

ought to be included. In mere illustrations, and in compositions designed to excite emotion, it is often proper to employ terms more particular than truth would warrant: but in purely scientific or didactic discourse, this should be avoided, because the more general expressions are the more extensive and concise.

Particular terms express more qualities of the thing meant; and hence their superiority for emotional and illustrative purposes. "Horse" expresses more qualities than "quadruped," and the latter, many more than "animal." The more species a term embraces, the fewer qualities it denotes, and conversely. The number of qualities a word expresses is termed its *comprehension*, and the number of individuals or species that it indicates, its *extension*. These two evidently vary inversely; the wider the one, the narrower the other.

2. *Our expressions should not be ambiguous.* We should reject expressions which admit of two or more significations, without violating any just rule of interpretation. Expressions are ambiguous, either where one or more of the terms bears several significations, and the context does not determine which is intended, or where a phrase is susceptible of various grammatical constructions.

3. *Our expressions should be perspicuous.* In order to this, we must avoid those expressions of which the exact signification cannot be detected without difficulty. We are liable to employ words of whose signification we have no clear notion, without ever being aware of our doing so, because they are familiar to our ears; and thus we are apt to overestimate our knowledge of the subject. One of the best means of testing the amount of this is, to write down what we know, in terms which are perfectly clear and unambiguous, and which will give any person of ordinary perspicacity a correct notion of what is meant. The application of this test will show that we are apt to mistake a familiarity with words for a knowledge of what they denote, two things widely different.

In order to secure perspicuity, we must not employ an expression without knowing precisely what we understand by it; and when we are obliged to use one that is obscure, ambiguous or not well known, we should clearly fix the sense in which we employ it, either by the form of the expression or by a definition.

Definitions are either *verbal* or *real*. The former point out what terms denote certain things, which they assume that we know; the latter unfold the nature of the things denoted by certain terms. If we are told that "*blue* is the color of the sky," this is a verbal definition, and teaches nothing unless we know what the color of the sky is. But if we are told that "a *dial* is an instrument for measuring time, by means of the different directions in which a shadow falls, throughout the day; and it consists of a flat surface, marked with the hours, and a style or axis, for projecting the shadow"—this is a real definition, which would give us a notion of what a dial is, if we had never seen or heard of such a thing before.

Definitions written for our own use exclusively will be verbal, since they are designed merely to aid us in remembering the meaning of words, or the sense we attach to them. We cannot give a real definition of a thing of which we are ignorant: and when we know the thing, the object of a definition for our own use can only be, to remind us of the term by which it is designated. Hence the subject of real definitions belongs to Education or Rhetoric, and not to Logic.

The following are the principal rules regarding verbal definitions.

1. *The definition should be free from the difficulties which it is designed to obviate or remove, and accord with the rules of correct expression.* If it is inaccurate, ambiguous, obscure, or couched in terms partly unknown, it is worse than useless, since it misleads, while it professes to guide. A common violation of this rule is, to employ defining terms which are as unintelligible as the word defined, so that the definition is tantamount to saying a thing is what it is. Thus, *life* has been defined "a system of vital forces," that is—"a system of forces pertaining to life." So, *justice* is defined, in the Institutes of Justinian, "a constant and perpetual desire to give every one his right," which is equivalent to saying that "justice is, a desire to do justice to every one," since a man's *right* requires to be defined as much as his *just due*.

2. *The thing defined should be clearly pointed out, and distinguished from all others.* In order to this, the definition must express the peculiar characteristic of the thing defined, and exclude that of every other being.

Consequently every negative definition violates the first part of this rule, since it never tells us what a thing is, but only what it is not; and every definition which applies equally to other things, violates the second part of the rule. It is not a proper definition of a *man*, to say that "he is not a fish," as this includes every being except fishes, while it wholly fails to tell us what man is. So, the definition "man is a rational being," violates the rule; for so are angels.

3. *The definition should comprise the whole of the thing defined, and not merely a part of it:* otherwise it is defective. It is not a proper definition of *ruminants*, to say that "they are sheep and oxen;" for goats, deer, antelopes and camels also belong to that order. Nor is it a proper definition of *swans* to say that "they are large, white birds, of the order *natatores* or swimmers;" for this excludes the black swans. This rule is very apt to be violated wherever our views of the subject defined are too narrow, since our attention is apt to be confined to that with which we are familiar, so that we overlook the rest, although this may possibly be the largest and most important part.

A proper definition requires both a good knowledge of the subject and care in forming it. Hence our definitions generally improve as our knowledge of the subject becomes more extensive, and further examination and experience gradually show the imperfections of former definitions.

Many definitions can generally be given of the same expression; and the best is, that which gives the most correct notion of its import to the party for whom it is intended. Hence definitions designed for our own use should be such as are most satisfactory to ourselves, although they might not be suitable for others. We should, however, depart from the ordinary signification of terms as little as a regard to clearness and precision will permit, since difficulty and error are very apt to arise from affixing peculiar meanings to terms. But when we are obliged to employ a vague term, we should either define or distinctly note the particular sense in which we employ it: otherwise obscurity or error will probably result.

When there is no term in the language to denote the meaning, without great risk of error, the best way is, to adopt a new word. This is generally desirable where a new name is required for something essentially different from anything which already has a name in the language: for, to attach the new meaning to some old term would lead to confounding the various significations, a thing which has been a fertile source of error.

Sometimes it is requisite only to borrow a word from a foreign language, and modify its form, if necessary. In other cases, we must coin a new term. It has been a common practice to form a compound from Latin or Greek roots, adhering to the proper rules of composition or derivation. The latter language is generally preferable, because it is more unlike the vernacular. Where the thing can be expressed exactly by vernacular roots, these should be adopted in preference, since the compounding terms directly suggest the signification of the word. Compounds of this kind are generally better for emotional purposes, as they affect the feelings more strongly: but they are frequently objectionable for didactic purposes, as they tend to mislead us regarding their exact signification.

The use of new words should be avoided, when there is no urgent occasion for them, since their introduction, in such cases, only clogs a language, like useless wheels in a machine. It also fosters error, owing to the common tendency to assume that different words must denote different things. Many new terms have sprung from pedantry, ignorance, or an affectation of originality, superior discernment, and refined feelings, which had no real existence.

§ 3. IMPERFECTIONS AND ABUSES OF LANGUAGE.—Origin of the Imperfections of Language.—(1) Idioms.—(2) Different admissible Constructions.—(3) Different Significations of Words.—(4) Terms expressing Nonentities.—(5) Deficiency of Words.—(6) Terms conveying false Impressions.—(7) Superfluous Words.—Pernicious Error.—Requisite in order to a right understanding of Terms.—Origin of abuses of Language.—(1) Using vague and obscure Words.—Common Instance.—Sources of this Abuse.—Means of guarding against it.—(2) Confounding different Significations.—How to be avoided.—(3) Conveying erroneous Meanings.—Instances.—How to be avoided.—Combination of Abuses.

Language represents, not the realities of nature, but the opinions of those who formed it; and as these were often erroneous, language exhibits corresponding imperfections, of which the following are the principal:

1. *Language abounds with idioms*, or peculiar phrases, the precise significations of which are not indicated by the terms composing them, but must be learned from observing the occasions on which they are used. Consequently they are very liable to be misunderstood by those who are not well acquainted with the language. These, again, may mislead others, by employing or defining those terms improperly.

2. *Many expressions admit of various constructions, every one of which gives a different sense*, while it may be difficult to ascertain the true construction; and consequently it is often missed.

3. Owing to the new significations superinduced on the primary, *many words denote things widely different, some of which are readily mistaken for others*, because the different significations are often connected in sense, and always, by the bond of the common name.

4. *Many expressions denote things which have no existence.* Such are *satyr*, *dryad*, *fairy*, *mermaid*, *sylph*, *griffin*, *empyrean*, and *primum mobile*. As the names exist, we are apt to think that they denote corresponding realities, which, however, are wholly imaginary.

5. *No word exists for many things which ought to be distinguished by a peculiar term.* Hence it becomes necessary, either to affix new significations to old words, or to introduce new terms, in order to denote such things; and this produces changes which render language either more vague or less intelligible.

6. Terms have been invented to express things regarding which vague or erroneous opinions prevailed at the time of their formation, so that *they convey false impressions regarding the reality*. Such are *animal spirits*, *humor*, *sensible species*, and *vitality*. Terms of this kind had no clear and correct signification originally; and subsequent alterations sometimes made matters still worse.

7. *Language contains many words so nearly resembling each other in signification that one or more of them might be discarded to advantage.* Such are *perhaps*, *peradventure* and *perchance*—*among*, *amongst*, *amid* and *amidst*—*nitrogen* and *azote*—and *between* and *betwixt*. Not only are slight modifications of the same thing often denoted by totally different terms, but the very same thing sometimes has several names; and hence we are apt to think that the things meant must be as distinct as

the words by which they are expressed. Wherever any of these modifications do not require to be frequently mentioned in discourse, as distinct from its class, the proper plan would be, to denote it by the general term, with some epithet indicative of its peculiarity, and to drop the special term altogether. This would render language more intelligible, precise, and easier of acquisition.

To overlook the preceding imperfections, and assume that language gives a faithful and intelligible representation of nature, is an ancient and still prevalent error, which has produced many fallacies. Some believe that there are fairies because we have a name for them; and most of the ancient astronomers never suspected that their "crystalline spheres" were wholly imaginary. Hence, also, has arisen the erroneous opinion that a knowledge of nature can be communicated by means of words alone, whereas we cannot be instructed by words which we do not understand, and the signification of a term can never be understood, unless we learn, by our own original comprehension, the primary elements of the thing which it denotes, just as no definition can give the blind a correct notion of color.

Owing to the fallible nature of our faculties, and their being often swayed by prejudices or bad motives, we are liable to use language improperly, even if it were perfect, while these abuses are increased by its imperfections. The most common of those abuses may be referred to one or other of the following heads:

1. *Language is used so vaguely or obscurely that it is very difficult to ascertain the sense, or whether any definite sense is conveyed.* A common instance of this abuse is, where figurative expressions are employed, so that we either cannot know what is meant, or we are unable to determine whether the expression is figurative or literal. When compositions which abound with this defect are carefully analysed, it will generally be found that the author had neither a clear conception of his own meaning, nor a good knowledge of his subject.

This abuse is of common occurrence, and frequently very pernicious in its consequences. It springs chiefly from ignorance, carelessness, an affectation of originality, learning or depth, and dishonest intentions.

A man cannot express himself clearly and accurately, in his own words, on a subject which he does not understand: and this appears to be the source of much of the obscurity and darkness which characterizes many treatises and compositions on mental science; for the language is frequently erroneous, whichever way it may be interpreted. But as the subjects of investigation are imperceptible to the senses, these faults are apt to escape the reader or hearer's attention; and he may thus admit as self-evident or fairly proved, a proposition really absurd, or incapable of valid proof.

It is easy to assume that everything is correct, where there is nothing manifestly false, while it is sometimes difficult to detect the true character of profound jargon or sublime twaddle; and hence these have often passed for being what they appear. It requires no severe thinking to employ such language, while it procures the admiration of those who believe that the depth of a discourse must be proportional to its darkness, and that what they readily and clearly understand, must be comparatively superficial, whereas the case is very frequently the exact reverse. This leads the shallow-minded and dishonest to use foggy, recondite and high-sounding expressions, when clear and ordinary terms would be much more appropriate.

Carelessness frequently leads to similar results. If a person is hurried, and inattentive to his words, he will often use terms which convey no clear or definite meaning to the party addressed, although his own views of the subject should be both accurate and profound. This is particularly apt to occur where he has paid little attention to the proper expressions for denoting his meaning.

A common temptation to indulge in this kind of composition is, the air of originality, profundity, and importance, which trite or puerile thoughts assume, in the eyes of many, when they are clothed in a new, vague, and pompous phraseology.

Another frequent source of this abuse of language is, a desire to mislead those addressed, or to shun the responsibility of telling the truth clearly, by employing obscure or ambiguous terms, while the speaker or writer escapes the charge of unequivocal lying, because his expressions are true in one sense, although he is well aware that they will be understood otherwise, or at least produce erroneous impressions.

Many terms denote things which have various forms, or are incessantly changing their import, so that when used without any qualifying or distinctive epithets, it may be impossible to ascertain the thing meant. Such are *wealth, capital, government, legislature, church,* and *polity.* A statement which holds true of one of the things denoted by such terms, may be quite false of another.

A careful attention to what is actually said, will enable us to estimate expressions at their true value, and prevent us from overvaluing puerility or absurdity, because they may be clothed in a novel or philosophic garb; and we shall learn that vague, obscure, and grandiloquent language generally covers error or inanity.

2. *The same expressions are employed in various significations, which are not distinguished,* so that we mistake one for another. The author uses the same words now in one sense, and then in another, while he gives no clear indication of the difference: but we are required, or at least expected, to assume that they are used in the same sense throughout. This abuse is of frequent occurrence, because it is very liable to escape the notice of both author and hearer or reader. We are so accustomed to use the same word in different senses, that the fact of this being done excites neither surprise nor suspicion; and hence we are very apt to overlook the abuse of doing so, while the argument, or object in hand, requires that they should be employed throughout in precisely the same sense.

To guard against this abuse, we must ascertain whether the expressions, while professing to mean only one thing, are not, in reality, employed to denote several things essentially different. This is done by ascertaining the precise sense in which the terms are employed, wherever they occur. In our own case, we must closely question ourselves regarding our meaning, ascertain whether we know the precise thing that we intend to express, and adhere to the rules of proper expression. In the case of others, we must also apply the proper rules of interpretation.

3. *Language is employed which conveys a clear and definite, but erroneous, signification:* and this may be done unconsciously, through ignorance and heedlessness, or wilfully, from some evil design. A common form of this

abuse is, to employ one term for another which it somewhat resembles in sense, while their proper significations are very different. Thus the words *impossible, inconceivable, incomprehensible,* and *highly improbable,* which properly denote four things essentially different, have often been employed indiscriminately, as if they all expressed the same thing.

Another common modification of this abuse is, erroneous definition. Thus, several of our older English lexicographers tell us that the *share* or *sock* of a plough is the *coulter* or *knife*, which is like saying that the hand is the foot. So the Hebrew, Greek and Latin term for *copper* is generally defined, in the dictionaries of these languages, by *brass*, an alloy which appears to have been wholly unknown to the ancients. This error is very apt to escape detection, because those who consult the definition are generally ignorant of the real import of the word. We should never rely on the definitions of persons who are not well informed regarding the thing defined, but either apply to a proper authority, or endeavor to acquire some knowledge of the thing defined, and then compare the reality with the definition.

A third common case of this abuse is, exaggeration and extenuation. On the one hand grandiloquent or hyperbolical language is applied to things which do not properly admit of it: and on the other hand, terms expressive of contempt are either overstrained or wholly misapplied. This course sometimes produces the desired effect, especially where it tallies with the prejudices of the party addressed. In other cases, the result is often the very reverse. Instead of being elevated, the language is only tumid or bombastic, in the one case: and compassion for its objects, with indignation against those who employ it, are the effects in the other.

This abuse is to be detected by ascertaining the character and circumstances of the author, and applying the proper criterions of testimony.

Not unfrequently two or three of the preceding abuses are combined, as where vague terms are employed to convey an erroneous meaning, while the same terms are improperly used in various senses. Here we should first try to ascertain what is said, and, if we succeed in this, endeavor to learn its character afterwards. If we fail in the former object, the second, of course, becomes impracticable.

§ 4. INTERPRETATION OF LANGUAGE.—Use and foundation of Rules of Interpretation.—Five classes of Expressions.—Three which require interpretation.—(1) Usual Meanings to be generally adopted. —On what this rule is based.—Principal Exceptions.—On what ground these are admitted.—(2) When Figurative Meanings are to be adopted.—Mode of determining when an expression is figurative.—(3) When special or technical senses are to be adopted.—Relations and places of the Figurative and the Technical.—Means of discovering the latter.—(4) Rule regarding antiquated Significations.—How these are to be ascertained.—(5) Rule regarding the intentional Sense.— Its Application.— Means of ascertaining this Sense.—(6) Cases in which the literal Signification must be adopted, and why.—(7) Usages to be observed.—Foundations of this Rule.—Means of ascertaining Usages.—Interpretation of Idioms.— Observations on Corresponding Terms, and those which gradually change their Significations.—Applications of the Rule.—Various things to be considered.— Caution.— Mannerism.— Influence of particular Pursuits and national Character.—(8) Rule where several meanings are admissible.— Its Foundations and Extent.— Means of application.—Common violation of it.—How to be avoided.—Remarks on cases to which it is inapplicable.—Latent Ambiguities.—Implications.—Aids where we depend on Translations. —Requisites to a proper application of the Rules of Interpretation. —Sources and Evils of Misinterpretation.

The difficulties which arise, in attempting to ascertain the precise meaning of expressions, are to be surmounted by a proper application of the rules of interpretation. These are founded on the structure and usages of language, so that their validity admits of no question, although people may differ occasionally regarding their applications.

With reference to interpretation, all expressions may be divided into the five following classes.(16)

1. *Those of which the signification is well known, clear, unchangeable and unambiguous.* This class includes most of the names of natural objects, and those words which denote simple external actions and direct comprehensions, such as *Sun, Moon, sky, tree, man, hand, foot, black, white, hot, cold, go, come, sit, eat, drink, see, hear,* &c. To all who possess an ordinary knowledge of the language, words of this kind require no interpretation; and it is chiefly by their aid that we can rightly interpret others.

2. *Those which convey no real meaning.* Here interpretation is needless, as it would be folly to search for a meaning where none exists.

3. *Those which are ambiguous,* or which admit of several interpretations.

4. *Those which are obscure,* or in which we cannot readily discover any certain meaning.

5. *Those of which the meaning is wholly unknown,* such as obsolete and unusual words, and foreign terms of which we have not yet, in any degree, ascertained the significations.

Thus we see that the ambiguous, the obscure, and the unknown comprise the whole field of interpretation: and the following are the principal rules for removing the difficulties which they occasion:

1. *The literal or ordinary signification of words is always to be adopted, except where there is some urgent reason for departing from it.* This rule is founded on the fact that the literal is the common or usual signification, to which all others are only exceptions.

The following are the principal cases in which we ought to depart from the literal signification: (1) *Where it renders the expression absurd,* or gives it a meaning which cannot possibly be true. (2) *Where it gives either a puerility or no meaning at all.* (3) *Where it gives the passage a meaning at variance with known truth, the whole scope and tenor of the discourse, or the views and opinions of the author.*

These exceptions to the general rule are admitted on the grounds that figurative and special meanings of words are common, that persons do not generally speak manifest absurdities and puerilities, that their expressions usually have a meaning, that they sometimes employ words which convey a sense different from what they intended, and that they seldom utter glaring falsehoods, or directly contradict themselves.

In all those exceptional cases, we must search for some other sense than the literal; and of this kind the *figurative,* the *special* or *technical,* the *antiquated* and the *intentional,* are the most common.

2. *A figurative rendering is to be adopted where it is admissible, and gives a good sense, which the literal does not give.* The various kinds of figures, and the circumstances under which they are employed, are so well known that it is generally easy to determine when a figurative meaning is to be adopted: and we are farther aided by the manifest difficulties attending a literal interpretation, in such cases.

The nature of the composition often indicates whether

an expression is figurative. In poetry, rhetorical compositions, and all impassioned discourse, figurative language abounds, while it is sparingly used in all scientific works, and in some it is hardly ever employed. As the oldest compositions mostly belong to the former class, they generally exhibit a free use of figurative language; and they are to be interpreted accordingly.

Those expressions which were originally figurative, but have entirely lost their old literal signification, are excepted from the preceding observation; for as the figurative has become the ordinary signification, they are properly classed with literal expressions.

3. *The special or technical signification is to be adopted where it is indicated by the subject or the context, and the ordinary signification is objectionable.* By the former terms is understood that signification which is peculiar to a certain art or science, or is employed only by a particular class of persons, in speaking of their distinctive usages, opinions, or doctrines.

Figurative and special expressions are generally in inverse proportion. The latter are extensively employed in purely didactic works, and very little in impassioned or emotional discourse. In the former, the special signification is the rule, and the ordinary, the exception, so far as the subject in hand is concerned.

The subject of discourse generally indicates the particular meaning which we are to expect. A sailor generally uses terms in their nautical, a soldier, in their military, a merchant, in their commercial, and a geometrician, in their mathematical signification.

The signification of the various terms is to be discovered from the oral or written testimonies of those conversant with the subject to which they refer, or from marking their usages and formal definitions; and much aid may be derived from good dictionaries of the particular subject, where such exist.

4. *Antiquated significations are to be adopted in ancient compositions, or such as treat of Antiquities.* The general subject, or the context, points out the cases in which antiquated meanings are admissible; and the principal thing to be avoided is, confounding the obsolete with the present meaning. The former may be ascertained from dictionaries which explain them, archæological treatises, and an extensive acquaintance with works

on the subject in question, written in the same age and country.

5. *The sense intended to be conveyed is to be adopted, though different from what the words ordinarily signify.* Speakers or writers sometimes inadvertently omit or insert words, or employ expressions which convey a sense quite different from what they intended, in all of which cases the latter is evidently the true interpretation.

In determining the applicability of this rule, the circumstances of the author are to be considered. If he wrote or spoke deliberately, and after careful consideration, the rule is not applicable, unless the intentional meaning is obvious and unquestionable, and certainly different from the literal. In other cases, this rule may safely be followed, wherever the words actually employed are such as might readily be used inadvertently, instead of those which would properly express the meaning supposed to be intended, while this harmonizes with the context, and the literal signification either contradicts it, or gives no sense at all, as when the negative particle, or the verb, is omitted. But the rule cannot safely be applied, where the supposed intended sense, though probable, cannot be clearly ascertained.

Wherever the intentional sense differs from the literal, it is generally to be ascertained from the context, or the author's other statements; and, not unfrequently, it is discovered from his acts, demeanor or circumstances.

6. *No meaning is to be attached to an expression, which it will not bear: and the ordinary signification is always to be adopted, where no other is admissible.* The rules of interpretation assume that the meaning which is to be attached to an expression, is one which it may possibly convey, and which is not wholly inconsistent with the usages of speech: for it is evidently absurd to attach any such meaning to words; and, therefore, we are always limited to admissible senses, whatever be the consequences.

7. *Expressions are to be interpreted according to the usages of the language, the age, the place, the subject, and the author.* This rule is based on the well-known fact that these usages widely differ. Every language abounds with idioms, or expressions which have a conventional meaning, well understood by those who are familiar with it, but frequently very obscure to all others: and a simi-

lar remark applies to different ages, places, subjects and authors.

Usage is learned from passages or expressions where circumstances render the exact meaning obvious, as when we observe the occasions on which the expression "How do you do?" is used. The meaning then becomes clear, whereas it is quite a mystery to a person who observes it for the first time in a book. In many cases, the sense is fixed by the immediate context, as if we notice the answer—"I am quite well," given to the preceding question: and when the usage is thus ascertained, it may be applied to remove the various difficulties which it occasions.

The usages of a particular author are best ascertained from his own works, although they are frequently pointed out in grammars and dictionaries: but much assistance may be derived from an acquaintance with his biography, and the circumstances in which he was placed, as well as the history of his age and country.

The peculiar usages of various countries and ages are best learned from their respective authors; and many of them are pointed out in grammatical and archæological works. But many are best learned from a good knowledge of the nation's public and private life and manners. The corresponding terms for artificial objects differ, of course, as much as the objects themselves. Thus, the words *writing, auger, plough, coat*, &c., have different shades of meaning, in different ages and countries. The corresponding words expressive of mental objects sometimes differ still more widely. Thus, among the ancient Romans, *pious* meant dutiful or well-behaved, and *religion* denoted only the rites and ceremonies pertaining to the worship of the gods.

Many words change their significations gradually, from age to age; and, therefore, the age, as well as the country, of the author must be noted, although the language may not have undergone any radical change. The significations of many words used by English authors of the fourteenth century, for example, differ widely from those which they bear at the present day.

In some cases we must consider, not only the age and country, but also the dialect, or local usage: and here we must interpret the expressions according to the peculiar dialect of their author.

The subject of discourse, also, requires attention: for many terms have, not only a general and a technical meaning, but several kinds of each; and the subject is frequently an important guide in determining which of these ought to be adopted. We should generally attribute to such words the peculiar sense which they usually bear, in the subject in which they occur, or among the class to which the author belongs.

The character, circumstances, objects, and pursuits, of an author, are often of great importance in determining the sense of his expressions. It is also necessary to note his general style. Some follow common usage so closely that their compositions present no peculiarities which can cause any difficulties of interpretation, while others follow an opposite course. The former are generally superior, in every important respect. Men of an original cast of mind frequently exhibit marked peculiarities of style: but the highest order of minds is characterized by the clearness and precision, as well as the force of their style.

A knowledge of individual peculiarities is occasionally of more consequence than that of an author's age, since some affect the style of former generations, while others study novelty of expression. But this has generally been done by inferior authors, who attempted to excite by their manner a degree of attention which could not be secured by their matter.

Individual peculiarities often affect language, by leading to the employment of terms in a technical sense, when they should be used in their more general signification; and persons often err in using the technical terms with which they are familiar, instead of the appropriate expressions. An old seaman's dialect generally shows many instances of this kind. Here we should look to the usage of the class to which the individual belongs, for an explanation of his meaning. Thus, ambiguities in contracts are often removed by referring to the usages of the trade or vocation to which they refer. Difficulties of this kind are sometimes surmounted by observing the conduct of the party, either when he employed the expressions or at some other time, just as we learn the signification of terms wholly unknown, by observing their applications.

The character of an author's nation ought not generally to be overlooked, in interpreting his language, as it frequently, to a great extent, controls individual peculiarities. Nations of ardent dispositions and undisciplined intellects are apt to use language stronger than truth warrants, while that of men of cooler dispositions and more discriminating judgements are to be taken much more literally. The lofty terms of the former often mean much less than the more sober expressions of the latter.

8. *Where several meanings are admissible, we should generally adopt that which best harmonizes with the context, the author's circumstances, views, and objects, and the character of those whom he addresses.* This rule is based on the assumptions that one part of a composition generally harmonizes with another, that an author will employ words in the sense best adapted to promote his objects, that his expressions have a meaning, and only one real meaning, and that he will endeavor to render himself intelligible to those whom he addresses. These assumptions generally hold true of the language of sensible men, who speak or write with proper objects in view, and on a subject which they understand. But when the author is a simpleton, or when he composes with other objects than to instruct, or deal fairly with those whom he addresses, this rule does not apply.

The proper application of this rule requires some knowledge of the general nature of the subject, the author's particular character and objects, and also the character of those whom he addresses. These may be learned partly from the general tenor of the language, and partly from other sources, such as the testimonies of contemporaries, acquaintances, or persons similarly circumstanced.

In the case of spoken discourse, the application of this rule is facilitated by observing the speaker's gestures, looks, and tones of voice. We are thus enabled, in many cases, to distinguish irony and other figures from literal speech, and to ascertain the force and purport of the language, without any difficulty. In written discourse, no such aid is generally available: but we may often ascertain the true meaning by referring to analogous expressions, or to the author's direct explanation of the same subject or the passage in question.

A frequent violation of this rule is, to interpret an obscure or ambiguous passage so as to make it clash with others which admit of only one interpretation, and then to force a meaning on the latter of which they do not admit, in order to make them harmonize with the false signification previously attached to the former passage. It is evidently absurd to force what is clear into conformity with the supposed signification of what is dark. The proper course is, to attach to words no signification which they will not fairly bear, and to interpret what is obscure or ambiguous so as to make it tally with what is free from any such difficulty, and rightly admits of only one meaning.

This rule is frequently inapplicable to the joint composition of different persons, who entertained conflicting views of the subject, such as state papers, laws, treaties, and contracts. As these often speak the language and opinions of different authors, in their various parts, they may be quite inconsistent with each other, while we may have no means of knowing the author of any particular part. The same remarks apply to the different compositions of a person who has changed his views in the interval, or who, at one time, expresses his own opinion, and at another time, those of other parties, whom he wishes to please.

Sometimes the same part of a joint composition has different meanings, because its authors understood it differently, or pretend that they did so, as in many cases of contracts.

With regard to all joint compositions, the safest course generally is, to adhere to the sense given by the other rules of interpretation, exclusive of the present, and, if these do not give a definite signification, to attach to the language no precise meaning whatever.

Sometimes expressions apparently very precise, are, in reality, ambiguous, owing to extrinsic circumstances, as where a man wills his large English Bible to his son John, and it turns out that he left two such Bibles. Here we should first ascertain, from extrinsic evidence, the facts affecting the sense, and then apply the ordinary rules of interpretation.

A frequent source of difficulty is, the uncertainty which exists as to what may be implied in the expressions: for we are left to infer the author's meaning from words which, of themselves, may convey no precise or unequivocal signification. The author's character, object, and circumstances, and the context of the discourse, are the principal guides for removing such difficulties; and where these fail to do so, the only safe course generally is, to let the expressions go for nothing.

Even where an inference may be necessarily deducible from a person's language, we are not always warranted in assuming that he intended to convey it: for he may either have overlooked it, or not discerned the necessary connection. We may generally infer that a person implies what one of his character and in his circumstances naturally would, but it is only where his character, or the immediate and palpable nature of the inferences excludes other suppositions, that we can safely assume he meant to express the inferences, unless we have some extraneous proof that such is the case.

Where we are obliged to rely on translations, we must look carefully to the context, and to similar or corresponding passages, either in the same or in other authors, wherever we encounter a serious difficulty; and much aid may be derived, in all such cases, from comparing independent translations of the same passage.

The proper application of the rules of interpretation require freedom from the influence of prejudice, and, in many instances, extensive and careful research and consideration. Hence misinterpretation has been a thing of very frequent occurrence, and a prolific source of pernicious errors. The interpreter strongly wished that a certain interpretation should be true, and consequently he adopted it; or he wished that the true meaning should be false; and therefore he rejected it.

There is a general prejudice in favor of interpreting the language of every one of whom we think favorably, so as to make it harmonize with our own views of what is true, right, or expedient. It is peculiarly apt to mislead us where we are not, in reality, well acquainted with the subject, and yet are not fully aware of the extent of our ignorance. We should remember that what *we* think or would have done, is no criterion of what others thought or did.

CHAPTER XII.

OF EVIDENCE.

§ 1. GENERAL PRINCIPLES OF EVIDENCE.—Two kinds of Signs.—Conclusive Signs.—How ascertainable.—Three classes of Probable Signs.—Mode of determining the value of a Sign.—On what the general credibility of Testimony is based.—Influence of Witness's moral Character.—Means of determining the combined force of several independent Evidences.—Circumstantial Evidence.—When satisfactory, and when not.—Criterion.—Principles of its Application.—Defects of Circumstantial Evidence.—Means of determining the degree of Probability.—Caution.

Signs are either *conclusive*, or only *probable*, evidence. They are conclusive when they are known, by induction or any other means, to be incompatible with the falsity of the proposition in question, or to be uniformly connected with it, and never to accompany any other state of things, except in some peculiar circumstances, which are absent in the case under consideration. Thus, the appearance of frost proves that the weather has been cold, and the mercury standing very high in the thermometer is a sure sign that it is warm.

The usual modes in which a phenomenon is shown to be a conclusive sign of something, is, by proving that they are connected as cause and effect, or that they are both effects of a known cause, or that they are connected as premise and conclusion.

Signs are only probable evidence in the three following cases.

1. *Where they are only generally, and not uniformly, connected with the thing in question.* The appearance of swallows is only a probable sign that spring has come, since they sometimes appear before that season; and the mercury standing very high in the barometer is only a probable indication of fair weather, since that phenomenon is sometimes speedily followed by a storm.

2. *Where they are sometimes connected with other things also.* A person's ceasing to breathe is only a probable sign of death, since that sometimes happens to one who is still alive.

3. *Where they are merely known to have accompanied it, in all the cases observed, but there is no proof that they will do so hereafter.* The appearance of comets, in ancient times, was followed by calamities; and hence it was inferred that they portended the latter,—whereas we now know that they have not the least connection with each other.

In judging of the value of a particular sign, as an evidence of the proposition under consideration, we should first ascertain its real character, and then attach to it the precise degree of weight which that warrants: otherwise we are very liable to be misled by prejudices and preconceived opinions.

The general credibility of *testimony* is based on the two following facts. 1. It requires an effort, for which some motive must exist, to invent a falsehood, or to conceive things differently from what actually occurred: for this is what naturally presents itself to our Memory or Apprehension. 2. The remembrance of a lie is always more or less painful, since we sympathize with the party deceived, and disrelish the thought that the falsehood may be detected, and possibly punished, either directly, or by loss of character for veracity, and withdrawing future confidence.

These principles operate even on the most immoral and unfeeling. Hence people will never lie, unless they have some object to effect by doing so; and, when there can be no such object, they will testify truly. The objects that will induce a person to lie, depend on his character. Some lie habitually, for very trifling objects, while others adhere to truth amidst the strongest temptations to the contrary. The latter are guided by certain moral principles, or fixed rules of conduct, which render their testimony faithful, under circumstances which would lead the former to falsify egregiously.

In many cases, there are various evidences, which all go to prove the same conclusion; but every one of them, taken separately, affords only a probability of its truth. Here we should consider whether the falsity of the conclusion is compatible with all the evidences, taken together. If so, there is only a probability that the conclusion is true. But if the reliable evidences are so numerous as to exclude the supposition of their being all fortuitous coincidences, the conclusion is established.

SEC. 1.]   CIRCUMSTANTIAL EVIDENCE.   211

Circumstantial evidence consists of signs or testimonies from which the proposition in question may possibly be legitimately inferred, but which do not directly prove it. It is sometimes more satisfactory than direct testimony, as being less liable to mislead us from the negligence or fraud of witnesses. It is frequently easy to state a simple direct falsehood which, if true, would decide the question, while it may be difficult to show the falsity of the statement, owing to its simplicity and conciseness; but it is extremely difficult to invent a series of circumstantial proofs the character of which cannot be easily detected by a careful examination of them, and comparing them with known truths.

This kind of evidence is always unsatisfactory where the circumstances are not so numerous and direct as to exclude the supposition of fortuitous connection, or where one circumstance disproves the conclusion drawn from another. The criterion of conclusiveness is, that when all the facts proved are certainly incompatible with any supposition but one, it is proved to be true. In all other cases, circumstantial evidence can prove nothing more than a probability.

Whether the circumstances of a particular case conform to the criterion or not, can be decided by no general rule, but only by testing them by the proper principles, and then fairly weighing the whole. If it does not appear that the point in question is a necessary inference from the facts established, it is not proved, although it may have been rendered highly probable.

It is further to be observed that we are liable to err in drawing the inferences from the circumstances proved, and that prejudices are apt to make the Imagination supply the links that are wanting in order to make the chain of evidence complete, while the facts are often made known to us by testimony, which is liable to deceive us, as in other cases. Sometimes, also, the apparent signs may have sprung from causes quite different from those by which they are usually produced, as where a stranger may have been handling a thermometer, unknown to the investigator, or stolen goods are secretly conveyed into the house of an innocent man, in order to avert suspicion from the thief. Sometimes, on the other hand, the usual signs are removed or concealed either from design or accident, as where a murderer hides the weapon of destruction, or a servant takes water from a rain-gauge, without its being known.

The degree of probability established by the circumstances, depends on their force, independence, consistency, and number. If several of them are dependent on one, or so connected with it that they must be true if it is true, all are tantamount to one only: and, in order to entitle the circumstances to any weight, they must not materially conflict with each other; else the inconsistent class will destroy the force of the rest.

This kind of evidence is susceptible of every degree of probability; and, therefore, we should beware of either receiving or rejecting it indiscriminately. Every case should stand on its own merits. The probability is often so strong that we should unhesitatingly act upon it, as if it were a certainty, while it is often so much the reverse that it is of no real value, except to suggest the course of further investigation.

§ 2. CRITERIONS OF TESTIMONY.—Importance of Testimony.—(1) Witness must have had means of knowing.—Usual Requisites.—General and Special Testimony.—Means of ascertaining witness's Credibility, on this head.—(2) He must have paid Attention.—Common sources of Inattention.—Partial Attention.—Where this point requires particular Consideration.—Means of determining the degree of attention given.—(3) Testimony must not be corrupted by Bias.—Kinds, influence, and sources of Bias.—How affected by individual Character.—Indications of its Absence.—Means of ascertaining its Existence, Character, and Influence.—(4) Testimony must not be a doubtful Inference.—Two classes of Inferences.—Various Sources of Error, and Means of avoiding them.—(5) Memory must not be in fault.—Influence of Bias on Remembrance.—Means of obviating its Defects.—Frequent Difficulty, and means of surmounting it.—(6) Witness must possess a competent Understanding. — Children's Testimony. — Indications of Defect. — (7) Testimony must be free.—Effects of Torture, Threats, Promises, and Suggestions.—Means of ascertaining whether they have operated.—(8) Testimony must be properly expressed, and faithfully transmitted.—How defects of this kind may be discovered.—Concurring Testimonies.—Nature of the Statements.—Common proof of Invalidity.—Caution.—Discrepancies.—Important Distinctions. —Sources and Character of apparent Discrepancies and minor Inaccuracies. — Real and material Discrepancies. — Means of surmounting Difficulties.—Particular use of Signs.—Probable Testimony.—Its Nature and Tests.—Caution.—Influence and Effects of Prejudices, in judging of Testimony.—Various kinds of Probabilities, and principles applicable to them. — Two futile Distinctions.

The subject of testimony is of the utmost consequence,

as by far the greatest and most important parts of human knowledge are based on it; and it is also frequently attended with difficulty. Hence its principles demand a diligent study and a careful application.

To render the testimony of an ordinary witness conclusive, as to the point which it professes to prove, it must generally possess the following characteristics.

1. *The witness must have possessed the means of knowing what he testifies.* The thing declared to have been comprehended must have been within the range of the proper faculties, in circumstances where they could act effectually; and these must have been in a sound condition, or at least not seriously deficient from disease or natural defect. The dim-sighted cannot see distinctly, nor can the deaf hear aright. In the case of sight, the object must not only be sufficiently near, but there must be enough of light, and not much more: men cannot see clearly in the dark; and a dazzling glare of light is equally unfavorable to proper vision. In the same way, sounds may be either too loud or too low for distinct hearing.

If the requisite organs are sound, defects of others do not generally impair the testimony. It is no objection to a witness's statement regarding what he saw, that he is deaf. Indeed a defect of one faculty frequently concentrates the attention so closely on the objects of that which is sound, as to render the comprehension and remembrance of them unusually distinct and vivid. It has often been observed that the blind mark and remember what they hear, and the deaf, what they see, better than those who labor under no such defect.

Testimony is of two kinds, *general* and *special*. The former alleges, not the particular things actually witnessed, but some generalization or inference from them: the latter states only what was directly comprehended. A person is incompetent to give a general testimony where he is not a judge of the subject: but such a restriction does not apply to special testimony. Thus, a man who is quite ignorant of seamanship, is incompetent to testify whether a ship of which he was aboard, was worked rightly at sea, during a tempest: but he may state the particular facts that he noticed, and aid mariners in forming a correct opinion.

The credibility of a witness, on this head, can be generally ascertained from a knowledge of his circumstances when he comprehended what he testifies.

2. *The witness must have sufficiently attended to what he testifies.* If his attention was absorbed by something else, or if he was indifferent, or so prejudiced that he did not fairly observe the object, his testimony is evidently unreliable. A partial degree of attention may enable a witness to testify correctly regarding the main facts: but in such cases little reliance can be placed on statements of details. In order to entitle these to credit, it must appear that the witness paid particular attention to the subject, and that he does not mistake for an apprehension what was only a conception, more especially if he belongs to that class of persons who are apt to be as much occupied with what they fancy as with what is present to their senses.

The degree of attention given can be generally learned, either from the witness's character and tastes, or from the nature of his statements. If these are full, and at the same time minute, attention may be inferred, while vagueness and confusedness of statement indicate inattention, in the absence of any design to conceal or mislead.

3. *The testimony must not be wilfully corrupted from bias.* In order to render a testimony quite reliable, the witness must be willing to relate the exact truth: otherwise it will be misrepresented by suppression, distortion or false additions. The bias may be either friendly or hostile. The former leads a witness to represent things more favorably for the party on whose behalf he testifies than truth warrants: the latter tends in the contrary direction.

The sources of bias are as numerous as the desires of the human mind: but the most prevalent are, self-interest, ambition, the love of ease, the love of the marvelous, vanity, and malice. Strong party feelings seldom lead to actual lying, where they are unaccompanied by baser desires; but they frequently produce concealment, exaggeration or distortion; and hence the testimony of partisans is justly liable to be suspected of such defects.

Where we are ignorant of a witness's character, we should not admit his unsupported testimony as conclusive, if there is room for the influence of bias: but where we know him to be strictly veracious, we may receive his statements without suspicion, wherever he could err only from conscious falsehood. In many instances, his veracity appears from the accuracy of his statements, as we learn from other sources, while it is sometimes established by the evidence contradicting his bias. Where a witness testifies against the prejudices or cherished views of himself or his party, and, more especially, where he exposes himself to pecuniary loss, suffering or general odium, by his testimony, without reaping from it any advantage that could lead him to falsify, we may consider his statements sufficiently free from the influence of bias. This conclusion is often corroborated by their exhibiting the artlessness, straightforwardness, candor, precision and minuteness which characterize faithful testimony, just as we often detect bias and fraud by the contrary qualities.

Where the testimony is otherwise unexceptionable, but it may possibly have been influenced by bias, we should receive it unhesitatingly only so far as it could not have been affected by the bias, and suspend any opinion as to the rest of it till we obtain further proof. Where the witness's character is known, we have only to determine whether there is any bias strong enough to have materially affected his testimony.

The existence and nature of the bias may be frequently learned from the witness's nation, profession, party, age, moral character, or peculiar relation to the testimony, as where he was himself concerned, and is anxious to show the best phase of his own conduct, or where some of his intimate friends or near relations are concerned. In other cases, these are learned from the sweeping or unqualified nature of the statements, laudatory or contemptuous expressions, and the manifest exaggeration, extenuation or coloring of the testimony. Sometimes they are detected by comparing the statements with those of an impartial person, or one of opposite bias, and by his evident desire to have us believe a particular thing.

Bias is sometimes discovered from the witness's hesitating before making his statements, for the manifest purpose of avoiding self-contradictions, and rendering his falsehoods plausible. At other times, it appears from the statements being vague, studied, evasive, impertinent or flippant, by the witness's pretending to forget what he could not but remember, by his avoiding definite assertions where he can be contradicted by other testimony, while he speaks positively and precisely where he cannot, or by his stating all that makes in favor of a certain conclusion, while he suppresses things of a contrary tendency, which he must well know. We should, however, distinguish the hesitation of bashfulness from that of bias, which can generally be done from observing the witness's whole manner, or his relation to the matter in question.

The extent to which bias has influenced the testimony, and the allowances to be made for it, may generally be learned from the circumstances of the case, or the testimonies of other witnesses; and, in many instances, it may be ascertained by comparing one part of the statements with another, or with what we previously knew of the subject.

4. *The thing asserted must not be a doubtful inference from what was actually witnessed.* Inferences are frequently drawn with such rapidity that we mistake them for comprehensions, while they are sometimes erroneous. Hence witnesses have thus frequently testified untruths. The inferences are generally drawn correctly, and, as was formerly explained, we may admit them without hesitation. But some extend such inferences beyond due bounds. They will say, for instance, that they saw or felt such things when they only apprehended something which they believed to be tantamount, but which was, in reality, very different. Biased and careless witnesses are especially liable to fall into this error, but it often results from a general illusion. Instances occur in the general belief regarding our seeing the heavenly bodies moving, and the distances of objects.

Sometimes a witness makes a statement as if upon his own personal observation, when, in fact, he derived his information wholly from others. Such testimony, at the best, directly proves only what the witness was told; and the accuracy of the transmission must be tested by the circumstances attending it. The manifest relation of the witness to the thing testified, or the nature of his statements, often enables us to ascertain whether he speaks from personal observation; and, in other cases, this can generally be learned either from the witness himself or from other parties.

5. *It must appear that there is no ground to suspect a failure of Memory.* When persons testify regarding things which they witnessed long ago, they are very apt

either to forget them altogether, and say they never comprehended such a thing, or to confound them with what they apprehended or imagined on some other occasion; and hence, where their testimony is unsupported by other evidence, it is seldom conclusive, except as to matters which they could neither forget nor mistake. The Memory of some, again, is so feeble, and their Imagination so active, that they can hardly distinguish what they remember from what they only conceived, after the lapse of a few days; and hence their uncorroborated testimony is of little value. Persons of this class are generally discovered by the palpable errors in their testimonies.

The bias and general character of a witness affect his remembrance, as well as his apprehension, because we remember what strongly interests us much better than what we view with indifference. Hence a witness frequently remembers well one part of what he apprehended, and wholly forgets the rest. Consequently, unless it appears that the witness felt equally interested in all that he may have apprehended, his having no recollection of one part affords no strong proof that it was not present.

The immediate defects of a witness's Memory may have been obviated in several ways. He may have accurately written down what he observed at the time of its occurrence, or before his recollection of it failed, and he may have a distinct remembrance that the writing is full and correct throughout, while he can identify it; or he may have perused a written statement made by another, while the occurrences or things witnessed were fresh in his Memory, and he can still testify to its accuracy from his recollection; or he may have faithfully related to another person what he witnessed, before he forgot it, and the latter may still have a perfect recollection of what was thus communicated to him, or may have written it down accurately, when he first heard it. In all cases of this kind, however, the various parts of the chain of evidence must be examined with care, in order to avoid false inferences.

6. *The witness must not be deficient in understanding.* The statements of an idiot, a maniac, or a young child, are generally as worthless as those of persons who labor under a defect of Memory or the organs of sense; and the testimony of one who labors under some hallucination or delusion relating to the matter in question, is

K

equally unreliable, although his intellect may be otherwise quite sound.

A child's testimony is sometimes satisfactory regarding simple and striking facts which fell under his observation; and it frequently possesses the advantage of being more free from bias than that of older persons: but the inexperience and volatile disposition of childhood, its great liability to imposition, and its strong tendency to mistake one thing for another, and draw false inferences, generally render its testimony unsatisfactory, and of little value except to corroborate other evidence.

Where we are not otherwise aware of the witness's character, testimonies of this kind are usually detected by their puerility, incoherence or absurdity.

7. *The testimony must be the free and spontaneous statement of the witness, unaffected by torture, force, threats, bribes, promises, or suggestions.* All these operate like natural bias, and, in many instances, still more strongly. It was formerly a common practice to torture witnesses who would not otherwise testify, in order to extort the truth: but experience showed, what might have been easily foreseen, that such applications rather led the witness to declare what he knew was required, in order to be released from the torture, even where his testimony subjected him to severe punishment.

Although this may be considered the extreme case, yet all similar methods of eliciting evidence are alike in principle, since they interfere with the fair statement of what the witness knows or remembers, and must generally lead to misrepresentation, if not to positive misstatement. Suggestions from others are equally objectionable, since they tend to make a witness confound what he remembers with what he only conceives, even where there are no suspicions of collusion, which suggestions of this kind naturally excite.

The relations, circumstances and conduct of the parties concerned, the character of the testimony, or the subsequent free statements of the witnesses, generally render it easy to ascertain whether motives of this kind have operated; and, in many cases, there is not even any ground to suspect them.

8. *The testimony must be properly expressed, and conveyed to us either directly, or substantially as it was originally delivered.* It would evidently mislead, either by

being stated in language which, when fairly interpreted, conveyed an erroneous signification, or by the original terms being essentially altered, through omissions, additions or substitutions. Defects of this kind are to be discovered by observing the witness's character and circumstances, and the mode in which the testimony is transmitted to us.

Where a testimony is unexceptionable on all the preceding heads, it is evidently conclusive. But, in many cases, we have not sufficient information on one or more of them; and hence it is necessary to determine the force of the testimony on other grounds.

The concurring testimony of several independent witnesses affords certainty, in many cases otherwise doubtful. Where the statements of one, although very unsatisfactory when taken alone, are corroborated by many others, while there is no reason to suspect collusion or unconscious error, we justly consider the evidence conclusive. For it is incredible that various unconnected persons, differently situated, should all invent the very same fictitious tale, or commit the very same mistake, unless there is some ground of illusion common to them all; and where this exists, it is generally obvious.

The absence of collusion or forgery is frequently evinced by various coincidences in the different testimonies, which are so indirect and recondite as to satisfy us that they could not have resulted from any kind of fraud. These frequently consist of incidental remarks or allusions, which are found to tally with the other testimonies only by examining them closely, and drawing various inferences, so that it would evidently baffle human ingenuity to forge such narratives, and yet escape real discrepancies. This conclusion is sometimes corroborated by the testimonies exhibiting several apparently glaring inconsistencies, which no one capable of forging such testimonies could have overlooked, but which can either be reconciled, or are of such a character as not to invalidate the substantial accuracy of the evidence.

The nature of the statements often entitles them to belief, independently of the witness's character, as where he states the results of recondite scientific investigations, on a subject of which he is ignorant, or plans and sentiments which are above his power of conception, if not above his comprehension. In cases of this kind, the testimony is credible, because the witness could not have forged it, while its consistency with what is known from other sources, may show that the facts are not materially misrepresented.

On the other hand, one of the most common means of ascertaining the invalidity of a testimony, is, its being at variance with what we know from other sources: for truth cannot be inconsistent with itself. Some statements are incredible, because they assert self-evident impossibilities; and others are so, because they are inconsistent with conclusive proof. But we must beware of hastily assuming that this is the case, especially where the witness's character is otherwise good: for we are very apt to be satisfied with fallacious proof of what we wish to be true; and cases are not rare in which conclusive testimony has been rejected, on the ground that it contradicted known truth, when, in fact, it contradicted only cherished errors.

In some instances, the worthlessness of the testimony is apparent from the fact that one part of it directly contradicts another, on points of great importance. This contradiction may exist between the different parts of one statement, or a witness's present allegations and former statements which he made on the same subject; or it may lie in the conflicting statements of different witnesses. Here we must distinguish between material discrepancies, which proceed from gross carelessness, forgetfulness or unfaithfulness, and those minor variations which generally exist in the testimonies of independent witnesses, or even between the different statements of the same witness. The former invalidate a witness's credibility, while the latter do not, but frequently establish it.

Mendacity, collusion and forgery are frequently detected by the absence of any palpable inconsistency with the belief of the person addressed, while a careful and extensive scrutiny will discover, in the different statements, serious violations of truth. A faithful witness, on the other hand, often makes statements which do not precisely tally with the opinions of those whom he addresses, or possibly even with all his own allegations. But their substantial accuracy is generally confirmed by subsequent investigations and discoveries. At the same time these detect a false testimony, since every error is necessarily inconsistent with some truth, which future researches or accidental discoveries generally bring to light. So numerous are the sources which gradually reveal error or falsehood, by unfolding some truth which is inconsistent with it, that it is very frequently exploded by the gradual disclosures of time.

Many discrepancies are only apparent, and arise from our misunderstanding or misinterpreting part of the testimony; some proceed from a witness's stating as a fact a false inference which he drew from what he comprehended; while others spring from one witness relating what another overlooked or forgot, every one generally remembering only what interested him, or attracted his attention. Some proceed from one witness suppressing part of what he apprehended, for reasons which can generally be discovered, without much difficulty, by considering his circumstances and character.

Many real discrepancies rather confirm than invalidate the testimony, because, while they are of very little consequence, they indicate the absence of studied and fraudulent harmony. So numerous are the sources of error, that the entire absence of any variations regarding such matters as dates, places and numbers, is not to be expected, under ordinary circumstances; and, therefore, such variations do not generally affect a witness's credibility regarding the main points of his testimony, especially when they are elicited by brow-beating bashful or timid witnesses. Discrepancies between different statements of the same witness often arise from a mere slip of the tongue or pen, or some other slight degree of inattention, or a failure of memory regarding points of no consequence, or misinformation derived from others, so that they do not in the least affect his veracity, and ought to have little influence on his credibility.

Where testimony is really contradictory or discrepant, on material points, we may either look for further evidence, or closely scrutinize the statements and character of the witnesses. We shall thus generally find an absence of the marks of conclusiveness on one side or the other. The defect is often so manifest as to render the testimony on that side worthy of no regard. The statements of a veracious and careful witness are not in the least invalidated by their being contradicted by those of a notorious liar or fabulist. So, when one witness of doubtful veracity states something which is omitted by another, of known credibility, and which the latter could have neither overlooked nor forgotten, the former statement should be disregarded.

If the difficulty cannot be removed by any of those means, we can obtain only a probability, and not knowledge, unless we have recourse to better evidence, which we should always do, if practicable, in every case of difficulty.

The credibility of a statement may often be ascertained from the witness's former conduct. The characters of men are generally so uniform and consistent, that one who lied yesterday is seldom found to be veracious to-day, and one who has been strictly veracious hitherto, will generally be found so hereafter. If a man has been found to adhere to truth, amid the strongest temptations to the contrary, and to have been constantly guided by rigid principles of veracity, we may safely believe his statements, provided he has not been misled. But if he is found to have generally yielded to strong temptations, although he may be veracious in ordinary circumstances, we can trust his unsupported statements only when there is no strong motive to falsify. If he has habitually lied, no dependence can be placed on his testimony except where there is evidently no possible motive for falsehood, which can seldom be safely assumed of such characters.

The credibility of a witness may sometimes be ascertained by observing how he states facts with which we are acquainted from other sources, and which form tests of his veracity, such as accounts of things discreditable to himself or his party, or things which militate against some of his favorite opinions or desires. But we must beware of trusting to doubtful testimony regarding a witness's character: for we sometimes receive the most contradictory accounts of a person's veracity, and it is folly to believe one stranger regarding the character of another, where he may have some motive for misrepresentation.

When a testimony amounts only to a probability, signs often supply the deficiency, and produce certainty. If a stranger asserts that there has been a severe earthquake, at a particular time and place, his testimony may be confirmed by palpable marks of its effects, such as rents in the ground and fallen houses. In such cases signs either corroborate or refute the testimony, precisely like the statements of additional witnesses. They may simply affect its degree of probability; or they may render it conclusive on the one hand, or entirely invalidate it on the other.

The probability of a testimony of this kind depends on the witness's circumstances and character. If he had good means of knowing the truth, and bears a fair reputation, there is a great probability that his testimony is true. But if his situation was unfavorable, and his veracity is very doubtful, his statements cannot be safely adopted, without corroborative evidence. If his statements have received a coloring from his prejudices, we should endeavor to ascertain its extent, and make due allowances.

If there are several independent witnesses, the probability will vary according to their number, their several characters and opportunities, the degree of attention they bestowed on the matter, the time that elapsed, and so forth. The probability may be so strong that we are justified in acting as if it were a truth; yet we should beware of ever assuming it to be such, and rejecting, without due examination, any alleged proof which seems to contradict or impugn it. Proof should never be rejected wherever we do not already possess certain knowledge: for it frequently happens that the highest probabilities are refuted by irrefragable proof.

Mankind are apt to be greatly influenced, in the reception of testimony, by their views regarding its extrinsic probability, or that which is based on the nature of the statements: but this is generally the effect of prejudice. The unlearned frequently measure everything by their personal experience; and the learned often view subjects in the light of their own theories. Hence conclusive testimony has often been rejected, simply because it contradicted the erroneous belief or strong desires of the party to whom it was addressed, while, owing to the same causes, worthless testimony was received as conclusive. Some travelers have been deemed liars, for faithfully recording what they saw, while others have related fictions, which were believed without doubt or hesitation.

Unexceptionable testimony is never to be rejected on account of the apparent improbability of the statement:

for testimony can prove anything comprehensible, which is not self-evidently impossible, or demonstrably inconsistent with known facts; and, for such propositions, no unexceptionable testimony ever can be given, for the simple reason that incompatibilities cannot co-exist. But we must beware of heedlessly assuming that anything professedly proved by reliable testimony, belongs to this class: for prejudice is apt to make us infer that a proposition is absurd or untenable, when, in truth, it is only distasteful to our wishes, or incompatible with our erroneous opinions.

Men have often admitted errors on inadequate testimony, because they appeared to them highly probable, while they rejected important truths, proved by conclusive testimony, simply because they appeared highly improbable, which was confounded with what is impossible. The proper course is, never to admit anything as a certain fact, whether probable or not, when the evidence is unsatisfactory, and never to reject anything which is proved by unexceptionable evidence, however contrary to our preconceived opinions or our wishes. To act otherwise is, to believe without any good ground on the one hand, and to disbelieve against conclusive proof on the other. The very improbability of a statement is often a strong indication of its truth, since a liar would invent something more likely to be believed by those whom he addressed, and an honest and careful witness is more apt to pay particular attention to an extraordinary phenomenon than to one of a different character.

Where, however, the testimony is not conclusive, and it only establishes a probability, we may set up another probability in opposition to it, and thus possibly nullify its whole force.

The *intrinsic* improbability of a testimony, or that which flows from the witness's circumstances and character, must be carefully distinguished from its *extrinsic* improbability, or that which is based on the nature of the thing testified. Where the former does not exist, the latter is entitled to no weight: otherwise this strengthens the former, or impugns an intrinsic probability, to a degree proportional to its force. Where the extrinsic improbability conflicts with the intrinsic probability, the resultant probability is equal to the excess of the weaker of these two elements over the stronger.

A distinction has been drawn between such things as are "contrary" and those which are "not conformable" to our experience; and it has been maintained that evidence of the latter is admissible, but not of the former.

To render this distinction valid, it is necessary to make the phrase "contrary to experience" mean that we were actually present, in the case alleged, and experienced no such thing as is affirmed, while we were so situated that we must have done so, if it actually occurred. But the distinction would now be irrelevant, since it amounted to saying that we should not receive another's testimony, to contradict what we ourselves properly witnessed, a doctrine which, though unquestionably sound, is evidently foreign to the point.

Another meaning of the dogma is, that we experienced nothing like what is alleged, while we were placed in circumstances where that might be reasonably expected. In this sense, it is evidently absurd. It would require a man, for instance, who had seen innumerable mountains, to reject all testimony as to the existence of volcanoes, because he had never seen a mountain pouring forth fire or smoke. So, it would require us to reject all testimony regarding the fall of showers of stones, because we have seen many showers fall, but never one of that kind.

The only other admissible meaning is, that the thing alleged is contrary to the experience of ourselves and all our friends and neighbours. This modification is not sounder than the former, as these persons' circumstances are generally the same as our own, so far as regards the matter in question. The fact that all our neighbours have uniformly seen the Sun in the south at noon, by no means proves that others do not see it in the north, at that time of day. The absurdity of the distinction, taken in this sense, is well illustrated by the case of the culprit, who offered to bring thirty witnesses, every one of whom could truly swear that they had never seen him commit the crime for which he was on trial, after three unexceptionable eye-witnesses had sworn to the contrary, the difference between the two classes being, that the latter were present, and the former absent.

The sense which would include, under the expression "*our* experience," that of every individual of mankind, is evidently inadmissible, as in that case the dogma would assume that the testimony is false, which was the very thing in question.

A similar distinction has been drawn between improbabilities based on previous experience, and those founded on the various ways in which a thing may happen. The former class, it has been said, are not legitimate subjects of testimony, but the latter are. It is sufficient to observe that History abounds with instances of things extremely different from what previous experience would have led us to expect; and, therefore, this distinction is as futile as the former.

§ 3. VARIOUS KINDS OF TESTIMONY, AND PECULIARITIES OF EACH.—(1) Explicit and Implicit Testimony.—Advantages of each.—(2) Immediate and Mediate Testimony.—Characteristics of the latter. —Why common Rumor is generally worthless.—When important. —Mode of testing the value of Mediate Testimony.—(3) Oral and Written Testimony.—Advantages and Disadvantages of the former. —Importance and Advantages of Written Testimony.—Effects of Lapse of Time.

1. Testimony is *explicit* when it expressly declares the very thing in question: it is *implicit* when it only states something which directly implies it, or is a sign of it. In the latter case, the witness is sometimes unaware of the implication: and hence it often elicits truths which could be discovered in no other way, owing to the witness's strong desire, or fixed determination, to conceal them: and such evidence here possesses the further advantage of excluding the influence of bias, in coloring the statements, even where there is no attempt at concealment. Yet, where there is no difficulty of this kind, explicit testimony possesses the great advantage of being free from the risk of false inferences, to which we are always more or less liable in the other case.

2. *Immediate* testimony is, where the party who addresses us personally comprehended what he relates: *mediate* is that given by one who only heard or read what he relates; and many parties may have intervened between the original witness and the last relater.

As men are generally liable to forget or misunderstand what they have read or heard, mediate testimony is seldom entitled to credit, especially where it has passed through many hands. It often contains a large admixture of truth: but it is frequently difficult to determine, with any great degree of certainty, which part is true and which is false. Even the most important part is sometimes entirely lost, or totally misrepresented. Hence the proverbial uncertainty of common rumor.

Mediate testimony is particularly worthless where, as frequently happens, some of the relaters labor under any of the difficulties which invalidate immediate testimony, as where an unprincipled man hears something disreputable about one whom he strongly dislikes. But the immediate relater is sometimes a person of a different character; and hence the muddy channel through which the statements have flowed is apt to be hidden from view.

Where the object is, merely to ascertain what common report says, such testimony is sometimes quite important, as where a man prosecuted for slandering another, shows that the latter's reputation, as to the matter in question, was previously very bad.

In estimating the value of mediate testimony, we must first determine the degree of credit due to the immediate relater, and afterwards examine every part of the chain of evidence. If we find a material defect in any link, it establishes, at the best, no more than a probability, which must be sufficiently corroborated by some other evidence, before it is entitled to be received as satisfactory.

Where a narrative includes both kinds of testimony, every part is to be tested by the proper criterion; and it may thus be found that some parts are quite credible, and others as much the reverse. Herodotus' account of the Persian wars is mostly credible, because he obtained his information from eye-witnesses: but this cannot be said of his history of earlier transactions, which is much less reliable.

3. *Oral* testimony is given by word of mouth : *written*, is that of a person who is not present, and whose written statements are read. The former possesses the advantage of allowing us to observe the witness's demeanor, and to obtain further information, or an explanation of difficulties, by means of questions. Suitable interrogations also enable us, in many instances, to test the witness's competency, and the character of his previous statements.

On the other hand, the presence and appearance of a witness not unfrequently excites prejudices which lead to his statements being estimated either above or below their real value, whereas a careful consideration of his testimony, by the light of other reliable evidence, and a knowledge of his character and peculiar circumstances, would obviate any such result. Immediate oral testimony, also, can be procured only during the original witness's life-time, after which the advantages of this kind of testimony are lost.

Written or printed testimony can be accurately communicated to distant times and places, while it obviates the numerous errors arising from failures of memory, and enables us to contemplate the statements as often and as deliberately as we please. It also enables us to have recourse to the original evidence, instead of relying on hearsay, compilations or abridgements, all of which are liable to produce numerous errors, and which frequently misrepresent or suppress something of importance.

There is generally less room for bias in written testimony, because the witness's prejudices are not so extensively concerned. An author frequently writes for distant times and places, or at least for unknown readers; and, therefore, he is less tempted to misrepresent than where his testimony has an evident bearing on immediate objects of desire, which may be secured by it.

Another advantage of written narratives is, that other evidences by which they may be tested, are more independent, and consequently more free from bias. The people of a neighbourhood are all frequently influenced by the same prejudices, which is rarely the case with witnesses who live in distant ages and countries. The lapse of time tends to confirm true statements and disprove others, by revealing things which confirm the former and disprove the latter, while these are so unconnected and varied, that collusion, or accidental errors of any great consequence, are inevitably detected. Thus, ancient testimonies, which are corroborated by the successive discoveries of distant ages and countries, become more and more irrefragable as they become older, while false statements are gradually seen in their true light.

While written testimony is of such great importance, it is sometimes attended with difficulties from which oral evidence is generally free. These we shall briefly discuss in the following section.

§ 4. MEANS OF ASCERTAINING THE ORIGIN AND CHARACTER OF WRITTEN TESTIMONY.—(1) External and Internal evidences of Authorship.—In what cases the former are reliable, and in what,

Sec. 4.] Evidence of Authorship. 229

not.—Principles of decision regarding the latter.—Means of ascertaining the Witness's Age and Country.—Important Requisite.—Means of distinguishing spurious from genuine Compositions.—When it is, and when it is not, of consequence who wrote a Composition.—Rules for ascertaining the Writer, in the former case.—(2) Sources of material Corruptions.—Rules for determining what parts are corrupted or spurious, and what, genuine.—Different Editions.—How Abridgements may be distinguished from Originals.—Means of ascertaining the Age and Country in which a Manuscript was written, or a Book printed.—(3) Origin of Various Readings.—Where one only, and where several, are genuine.—Comparative purity of Manuscript and Printed Copies.—Effects of Time, in removing Errors.—Influence of Printing.—Rules regarding Various Readings.—General Character of these Readings.—(4) Means of distinguishing authentic from fictitious narratives.—Applications of this section.

I. The evidences of the authorship of a composition are either *external* or *internal*. The former consists of the testimony of contemporaries or persons who lived near the period: the latter is, that which is afforded by the matter and style of the production.

External evidence is to be examined like any other testimony. In the case of contemporary or very recent writings, it is generally conclusive, and free from any serious difficulty, as there is direct credible testimony: but, in regard to ancient compositions, it is often worthless, owing either to the witness not living sufficiently near the time when the work was composed, or to the testimony being spurious, or to the known mendacity or inaccuracy of its author.

This testimony may consist either of direct statements, or of allusions, quotations or translations. As a book cannot be quoted or translated till it exists, the fact of such references being found proves that the composition is older than the one which contains the references. But these should be unequivocal; and the age and origin of the work which contains them should be well known: otherwise they will rather mislead than enlighten.

The supposed author himself sometimes testifies regarding a composition attributed to him: but his evidence should be received with caution: for some dishonest men have claimed the works of others as their own, while some have denied their real compositions. If the author's veracity is unimpeachable, however, such testimony is conclusive.

The internal evidence is based on the various peculiarities of thought and expression which distinguish different ages, countries, and individuals. A work which bears the characteristic and peculiar indications of a certain age and country, may fairly be inferred to belong to them, while one which lacks these, may be safely taken to belong to some other time and place.

The evidence afforded by particular allusions is often conclusive, as to the time after which a work must have been written. A treatise on the Trojan War, that alluded to the Crusades, must have been written after the tenth century; and one which mentions the presidents of the United States of North America, cannot be older than the end of the eighteenth. It is possible, however, that a work may have been altered, so as to contain allusions to things much more recent than the original treatise. If the part containing the allusion is genuine, its evidence is conclusive: if not, the age of the composition must be ascertained by other means.

The author's country may often be determined in a similar manner, because some nations were ignorant of things well known to others. A history of the Crusades could not have been written by a Japanese contemporary; nor could a Roman history have proceeded from an ancient Hindoo.

The application of the preceding principles requires an accurate knowledge of the real peculiarities of the different ages and countries, which were not always such as they have been represented.

Compositions of which the authors are doubtful or unknown, have been mostly assigned to persons of whom we possess some genuine works: and, therefore, we can compare the composition in question with the latter. But, in so doing, we should beware of assuming that all the works of an author must be precisely alike, in every respect. His subject, time of life, and varying circumstances and opinions, often produce marked diversities in an author's different works, although certain peculiarities are generally common to them all. We should, therefore, observe whether the differences in question are greater than may reasonably be attributed to those sources.

The difficulty of detecting some spurious productions is increased by their having been designedly composed in imitation of the alleged author's real works, and their consequently adhering closely to his style and sentiments. But a close inspection will generally show their different origin. It is more difficult to imitate a person's style exactly, throughout a composition of any length, than to mimic his voice. The imitation will be either a spiritless copy of the original, as unlike as a corpse is to the living person; or it will exaggerate his peculiarities, so as to resemble a caricature.

If the imitator is incessantly on his guard against any expression which would betray him, his composition will necessarily exhibit a constrained and affected mode of speaking, reminding us of a person walking on stilts, with an absence of originality and the ease which distinguishes one who uses his own natural style. If, on the other hand, he should venture on new ground, and adopt some freedom of expression, his own peculiarities will appear, and some words will escape him which will at once betray the origin of the composition.

The question who actually wrote a composition is frequently of no consequence. For, even where the author employed an amanuensis or one who wrote from his dictation, we may fairly presume that he either perused it himself or that it was read over to him, and all serious errors corrected, before it passed out of his hands. But this question becomes very important wherever the writer must have been the author, as it then becomes identical with that of authorship, which it may be the principal means of ascertaining. In such cases, there are the four following ways of ascertaining the writer.

1. His own acknowledgement, which is quite satisfactory wherever his character and motives are above suspicion, and he gives his evidence unambiguously; but, in other cases, it may be entitled to very little weight.

2. The evidence of a person who saw him write it, which is to be examined like other cases of testimony.

3. The opinion of one who is familiar with the handwriting. As various persons often write like each other, and the same person often writes in different hands, either from hurry, design, or gradual change, this evidence is seldom quite satisfactory; and it generally furnishes only a probability.

4. The opinion of one who is a judge of handwritings, and who compares the writing in question with others, the writers of which are supposed to be known. This

method is still less reliable than the last, since we are apt to err regarding the origin of the writings used for comparison; and the opinion of one who is not familiar with the party's handwriting is less reliable than that of a person who knows it well.

II. Written testimony has sometimes been materially corrupted from design. In some instances, the corrupter believed that he was correcting the testimony, and removing errors, while he was only falsifying it: in other cases, he acted wilfully, from a desire to make it conformable to his wishes. Here parts which exhibit any of the three following characteristics are to be rejected as interpolations.

1. Where they do not exist in any good manuscript, or in any printed edition superintended by the author.

2. Where the statements could not possibly have been made by the author of the work, on account either of his circumstances or his character.

3. Where we have conclusive external evidence that the parts are either wholly spurious or materially altered. Such evidence may be found in the testimony either of the author himself or of persons who derived their information directly from him, or examined a copy of the work known to be correct.

A part is to be deemed genuine in the following cases, provided there is no conclusive proof to the contrary.

1. Where it is found in all good manuscripts, or in a manuscript or printed copy examined by the author.

2. Where there is conclusive external evidence of its genuineness. This may be found in the author's acknowledgement, either direct or indirect, (as by quoting it or alluding to it) or in the statements of parties who read or heard the original or obtained reliable information from those who knew.

3. Where it forms an essential part of the composition, which would be rendered unmeaning or absurd by its removal.

Some variations are possibly alterations made by the author, in successive issues, in which case widely different copies may all be equally genuine, although the latest may be fairly presumed to be the best. In printed books, the date, place, printer, and preface, often enable us easily to distinguish the successive variations, which are occasionally so great as to render the various edi-

SEC. 4.]   MATERIAL CORRUPTIONS.   233

tions, in reality, different works. But we seldom have such aids in the case of manuscripts. Yet the circumstances and habits of ancient writers were such, that different editions of a work can rarely be presumed to have emanated from the author, while unequivocal proof of such a thing can hardly be found; and consequently we may presume that there is only one genuine version.

It is sometimes doubtful whether a work is an original composition, or only an abridgement, executed probably by some other person than the author. Here we must generally rely on external evidence, as the question usually arises only where the original has perished. In other cases, differences of style and expression may clearly indicate that the abridgement was not made by the original author.

In investigating the character of a particular copy, it is frequently important to know the age or country in which it was written or printed. The former can generally be determined, to a great degree of accuracy, by observing the language, orthography, divisions of lines and words, punctuation, contractions, abbreviations and diacritical marks, embellishments and flourishes, material, and ink. These all varied so much, from age to age, that a careful examination of them will generally enable a person who has studied the subject, to determine the age of a manuscript within half a century. The country in which a manuscript was written, can frequently be ascertained by the same means, since it is found that different countries adopted different forms and methods.

The characteristics of the several ages and countries are ascertained from writings of which the origin is known, either from their nature or from immediate testimony. To the former class belong original deeds, charters, letters, proclamations and public records. Some of these, however, are counterfeit; and, therefore, care is required to distinguish them from the genuine.

Printed works generally indicate the place, printer, and year, on the title-page and its reverse, or at the end. But some books lack these, or give fictitious ones, in which case recourse must be had to the preceding methods. The character of the materials, and the appearance of the type, ink, and binding, often indicate their age and origin, while the printer is frequently known from his colophon, or peculiar device, at the end of the volume, or in the title-page.

III. A third difficulty, which frequently attends written testimony is, that there are various readings, or slightly different versions of it. These arose from various causes. The old copyists, who supplied the place of printing, before the use of that art became general, did not always detect or correct the accidental mistakes which occurred in their manuscripts, since this both imposed labor and disfigured their appearance. Other errors arose from the same sources as spurious passages, and some proceeded from the marginal glosses of preceding copyists or readers being mistaken for omissions, and consequently inserted in the text. Others were caused by substituting a phrase which was better known for one which was becoming obsolete or belonged to another dialect. Many arose from misreading or forgetting the original, and then substituting a similar but different expression.

It is also very possible that there may be several genuine readings of a passage, just as there may different genuine editions; and the remarks already made regarding the latter, apply equally to the former. In modern authors, they are of common occurrence, and easily ascertained: but the case is frequently otherwise with older compositions, in which we have rarely any proof that more than one reading ever proceeded from the author.

Printed copies are generally more accurate than ancient manuscripts, because various errors are corrected, by means of proof sheets, before the impression is struck off; and they are also less subject to omissions or additions, either accidental or designed, while they render impracticable any subsequent alterations of all the copies of an edition; and although printed copies are by no means exempt from errors, yet the critical comparison of good manuscripts or copies, and more care in printing, tend to produce more and more accurate versions, till a text is formed free from any serious error.

Although a single copy or impression lasts only for a definite period, yet the arts of printing, photography, engraving and stereotyping enable us to multiply accurate copies, or facsimiles of manuscripts indefinitely, so that the testimony can be preserved unchanged, to the remotest times.

The rules already given, regarding the genuineness of a composition, are equally applicable to this subject; and the following are the principal additional rules, for ascertaining the true reading:

1. The oldest manuscript of an ancient author is to be preferred when other things are the same. But the rule does not apply where a later manuscript has been more directly derived from the original, or where a manuscript is found to have been either carelessly written at first, or afterwards corrupted.

2. Where several genuine readings exist, the most recent is to be deemed the best. This rule assumes that the last reading expresses the author's most mature thoughts; and, therefore, where the case is otherwise, it does not apply, as if a man in his dotage or in a fit of insanity should undertake to alter the productions of his better days, or should wilfully corrupt them, from some bad motive.

3. The reading of the great majority of copies is to be preferred, unless there is conclusive proof that it is erroneous.

4. A reading which is conformable to the known views, sentiments or style of the author, is preferable to one which is not, unless the former is demonstrably erroneous.

5. A reading which gives a good or correct sense, is better than one which gives either nonsense or an erroneous statement. This rule applies only to the testimonies of persons who are not destitute of sense or honesty.

6. A reading which violates the idiom or grammar of the language, is to be rejected for one which does not. It is assumed that the author wrote with care, and knew the language: else this rule does not apply.

7. Of two admissible readings, that which was most liable to be changed, is to be preferred. Thus, a reading at variance with the opinion of the copyists, or conflicting with some strong prejudice of the party addressed, is preferable to one which conforms to it.

8. Of several readings otherwise equally probable, that which best agrees with the context, is to be preferred.

9. A reading which is a manifest blunder or corruption, is to be disregarded.

10. The reading of a good is preferable to that of a bad copy, where there is no proof to the contrary.

11. A good copy is known by its general accuracy, and its containing few manifest errors, or by the testi-

mony of persons who knew the original, or copies directly and carefully made from it.

12. An inaccurate copy is known by its abounding with unquestionable errors, its being evidently made without care, or by the testimony of those who know its real character.

13. A printed copy of which the author corrected the proof sheets, is good: but if he did so carelessly, a copy diligently corrected from his original manuscript, may be better.

14. Where all the copies of an early translation show no variations of a passage, its reading is equivalent to a copy of the age in which it was made. The rule assumes that there is no room to doubt the genuineness of the passage, or the mode in which the translator read his original: otherwise it is inapplicable.

15. The preceding rule is applicable to quotations, if not made from memory. It is observable, however, that quotations are often made from memory; and, in such cases, they are of little use in ascertaining the true reading.

16. In modern compositions, the true reading may frequently be ascertained by referring to the original manuscript, or to some person who directly knows that reading. But we must beware of mistaking an inaccurate copy, or the rough draught, for the original manuscript. The true original is the finished composition, and not the first copy.

17. Conjectural emendations are allowable only where the text is certainly corrupt, while the proposed emendation harmonizes with the context, and has every appearance of having been the original reading. The obscurity of a passage does not justify the admission of such emendations, since it may arise from our ignorance, and disappear after we have become better acquainted with the language and the subject of discourse. Hence the text might be corrupted, instead of being improved, by a too ready admission of emendations of this kind.

Most various readings are of little importance, because they do not materially affect the sense. But some of them are serious; and these abound in works which have been copied very carelessly.

IV. The fourth difficulty which sometimes attends written testimony, is, that we are left to doubt whether the narrative is authentic or fictitious. Here we must examine the external and internal evidences, as when we are investigating the authorship, the determination of which will generally remove this difficulty also.

The character of the composition may possibly be learned readily from testimony. Thus we learn that "The Life and Surprising Adventures of Robinson Crusoe" is a fictitious autobiography, written by Daniel Defoe, though partly based on the real adventures of Alexander Selkirk. So we learn that Julius Cæsar's "Commentaries on the Gallic War" is an authentic narrative, regarding real personages and events.

Fiction generally betrays its character, either by its absurdity or its inconsistency with what we know from other reliable sources. If the writer gives dates, and the names of persons and places, we may compare his statements with what is otherwise known regarding them, when we shall generally ascertain the character of the narrative, with little difficulty. If he gives only fictitious names, his testimony is not entitled to belief, unless we possess other conclusive evidence of its authenticity.

The character of a narrative is generally unfolded by subsequent researches or discoveries, regarding the subjects of which it treats, as in other cases of testimony. Thus, we learn that the accounts which several ancient authors give of the Pygmies, though long believed, are quite fabulous, while, on the other hand, numerous statements of ancient historians, which were long regarded as fictitious, have been confirmed by recent discoveries.

This section is mostly applicable to all kinds of written composition, as well as to testimony in its narrower signification, or that which regards only the witness's immediate comprehension, since the general modes of ascertaining the origin and character of writings are the same, to whatever class they belong: and, so far as concerns our present purpose, everything which a person asserts, may be termed his testimony.

CHAPTER XIII.

OF CLASSIFICATION.

§ 1. NATURE AND USES OF CLASSIFICATION.—Definition of Classification.—Two kinds of it.—Naming essential.—How distinguished from Generalization.—Objects susceptible of Physical Classification.—Why this is preceded by the Mental Process.—Five main Objects of the latter, with Observations.—Division.—Important Distinction.

Classification is, arranging together such things as resemble each other, and separating them from such as are unlike: and it is either *mental* or *physical*. The former consists in determining what things are alike in some respects, and unlike others in the same respects, and distinguishing them by a suitable name. The latter consists in actually arranging or assorting together physical objects, which have been formed into classes by the mental process. If a man has a lot of shells, he may first determine how many kinds there are, with the peculiarities of each, and give every class a name. This is the mental classification. When he places together all the shells that belong to the same class, and removes those which do not, this is physical classification.

Naming the various classes is an essential part of classification: for it is requisite to a proper remembrance or description of their several peculiarities. Hence a good system of nomenclature greatly facilitates induction; and it is also requisite in order to secure the chief objects of classification.

Classification differs from generalization in comparing like with unlike, separating the two, and distinguishing every class by a suitable name. We might generalize if we had only one species before us: but classification requires us to compare individuals of several species. Generalization necessarily precedes classification, because we cannot classify objects till we have first ascertained their common resemblances, as well as their common differences.

Physical or material objects alone are susceptible of a physical classification: for, although we may classify mental phenomena, or the peculiarities of individual character, it is impracticable to arrange such things together.

As physical classification is only an actual assorting of things, according to the dictates of the mental process, the latter must always precede it.

The main objects of mental classification are the five following.

1. *To aid the Memory.* The simple objects in nature are so numerous, and frequently so unlike each other, that we cannot remember the properties of any great portion of them, without arranging them in such a way that we can consider those which are alike consecutively, and thus obviate the necessity of running continually from one object to another extremely unlike it. This can evidently be effected only by means of classification.

2. *To facilitate induction.* It is only after those things which are obviously alike have been arranged together, and separated from those which are unlike, that we can advantageously proceed to investigate their general properties. We first ascertain individual facts, and then extend our observations, till we establish some kind of generalization. This is followed by a corresponding classification, which is usually succeeded by an induction: and the latter often leads to an improved classification, which may prepare the way for a more extensive induction.

3. *To assist us in determining the character of an individual.* By simply ascertaining that it exhibits some characteristic mark of a class, we know at once that it belongs to that class, and possesses all its peculiarities, a thing which might otherwise be very difficult, or even impracticable. Thus, a zoologist can ascertain the general structure and habits of a quadruped from simply inspecting the bones of the foreleg.

4. *To prepare for physical arrangement.* Where this preliminary process does not receive due attention, much confusion and loss of time are apt to result, because a thing cannot be found when it is wanted, or it cannot be ascertained what is wanted and what is superfluous. Proper physical arrangement of tools, apparatus and materials is generally requisite to success in the arts, and also in many scientific investigations.

5. *To facilitate the communication of knowledge to others.* This is sometimes termed *division;* and it belongs to Rhetoric, rather than to Logic. Its leading principle

is, that things should be so arranged as to render an understanding of the subject as easy, and the proofs as clear and conclusive, as possible. The principle that things the most like should be classed together, which is generally supreme in Logic, should here be subordinated to the leading principle; and many didactic treatises have been seriously injured by the author's confounding rhetorical with logical classification, two things which often differ widely.

§ 2. PRINCIPLES AND METHODS OF CLASSIFICATION.—General Principle of Classification.—Why there are many Special Principles.—Primary and Subordinate Principles.—What determines the former.—Seven Rules of Classification.—Essential and Non-essential Properties.—Classification of Organic Bodies.—Limits of Species.—Method of forming Genera, Families, Orders, and Classes.—Its Advantages.—Modes of naming Organic Divisions.—Application and Improvement of Principles.—Influence of Prejudices, and how avoidable.

Similarity and diversity form the bases of all classification, the general principle being, that *things which are similar are to be classed together, and separated from such as differ.* But as things are like and unlike each other in various respects, there are many special principles of classification.

The *primary* principle of a classification is, that according to which things are grouped in the first instance, without allowing any other to interfere. A *subordinate* principle is, that according to which these groups are subdivided.

The particular object of the arrangement determines which of several possible primary principles is to be adopted. The tax-gatherer classes men according to their residences and amount of property—the physiologist, according to their corporal peculiarities—and the moralist, according to their ethical principles and conduct.

The following are the principal rules of classification:

1. Every classification should be made according to a *certain principle*, with which no other should interfere. This rule is evidently requisite to prevent confusion. If a librarian, in first classifying his books, should at one time arrange them according to their subjects, and at another, according to the language in which they are written, volumes which treated of different subjects would

stand together in one place, and some which were composed in different languages would be found in juxtaposition elsewhere, so that both principles would be violated.

2. The principle should be *definite*, and not difficult of application. Otherwise it would be doubtful in which class a certain thing should be placed, and it might happen to belong equally to two co-ordinate classes. The principle that all those plants are to be classed together which are essentially alike, although good as a general rule, is too indefinite as a general principle, because it leaves doubtful what essential likeness is.

3. The principle should *apply equally to all the things to be classified.* Otherwise some part may be entirely omitted in the classification. The primary principle that all animals are to be classed according to the structure of the nervous system, is objectionable, because many animals have no such system.

4. The principle should *bring together* those things which our object requires us to unite, and *separate* them from all others: else the classification would not answer the purpose. The principle that animals should be classed according to their apparent affinities, is bad, because these often differ widely from the real affinities.

5. We should *commence with the highest*, and proceed gradually to the lowest divisions of our subject. As the former are separated by the greatest and most striking differences, they are generally the most easily made: and it is necessary to survey all that is to be classified, and form the higher divisions, before we can rightly fix the limits of the lower. Thus, in classifying all animals, we should begin with what are termed the sub-kingdoms, then proceed to the classes, thence to the orders, thence to the families and genera, and, lastly, to the species: otherwise it would be impossible to effect an harmonious and satisfactory classification.

6. In all the sciences which treat of organic beings, as such, those things are generally to be classed together which are *most like*, or resemble each other in the greatest number of particulars. For the properties of such bodies are best understood and remembered, when everything is arranged beside that which it most resembles, because this renders the points of similarity most apparent, and the mind naturally associates things according

L

to their similarities. Hence plants should not be classed solely according to the structure of the organs of fructification, because many plants have no such organs, and, in the case of others, the general structure is frequently unconformable to that of those organs.

Essential and *non-essential* peculiarites should be distinguished. The former are those which determine a thing to be what it is, and which cannot be altered without subjecting it to a complete structural change: the latter are such as may alter without producing any structural change. In animals, the structure of the organs of nutrition and motion are essential properties, while the size, the color, and the appearance of the covering, are non-essential. Thus, a dog may be larger or smaller, black or white, straight-haired, curly-haired, or wholly hairless: but he could not have teeth and feet like a sheep's without ceasing to be a dog. This distinction leads to the following principle, which qualifies the sixth.

7. Organic beings which resemble each other in *essential properties* are to be classed together, although they may differ in others. Grayhounds must be classed with dogs, and not with hyænas, because they exhibit the essential peculiarities of dogs, although, in several respects, they resemble hyænas more than they do poodles.

In classifying organic beings, all those individuals are considered to be of the same species which differ only in such peculiarities as are found to vary in individuals known to have sprung from the same parentage; and where two individuals exhibit differences that are constant, and resist change, in those which have a common origin, under every variety of external circumstances, they are considered of different species.

In forming species into genera, a different course must be adopted: for the latter, unlike the former, are marked by no definite natural boundaries. The following method appears to be the best which has been hitherto proposed.

A particular species which exhibits, in a marked degree, the chief peculiarities common to several very similar species, is taken as the *type*, or best representative, of the whole group; and all those species which resemble this type more than they do any kindred one, are classified with it, as being of the same genus.

The number of types to be adopted, depends on the number of groups of species which exhibit such differences that it would not comport with the object of the classification to arrange them in the same genus.

The higher divisions of families, orders and classes may be formed in the same way, by adopting one of the subordinate divisions as a type.

This method facilitates both classification and study. For we have only to compare a new or unclassed species with the types which it closely resembles, in order to determine its proper place, while the type gives the student a good view of the characteristics of the group, and the peculiarities of every species are easily acquired afterwards. So, the typical genus may represent its order, and the typical order, its class.

A genus is distinguished by a generic name, and this is prefixed to a specific designation, to form that of the species, which thus indicates both the genus and the species. Thus, *bos* is the generic name of the ox tribe; *Bos taurus* is the common ox, and *Bos Americanus*, the American ox, frequently miscalled the buffalo. So, *Quercus* is the generic name of the oaks: *Quercus pedunculata* is the European white oak, and *Quercus alba*, the American white oak.

The orders are named from some striking peculiarity or some well-known genus. Thus, *Ruminantia* (cud-chewers) is the name of that large and important order of mammals which possesses four stomachs and ruminates; and the rose gives name to the extensive vegetable order of the *Rosaceæ*, including the rose, apple, pear, quince, plum, peach, cherry, currant, blackberry, gooseberry, raspberry, and strawberry, which are all characterized by alternate leaves, several sepals, regular petals, distinct stamens, and separate carpels.

Families or sub-orders are generally named by modifying the name of some well-known genus. Thus, *Canidæ* denotes the canine family, from *canis*, the term for the dog genus; and *Bovidæ* expresses the cattle family, from *bos*, the name of the ox genus.

The classes, which are comparatively few in number, are designated either from one of their chief characteristics or from their common collective name. Thus, the class *Mammalia* is named from the peculiarity of suckling their young, and the birds are termed *Aves*, which is only their ordinary Latin name. So, in Botany, one great class is termed *Exogens* (out-growers), from their growth being formed by new layers around that of the preceding season, such as the common forest trees of temperate regions; and another is termed *Indogens* (ingrowers), from the new growth being deposited among and within the old, such as the grasses, lilies, and palms.(17)

In effecting a classification of many objects and kinds, we must first lay down proper primary and subordinate principles, then arrange the objects according to the dictates of our primary principle, then subdivide every division, according to the requirements of our most general subordinate principle, and so on, till we come to the lowest subdivision.

When we have to class a newly discovered or invented object in its proper place, we must ascertain the division to which it properly belongs, from its classic, ordetic, generic, and specific characteristics, or those which distinguish the class, order, genus, or species from its coordinate divisions. We begin with those of the highest division and proceed gradually to those of the lowest, arrange the object accordingly, and distinguish it by a suitable name. Thus, if we had just discovered the musk ox of Arctic America, the first glance would show that it is of the mammal *class*, while its horns and teeth characterized it as of the ruminant *order:* but its generic character is not so easily determined. Its horns and general character indicate that it is an ox, while its long, woolly hair, its short legs, its nose and its face, ally it to the sheep tribe. Hence it is classed as an intermediate genus, under the term *Ovibos* (sheep-ox), while the species is designated *Moschatus* (musky), from its rank musky smell.

In effecting a physical classification of objects, we first select and arrange together those which exhibit the characteristic marks of the various divisions immediately below that to which they all belong, and then subdivide the groups thus formed, by placing together, apart from the rest, all those objects which possess the characteristics of the various groups of the next lower subdivision, and so on, till we arrive at the lowest.

Wherever it is found that our principles are objectionable, because they fail to effect the desired result, we should modify or alter them, so as to remove the difficulty, by which means we may gradually arrive at perfect principles. Thus, if a librarian attempts to arrange all his books on the primary principle that those treating of the various subjects were to be separated, he would soon find that his principle is objectionable, because many books treat of different subjects. This would lead him to the better primary principle, that books which treated of several distinct subjects, were to be separated from such as treated of only one subject; and then he could properly apply his original principle to the latter.

Prejudices may injuriously affect this subject, as well as every other, because we are strongly tempted to place a favorite individual, and especially ourselves, in a class better than that to which he really belongs, and to reverse this error, in the case of a being that we dislike. The proper course is, to ascertain the individual's real characteristics, and then set him down in that class to which he is unequivocally attributed by the proper principle.

CHAPTER XIV.

TABULAR VIEW OF THE MEANS OF ACQUIRING KNOWLEDGE.

I. *Necessary and Universal* Truths are learned by

1. Direct Intuition, including
   - (1.) Truisms.
   - (2.) Self-evident properties of Time and Space.
   - (3.) " Abstract Quantity.
   - (4.) " Substantial Beings.
   - (5.) " Relations of Things.

2. All the intellectual faculties, aided by signs and symbols, including
   - (1.) The necessary properties of Space and Time, which are not self-evident.
   - (2.) The necessary properties of Abstract Quantity, which are not self-evident.
   - (3.) The necessary properties of Substantial Beings, which are not self-evident.
   - (4.) The necessary properties or Relations of Things, which are not self-evident.

II. *Particular Contingent* Truths are known by

1. Direct Comprehension, including our present
   - (1.) Perceptions.
   - (2.) Sensations.
   - (3.) Ideas and phantasms.
   - (4.) Emotions and desires.

2. Comprehension, Intuition, and Memory, often aided by Testimony and numerous external appliances, whence we know
   - (1.) Our own existence.
   - (2.) That of other substantial beings around us.
   - (3.) The obvious properties of us and them.
   - (4.) Our own past experience.
   - (5.) That of others, including History and Biography.

III. *General Contingent* Truths are known by

1. Comprehension and Abstraction, whence we learn empiricisms regarding the present.
2. Comprehension, Abstraction, and Memory, aided by audible and visible signs, including Testimony, whence we know empiricisms regarding the past.
3. Comprehension, Abstraction, Memory, and Intuition, generally aided by signs and apparatus, whence we learn obvious inductions regarding the past, present, and future.
4. All the intellectual faculties, aided as in the preceding case, whence we learn recondite inductions and inventive truths.

IV. *Hypothetical* Truths and *Probabilities* are learned by the same means as the third class.

PART III.
OF FALLACIES.

CHAPTER XV.

NATURE AND CLASSIFICATION OF FALLACIES.

§ 1. NATURE OF FALLACIES.—Definition and Operation of Fallacies.
—Why they frequently produce Belief.—Their Number.—Two
Evils resulting from Fallacies.

A *fallacy* is, any thought, expression, or argument, which tends to produce erroneous belief. It has always some semblance of proving a certain conclusion, without which it could have no such tendency; and we are frequently liable to adopt the error, without ever suspecting its existence. When a fallacy is clearly exposed or understood, its worthlessness as a proof appears so glaring, that we are apt to think it could never impose on anybody; yet, when presented in the disguised and indirect form in which fallacies usually come before the mind, it often produces conviction, especially when favored by prejudices; and it cannot sometimes be detected without close attention and great care.

The difficulty is frequently increased, and the reception of the fallacy facilitated, by the argument's containing a combination of fallacies, in which one conceals, and consequently strengthens, another, and its being believed by the party who propounded it, and who is consequently sincere in his advocacy of it. We are also liable to overlook fallacies which originate wholly with ourselves, because we do not suspect their existence, and are unwilling to believe that we have been misled. Many fallacies, again, may be detected by a very brief close examination of them; yet, owing to ignorance of their nature, and attention not being called to the defect, no such examination is ever made.

Fallacies have no definite limit as to number; and they admit of endless modifications and combinations, so that a complete enumeration or precise definition of every fallacy is impracticable: yet, when we understand the nature and operation of those which most usually occur, we need have little difficulty in detecting all others.

Fallacies hide truth, and substitute positive error in

L 2

its place. As conscious ignorance is preferable to delusion, it were better not to engage in the pursuit of truth at all, than to do so, and fall into error, which not only misleads us directly, but prevents further investigation, and leads to the rejection of every incompatible truth.

§ 2. CLASSIFICATION OF FALLACIES.—Three Classes of Fallacies.—(1) Paralogisms.—(2) Sophisms.—(3) Aberrancies.—Universal defect in Fallacies.—Why these three include all possible Fallacies.—Their independence of each other.—What invalidates an Argument.—Practical Application and Illustration.

Fallacies may be divided into the three following classes, corresponding to the three parts of a syllogism.

1. A comprehension exhibits the appearance of something different from the reality, or something appears to be self-evident or known by satisfactory proof when it is neither; and thus we are led to infer that we discern or learn something, which, in truth, we do not. This class we term *paralogisms*, or fallacies of primary assumption. A paralogism may, therefore, be defined a delusive representation, leading directly to an erroneous belief.

2. Something is inferred from a premise which is not, in reality, implied in it. This class we term *sophisms*, or fallacies of intermediate reasoning. A sophism may, therefore, be defined a syllogism or argument in which the inference or conclusion is not implied in the premises.

3. The actual conclusion is not the proposition which ought to have been proved, but one essentially different, which forms an *aberrancy*, or fallacy of irrelevancy. It is, therefore, an argument or syllogism, the conclusion of which is not the question, or the conclusion which ought to have been proved, but one essentially different.

There is a false inference in every fallacy: but in paralogisms it regards primary assumptions—in sophisms, intermediate reasoning—and in aberrancies, a final conclusion. Hence, the same fallacy which is a sophism when it is employed to prove a further inference, may become a paralogism when it is used in the final syllogism of an argument.

The above three classes include all possible fallacies; for, if all our primary assumptions are sound, our inferences from them legitimate, and our final conclusions the very things to be established, there is evidently no possibility of error. Hence all fallacies must regard our primary assumptions, our reasonings from these, or our final conclusions.

Those three classes of fallacies are all distinct and independent. We may reason correctly from unsound premises; or we may reason sophistically from unobjectionable premises; or both premises and reasoning may avail nothing, since they are beside the real question: and an argument is rendered worthless by involving any one of those fallacies. One material defect invalidates the whole, as much as if no part was sound. It may possibly contain two or three kinds of fallacy: yet, in examining its validity, if we find one fallacy, it is quite unnecessary to search for a second.

If the premises require proof, and none is given, we need not search to see whether the conclusion is implied in them, or whether it is relevant. The whole is like a fabric built on sand; and it is not stable, however strong the superstructure. It is not requisite to show that the premises are false, in order to refute an argument based on them: it is sufficient to show that, for any thing which appears to the contrary, they *may* be false. If they are doubtful, an argument which, in order to be valid, requires that they must be known truths, is as worthless as if they had been proved to be false.

If it appears that the conclusion is not implied in the premises, it is unnecessary to test their soundness, or the relevancy of the conclusion. Here the fabric has no strength in itself; and, therefore, the firmness of its foundation avails nothing.

If the conclusion is irrelevant, all that precedes is tantamount to nothing: for here the fabric is erected on a wrong foundation; and, therefore, it leaves the place where it ought to have been built still vacant.

CHAPTER XVI.

SOURCES OF FALLACIES, AND MEANS OF GUARDING AGAINST THEM.

§ 1. SOURCES OF FALLACIES.—Twofold Source of Fallacies.—Intrinsic Sources.—Extrinsic Sources.—Immediate extrinsic cause of all Fallacies.—How it operates.—Seven causes of Inattention, with Remarks.

THE sources of fallacies are either *intrinsic* or *extrinsic*. The former lie in the subject, and the latter in the investigator.

All intrinsic sources of fallacy lie in one thing's resembling another so much as to be readily mistaken for it, while they are materially different, or in two things differing so much that we are apt to think they differ altogether, while they are virtually alike, so far as concerns the matter in hand.

The intrinsic source of paralogisms is found in a delusive representation, which is apt to lead directly to the inference that we know or learn something which, in fact, we do not.

The intrinsic cause of sophisms lies in one thought or expression's being so like another as to be readily mistaken for it, while the two are radically different, whence we are apt to infer that a premise implies an inference which, in reality, it does not.

The intrinsic source of aberrancies consists in the question at issue bearing some resemblance to another, which is essentially different, whence it is apt to be inferred that the former is proved, when the latter only may have been proved.(18)

Inattention is the immediate extrinsic cause of all fallacies: for, if we carefully attend to a subject, we shall know whether the premises are sound, whether they imply the conclusion, and whether the one established is the right conclusion. Fallacious phenomena often lead to false inferences: yet the appearances are real; and no false inferences are drawn regarding them, wherever we give them sufficient attention.

Inattention operates by leading us, either to overlook something altogether, or to draw immaterial distinctions, or to overlook characteristic differences, so that we mistake one thing for another.

In order to secure attention, we must ascertain the causes of inattention, of which the following are the principal.

1. *The painfulness of close and continued attention.* This generally requires a strong and unpleasant effort: for the mind naturally tends to run off from the point under consideration to something else, which it suggests, and which it is less irksome to contemplate, while we are glad to avoid the painful feeling of exhaustion which such attention produces. Hence investigation is often hurried; and various points are considered inattentively, as the mind wanders from the subject, before we have obtained either an accurate or an extensive insight into it.

2. *The sacrifice of present enjoyment often involved in continued attention.* The earnest investigator must often deny himself many pleasures which are freely enjoyed by others, as when he sits in silence, pondering over difficulties, or when he is busy amid objects offensive to his senses, while his friends are enjoying the pleasures of social intercourse, or the beauties of external nature.

The two preceding causes generally operate simultaneously on the individual, and, owing to their combined influence, the attention often wanders from the subject; things are viewed hurriedly and inaccurately; conclusions are adopted or rejected without proper investigation; and erroneous opinions result.

3. *Thinking of too many things at once.* The attention to any point becomes less and less, as the number of objects to which it is directed increases, supposing that none of them excite it strongly. Hence a person who thinks at once of several things, each of which requires close attention, is very liable to err regarding one or more of them.

4. *Strong sensations or perceptions.* When a person endures acute pain, or has his attention forced on some strong perception, such as the discharge of a gun very near him, it is generally impossible to attend either closely or continuously to the matter in hand. Even moderate pain, if continuous, such as hunger and thirst, distract the attention, in ordinary circumstances, to such a de-

gree as to render the investigation quite unreliable. Strong pleasing sensations are equally unfavorable to the discovery of truth. A person whose ear is delighted with sweet music, or who views a very striking and beautiful scene, for the first time, need not hope to solve any difficult problem. In all such cases, the attention is drawn so powerfully to the apprehension, that we cannot examine another subject with any degree of care.

5. *Obtuseness of the faculties, arising from fatigue, disorder, or some permanent defect.* A person who is sleepy, or suffering under the influence of a narcotic drug, cannot pay close attention to anything which does not produce unusually strong feelings; and the same remark applies to one who is prostrated by strong previous excitement, or whose faculties are naturally very dull and torpid.

6. *Strong emotions, unconnected with the subject of investigation.* These distract the attention quite as much as strong apprehensions, and operate precisely in the same way. A man who is excited with wrath, or in a paroxysm of joy, is as unfit for investigation as if he were laboring under a burning fever, although his passion should be entirely unconnected with the subject of inquiry.

7. *Prejudices*, or emotions connected with the subject of investigation. So much often depends on the conclusion at which we may arrive, that strong feelings are excited, which mislead us like other emotions. But here we are less apt to be sensible of their influence, while, at the same time, this is more constant, and not so easily avoided. Hence prejudices are the most fertile of all sources of error.

§ 2. OF PREJUDICES.—Modes in which Prejudices operate.—Causes of their power.—In what cases they exist.—Five classes of Prejudices, with Remarks.—Combination of Prejudices.—Source of the power of Error.—Why we readily take for granted what we are taught.

While prejudices distract the attention, like other emotions, they also tend to withdraw it from whatever leads to a disagreeable conclusion, and to concentrate it on those of a contrary kind, because we naturally turn away from painful feelings, and fix the attention on such as are agreeable, just as we turn from loathsome sights or odors, to such as are pleasant. Hence we are disposed, not only to overlook that side of the subject which makes against the conclusion we desire to be true, but also to view the other side as being brighter or stronger than the reality, because we overlook all the weak and hostile parts of it, and confine our attention to the others.

The case of persons who take a more gloomy view of a subject than truth warrants, is no exception to this remark: for security against some anticipated evil, or a deliverance from present pain, appears to them extremely desirable, as the amount of evil is generally exaggerated, from some mistaken opinion, or a timid and oversensitive disposition.

The attention is thus apt to be confined to those things which make either for or against the conclusion, while the rest of the subject is either wholly overlooked, or not considered with any degree of care. Hence futile distinctions are taken for essential differences, while the latter are either overlooked or treated as immaterial, and one thing is mistaken for another, of a widely different character, so that fallacious arguments are adopted as conclusive, while irrefragable proofs are rejected as unsatisfactory.

Not unfrequently the Imagination rivets the error, by drawing unfaithful pictures, which are taken to be correct, because the attention is so absorbed that their true character is not perceived. The bright parts are exaggerated, because their worse aspects are overlooked, and imaginary excellence is superadded, while the dark parts are equally misrepresented, because we do not attend to their better aspects, and we attribute to them imaginary evils. At the same time the emotions excited prevent us from adverting to the fact that the picture is only imaginary, and lacks proof.

The influence of prejudices is very apt to escape our notice, because they operate rapidly, quietly, and without any noticeable effort, whence their power is increased, because it is unperceived, or even unsuspected. Thus the Judgement becomes the dupe of the feelings, and we unconsciously form very false opinions, under the partial and erroneous views of a subject thus produced. Such opinions are often held with great tenacity, because we are quite unaware of the deception which we practice upon ourselves; and they are more numerous than the fallacies arising from other emotions, because prejudices operate secretly and incessantly. Every important subject generally excites some prejudice, which arises quietly and operates, in a great measure, undiscernibly, because there is no palpable indication of its presence, whereas a man under the influence of violent wrath, or deep grief, is apt to allow his emotion to subside, before he undertakes to investigate any subject of consequence; and he has, at all events, a distinct knowledge of the disadvantage under which he labors.

The influence of prejudices is strengthened by our being very unwilling to believe its existence, or to examine their operation aright. It is mortifying to our self-esteem to think that we have been deceived by our wishes; nor is it easy to understand how this happens. Hence many never acquire that knowledge of the nature and operation of prejudices which is requisite to guard us fully against their influence.

Prejudice exists wherever strong emotions are excited by the contemplation of the results at which we may arrive: for these feelings excite a strong desire that the proposition under consideration should be true or false, owing to our belief regarding the good or evil connected with the alternatives.

As all prejudices operate in the same way, and their number is indefinitely great, a complete enumeration of them is neither desirable nor practicable. The following are the most common.

1. *Prejudices of self-love in general.* Every one necessarily desires his own welfare, and is, therefore, prone to believe whatever tends towards that object, and unwilling to believe the reverse. Hence the readiness with which youth expects a happy manhood, and the latter believes that visionary projects are highly eligible. Hence, also, the readiness with which many listen to flattery, or fish for applause, or foster pride, by believing that they and their kindred are superior to others. Hence our unwillingness to listen to reproof and bitter truths, and our readiness to believe agreeable falsehoods, to interpret the language of others so as to square with our views and wishes, and to accept or reject their testimonies, according as it is agreeable or the reverse. Hence we are apt to form too high an opinion of ourselves and what concerns us, and too low an estimate of

others. Hence, also, the undue value we often attach to what affects our own welfare, and the underestimate we make of what concerns our neighbour's; and hence our unwillingness to believe that our own characters, views, opinions and prospects are bad.

2. *Prejudices of the ruling desire.* A great portion of mankind consider some particular thing much more valuable than any other good; and hence they desire it with corresponding earnestness and constancy. Everything is viewed under the influence of this prejudice; and nothing is patiently considered which tends to prove that the object of pursuit is unworthy of such anxieties and toils, or that there is something else of much more consequence. A miser will not consider the evils incident to an inordinate pursuit of gain; and a vain person is unwilling to believe that his flatterers are either mockers or designing knaves, although this may be very apparent to others. The man who makes reputation his chief good, rejects all arguments to prove that he is pursuing a shadow; and the sensualist will listen to nothing which goes to prove that there is anything within his reach of vastly more importance than sensual gratifications.

3. *Prejudices arising from the love of present ease or enjoyment, and an aversion to present toil or suffering.* What is present, or in immediate view, is easily appreciated, and if it be deemed of much consequence, it usually excites a strong desire to secure or avoid it. But the case is far otherwise with the distant future. This generally requires careful consideration, in order to be properly estimated, because remote objects appear indistinct and smaller than the reality to mental, as they do to ocular, vision; and, in many instances, the real character of the future can be ascertained only by means of continued and painful efforts, which all dislike, and which an examination of the present does not involve. At the same time the attention is naturally drawn first to that which is near, as being the most striking, and often the most urgent; and the anticipation of speedy pleasure or pain withdraws it from a careful consideration of the future, which is also less vividly pictured by Conception. Hence, when the choice lay between the present and the future, mankind have frequently preferred the former, even where the latter was incomparably more important.

From this prejudice originates the practice of taking up a subject, and studying it only till its novelty has ceased to tickle the fancy, when it gives place to some other, which soon makes way for a third, and so to the end of the cycle. In this way many have become proud of their attainments, when, in truth, they never mastered the rudiments of a single important study. A similar effect of this prejudice is, the common predilection for such methods as profess to furnish a short and easy road to knowledge, and the dislike for such as require a great degree of labor, and also the predilection for those studies which either please at the time or promise the speedy gratification of some favorite desire, in preference to those of a contrary character, though the latter are generally by far the most important.

From the same source has originated many false maxims, such as that all valuable truth is easily acquired, that whatever is unknown must be of little value, that our faculties are very blind and weak, that we can know very little at the best, and that all real knowledge is very simple. Although it is easily seen that all such dogmas are false, yet, with their aid, many contrive to keep themselves very ignorant of various truths which it deeply concerns them to know, while they adopt errors in their stead, without ever suspecting that all is not right.

Another instance of the effects of this prejudice is, the dislike which men swayed by evil desires or groveling appetites bear to true religion and sound morality. These militate against their present enjoyments. Hence they generally fly to religious scepticism, fanaticism, or superstition, any of which is more conformable to their wishes than truth; and consequently the same person often swings repeatedly from one to another of these errors.

Another frequent instance is, the devotion to that course which promises most money. This commands various immediate pleasures, while it secures men against various present pains, and hence the avidity with which it has been sought by the countless votaries of Mammon. The preference of sensual to mental gratifications is an equally common instance. The former can be enjoyed without any previous study or self-denial.

From this prejudice also springs that fondness which many show, for adhering uniformly to general rules or modes of acting, however requisite modifications or deviations may be, in certain instances. Hence they frequently run from one extreme to another. If they get into serious trouble, from following their usual course, they are apt to discard all rules; and if they find that some of their opinions are not wholly true, they will probably reject them altogether, and adopt the contrary, which may be more erroneous than the former.

4. *Prejudices of education and profession.* The habitual consideration of some objects, or parts of a subject, to the exclusion of others, tends to produce very erroneous opinions, since we attach little importance to that of which we are ignorant. Hence those objects often excite emotions and desires extremely disproportionate to their real importance. The subjects which we have been taught in youth, and those things with which our vocation makes us familiar, are frequently viewed through this distorting medium, and seem very different from the reality, while we test everything by that with which we are familiar, however inapplicable. Every bigot thinks that saving truth is found with his creed alone: the artisan is apt to think that his art is the most important, ingenious or beautiful in the world: the merchant is inclined to apply figures to everything, and to estimate the character and prosperity of a nation by its exports and imports; and the mathematician sometimes attempts to improve mental science by very rigid arguments, based on specious, but inaccurate, definitions, and plausible, but erroneous, assumptions, leading to irrelevant conclusions. So every nation and class are apt to think more favorably of their own institutions, laws, customs, and manners than would be warranted by an impartial judgement.

5. *Prejudices of association.* It often happens that things are associated in our minds with excellences or defects with which they have no necessary and uniform connection, whence the contemplation of them excites emotions which properly belong to the latter, but which are referred to the former, as their cause. Consequently they are judged according to those emotions, and very erroneous opinions are formed regarding their real nature. A person, for instance, flatters those whom he addresses. This pleases them; and they attribute the pleasure to his superior judgement, the excellence of his arguments, or the goodness of his cause. Hence flattery has

ever been a powerful instrument of deception. One man has a repulsive aspect, and hence a prejudice against his sentiments: another appears benevolent, and hence a prejudice in favor of his opinions.

This class of prejudices has produced much evil throughout the world. The heathen has often clung to his gods, and rejected the clearest proofs that they are only imaginary monsters, because their worship was associated with his domestic joys, and they did not prohibit wicked practices to which he was strongly attached. So the child, whose vicious propensities were curbed by his parents and teachers, has often been prejudiced, on that account, against the truths which they inculcated, while another, whose wicked inclinations were freely indulged, was strongly prejudiced in favor of the false teachings of those who so indulged him. Men are thus very apt to be prejudiced in favor of false opinions which accord with their vices; and here lies the power of error.

Prejudices often derive force from the combination of several, to produce the same result. A man, for instance, is strongly prejudiced in favor of his own views or opinions, because it is humiliating to think that he has been mistaken; he has been so taught by his parents or teachers, whom he venerates; if he should be in error, he must undergo the labor of re-examination; and possibly he may arrive at conclusions repugnant to his ruling desires, and tending to render him less acceptable to persons whose good opinion he wishes to possess.

From such sources originates the common practice of taking for true what we have been taught or have hitherto believed, and looking to the mere opinions of others, instead of searching for conclusive proof. It is so much more pleasant and easy to accept current opinions, which tally with our wishes, and to compare and criticise written statements, than to investigate the subject properly for ourselves, that we need not wonder a large portion of mankind have adopted, from the earliest times, error which a very moderate degree of independent and careful observation and reasoning would have completely exploded.

§ 3. MEANS OF AVOIDING FALLACIES.—Requisites in order to avoid Fallacies.—Means of obviating the influence of Prejudices.—Two important means of avoiding Error.—Why Truth is often undervalued.—Consequences.—Proper Course.—Dangerous Practice.—

Evil results of one fundamental Error, and of they will readily—How important Conclusions should be tested.—ent wherever guments which prove too much.—Means of counterac.. -equired fects of Prejudices.—Refuting and proving Arguments.—A tendency of the Mind, and consequent Caution.—Exception en Common Error.

In order to avoid erroneous opinions, we must attentively examine everything requisite to obtain a correct view of the subject under consideration; and this requires that we avoid the various sources of inattention. Hence we should never study or investigate any subject while we are influenced by a feeling or perception so strong that we cannot readily concentrate our attention on any point at will, or while our faculties are blunted by sleepiness, languor, stupor, fatigue, exhaustion, or bodily disorder, so that we cannot pay close attention to the subject; nor should any confidence be placed in conclusions formed under such circumstances.

We should guard against strong apprehensions and emotions, by keeping at a distance from their various sources, such as exposure to great heat or cold, sensual habits, loud noises, and exciting scenes, conversation or reading. Sensual habits are particularly injurious, because they produce unnaturally powerful appetites for particular indulgences, which greatly distract attention, and at the same time cause a languid and irritable state of mind, very favorable to the adoption of error. Passions operate in the same way: but their influence is much less extensive, because they are much less permanent in their effects.

We should also beware of having our attention distracted by considering several things at once: and hence an investigation should be subdivided into parts, every one of which can be closely examined, and its character determined, without paying any attention to the rest, during its examination.

No subject should ever be considered negligently or inattentively. If it does not deserve careful attention, it were better let alone: for otherwise we shall be apt to form a habit of inattentive examination, which will cleave to us in investigating the most important subjects.

The principal means of obviating the influence of prejudices is, to study their mode of operating, once for all, until we see the inevitable consequences of yielding ever been a power, to mark the nature and tendency of prejudices regarding the immediate subject of consideration, and to cherish the dispositions and prej.ts requisite for proper study and investigation. If we allow our minds to be controlled by prejudice, the flood-gates of error are thrown open, and we readily adopt it instead of truth, without ever suspecting the deception. We should distinguish the question before us from those which prejudice leads us to substitute for it, and to examine it aright, on its own merits, regardless of the real or fancied consequences.

It is frequently impossible to eradicate a prejudice: yet its mere existence need not prevent us from adhering to all the principles of proper investigation: for where conclusive proof is fairly understood, it necessarily produces conviction, however disagreeable the conclusion, as when a man hears very bad news, and the testimony leaves no room to doubt its truth, or opens his purse, thinking it contains money to pay a demand made on him, and finds it empty. Prejudice can mislead us only by withdrawing or distracting the attention: and, therefore, if this can be sufficiently secured, the subject will be seen in its true light, whether pleasant or the reverse.

We shall be greatly aided in overcoming the influence of prejudice, by forming a proper estimate of the value of truth, and habituating ourselves to endure a little present evil, for the sake of a great future good: for prejudices derive their force chiefly from the inducements which they hold out to sacrifice the long future for the present moment. The discovery of truth may hurt our feelings for a little while: but, like a surgical operation, it has a healing effect.

Prejudices are often on the side of truth: but, even here, they are apt to act injuriously, by preventing a proper examination of the subject. Thus, they may prevent us from knowing the real grounds of our opinions, and the futility of plausible objections and counter arguments, so that when the advocates of error afterwards assail us with these weapons, we are liable to be much perplexed, if not permanently misled.

A knowledge of the great value of truth and the vast importance of the future, forms one of the most important means of avoiding error. Mankind do and suffer much for the sake of gain; and, as truth is of incomparably more importance than money, they will readily make much greater sacrifices for its attainment wherever they understand its worth, while those actually required are much less. The self-denial, toil and suffering often undergone for the sake of gold, greatly exceed anything which the proper pursuit of truth demands, although, like every real good, it requires sacrifices.

In order, however, to know the value of truth, it is often necessary to look steadily into the future; and this is strongly opposed by the desire of immediate enjoyment, and the aversion to present toil or suffering. Hence truth is often exceedingly undervalued; and men fly to false maxims or opinions, to justify themselves in neglecting to search for it aright. It is assumed, for instance, that all important truth is easily discovered; and then the dictates of prejudice, the illusions of the senses, or the deceptions of knavery, are received as truths that require no further examination, and ought not to be doubted. Or, it is assumed that nothing can be certainly known, and, therefore, it matters little what we believe.

Errors of this kind should be carefully avoided: for they undermine the very foundations of knowledge. There is no more important requisite for right investigation than correct views of its value; and the absence of it is a most fruitful source of error. False maxims and opinions, such as those just adverted to, are so frequently echoed from man to man, and chime in so exactly with our wishes, that they have imposed on the great majority of mankind, from the earliest times, and, by preventing due investigation, led them to adopt error in the place of truth.

In order to understand any important subject correctly, we must always labor on things which are intrinsically uninviting, if not unpleasant, but which are interesting as necessary means to future results. We should beware of the childish practice of looking only at what is near, and overlooking all beyond as of little consequence: else we shall allow things comparatively trifling to set aside the consideration of the most momentous subjects, and resemble a man journeying on very important business, who lost his object, by turning aside to gather flowers. Everything should be estimated at its real value, without making any distinction between the present and the future, since the former is incessantly vanishing and giving place to the latter.

Some are quite satisfied to believe like their friends; and these, in their turn, think they are safe in imitating the former, when possibly all are equally in error, and they are mutually confirming each other in false opinions.

The importance of care, at every step of our investigations, appears from the facts, that one error may lead to many others, as sure consequences, and that a mistake on a single point may mislead us regarding the whole subject. It is often impossible to foresee the future effects of a certain conclusion; and hence we are liable to attach little importance to points which may be of the utmost consequence. If a man radically err regarding the doctrines of causation and free agency, for example, all his opinions regarding God and duty will be little more than a tissue of errors, in which one fallacy supports another; and a man who believes that we cannot acquire any certain knowledge, will never be successful in its pursuit.

Mere feebleness of intellect rather fosters ignorance than leads to positive error, since there is inaction, and not misdirected action. Hence idiots never adopt many errors which have been held by men of great abilities, into which they were led by prejudices and ill-regulated feelings.

Every conclusion should be carefully traced, step by step, to its foundations; and none should be admitted as true unless the proof is found conclusive. In cases of difficulty and importance, it is proper to write a synopsis of the whole proof, from the primary premises to the conclusion, omitting all unessential matter, and laying down every part in its proper place. The synopsis should then be scrutinized till we know that we have carefully considered every part, and ascertained its true character.

Fallacious arguments frequently pass for sound reasoning, because they are interlarded with various illustrations and irrelevant matters: and, therefore, it is desirable to separate the essential parts from the rest, and exhibit the argument in its naked and most concise form, when its true character can be generally ascertained by a little close attention to the exact nature and connection of its various parts.

Wherever it is alleged that an argument proves too much, or that it leads to some false conclusion, we ought to examine the subject till we have certainly ascertained whether it does so or not. If it do, it is unsound. If it do not, we must not infer that it is conclusive, as it may possibly be refutable by other means; but it should be tested by a proper examination. Men frequently allege that an argument which establishes a conclusion that conflicts with their wishes or opinions, proves too much when such is by no means the case.

Where prejudices are possibly concerned, we should compare our own conclusions with those of persons of opposite prejudices, if such there be, and then examine and compare the arguments on each side, till we ascertain the source of any discrepancy which appears in the conclusions. In order to *know* that a conclusion is true, it is not sufficient that we are satisfied and firmly believe it is so: we must have the evidence of Consciousness, at every step, as was formerly stated; and where we have obtained such evidence, the conclusions are no longer opinions, but cognitions.

We should never slight an argument which professes to prove a disagreeable conclusion, or one that runs counter to our prejudices. If we are inclined to do so, we should try whether we can refute it, or demonstrate its inconclusiveness beyond all possibility of error, a course which will sometimes show us that it is irrefragable. On the other hand, we should most rigidly test the soundness of arguments which prove a conclusion that we desire to be true, and not rest satisfied that it is conclusive till there is no possible room for any fallacy. We shall thus be likely to escape the numerous errors into which we are liable to be led by prejudices.

The usual tendency of the mind is, to believe too readily what it wishes to be true, and to disbelieve too readily what it wishes to be false. Hence we should be particularly circumspect in examining proofs, in all cases of this kind. A person sometimes hesitates in believing conclusive evidence of an agreeable truth, or in rejecting similar proof of the falsity of a disagreeable proposition: but this is only an exception to the general rule, arising partly from the antecedent improbability of the fact, and partly from the confusion caused by the strong emotion excited. Consequently the disbelief generally disappears as soon as the feeling has subsided, and the matter is viewed more calmly.

As the sources of fallacy are very numerous, some

have concluded that we can be positively certain of nothing. But the fact that we sometimes adopt erroneous opinions, by no means proves that we cannot possess certain knowledge. A proposition which conforms to the criterions of truth, cannot be false, although we sometimes admit those of a different character as true. Our liability to err should teach us caution in forming conclusions, and distrust of our opinions on subjects that we have never properly examined: but it furnishes no good ground for scepticism regarding matters where no fallacy can exist.

CHAPTER XVII.

OF PARALOGISMS, OR FALLACIES OF PRIMARY ASSUMPTION.

§ 1. PARALOGISMS OF INTUITION.—Intuitional Assumption.—Intuitional Rejection.—Sources of these Fallacies.—Test of Intuitions.

Intuitional assumption is, assuming a proposition as a premise, on the ground that it is self-evident, when, in reality, it is not.

Intuitional rejection occurs where a self-evident premise is rejected, on the ground that it is untrue, or wants proof.

The former of these paralogisms arises chiefly from mistaking very familiar propositions for intuitions, from a desire to extend the sphere of universal truths, or from a bias in favor of a conclusion which the supposed intuition tends to establish. We are disposed to assume as self-evident what we have always considered such, especially where the proposition proves a favorite conclusion, or saves the labor of close and continued observation. Hence propositions have been taken for intuitions which are self-evidently impossible, such as that we see distance.

Intuitional rejection generally proceeds from the proposition's conflicting with some cherished belief, or its appearing novel or strange. We are as much disposed to reject what militates against the truth of favorite opinions as we are inclined to receive, without due caution, things of a contrary character. Many persons, also, make their own views the standard of truth, and unhesitatingly reject, as unworthy of being deemed self-evident, anything which never appeared to them in that light. Instances of this occur in the repeated rejection of some intuitions regarding volitions and motives.

To determine whether a proposition is self-evident, we must ascertain exactly what it is, and then attentively consider the thing meant, apart from the words in which it may have been expressed. If it is self-evident, its truth will then appear clearly and irresistibly, accompanied with a discernment that it cannot by possibility be false, and that its contradictory is an utter impossibility. The criterion of intuitions is, as the name implies, that the attentive mind discerns they are necessarily and universally true, and cannot possibly be false, in a single instance.

The self-evident nature of intuitions is evinced by all persons believing and acting on such as are clearly brought under the notice of their understandings, however much some may profess to repudiate them. When a strange event occurs, without any known cause, those who maintain that causes are not necessary antecedents of changes, believe there must be a cause, while there is no other evidence of this except intuition. Experience does not show that every event has had a cause; for we do not know by experience that many events ever had a cause; and even if we did, this would not prove anything regarding the future. The experience of persons in the various zones, regarding the apparent positions and the risings and settings of the heavenly bodies, are uniform, and yet quite different from their neighbours'.

§ 2. PARALOGISMS OF ASSUMING WHAT IS ATTEMPTED TO BE PROVED.—Nature of these Fallacies.—(1) Assuming the Question.—Its Criterion.—(2) Reasoning in a Circle.—Its Criterion.—Origin of this class of Fallacies.

In this class of fallacies, something which requires proof is assumed as a primary premise, while it is either virtually identical with the conclusion or is inferred from it, although the reasoning subsequent to the assumption may be quite valid. It includes the two following fallacies.

1. *Assuming* (termed also *begging*) *the question* is, where we assume, as a primary premise, a proposition

which requires proof, and which is virtually identical with the conclusion, although it may be differently expressed.

The criterion of this fallacy is, not merely the identity of the primary premise with the conclusion, but also that it requires proof, which is not given: for, if the primary premise requires no proof, there can be no fallacy of primary assumption. Nor would the argument always be useless, since we may rightly assume the primary premise without seeing that it is, in effect, identical with the conclusion, because the latter is differently expressed. A person often admits a proposition when it is expressed in one form, while he will deny it under another form.

Instances of this fallacy occur in those cases where the premise is only the conclusion expressed in learned phrase, as where it is attempted to be proved that we ought to do a thing, on the assumption that we are under a moral obligation to do so, or that a man must be honest, since his conduct is always guided by principles of moral rectitude, or that a proposed law is good, since it will ameliorate the condition of the people, or that a politician cherishes no selfish views, since he is a disinterested patriot, or that *our* views are right, since they are orthodox, and our opponents wrong, since they are heterodox.

2. *Reasoning in a circle* (termed also *the vicious circle*) is, where we either assume, as a primary premise, an inference from the conclusion and provable only by means of it, or attempt to deduce our first assumption from the conclusion inferred from it. In the former case, we first assume the inference, and then attempt to deduce the conclusion from it, while, in reality, the assumption requires proof, and can be proved only by means of the conclusion which it is employed to establish. In the latter case, the conclusion is first proved by means of the premise, and the latter is then proved by means of the former. Both processes may be compared to an attempt to make the top of a house serve for its foundation.

The criterion of this fallacy is, that something is assumed which requires to be proved by means of the conclusion inferred from it. A frequent instance of it is, assuming a proposition generally, and then attempting to prove a particular case of it, by means of this assumption, when, in fact, the former assumption requires proof, which cannot be given otherwise than by means of the latter. It is assumed, for instance, that all men are mortal, and from this it is very easy to prove that A, B, or C is mortal. The principal difficulty lies in proving the assumption, which cannot be rigidly done at all, and which can be even proved to hold true generally only by assuming that some men are mortal.

Another common instance is, where we take the word of a stranger, or a person suspected of being a liar, for his own integrity or veracity, or where one man's veracity is vouched by an unknown neighbour, and the man himself then vouches for the latter's veracity. This has often been done, where all the parties were rascals, colleagued with each other.

Both these kinds of paralogism are frequently derived from false systems of philosophy or science: but they may also proceed from inattention in our own investigations, in which we should guard against them, by attending closely to the nature of our primary assumptions.

§ 3. PARALOGISMS OF COMPREHENSION. —(1) Miscomprehension.— (2) Incomprehension.—Influence of Intuitional Assumptions.

1. *Miscomprehension* occurs where, owing to a want of due care and attention, we overlook or mistake characteristic marks and distinctions, whence we mistake one thing for another, or attribute to a subject something which does not, in reality, belong to it. A common instance of it is, where we attribute to one thing what really belongs to another, with which it is associated, as when we erroneously fancy that certain forms and colors are beautiful because they are associated with something that is so, or think that the appearance of a person who has wounded our feelings is disagreeable, when it is the reverse, or that some insipid substance has an agreeable taste because it is fragrant. In the same way we sometimes falsely attribute a certain character to a person, because there is something in his dress or appearance which we have erroneously associated with that character. Another common instance is, where we mistake one person for another, because we have not sufficiently attended to their distinctive differences.

Strong emotions, especially when the organs are unfavorably situated, often produce this fallacy. A timid man, in a lonely forest at night, is very apt to take every black log for a bear, and every gray stone for a wolf. His fear causes vivid ideas of these animals, to which the things apprehended afford a basis of reality. So, when a person is very anxious to see a particular thing, or when it is suggested to him by others, a vivid similitude of it arises before his mind, and if there is anything visible to which it bears the least resemblance, on which it can be superinduced, he blends the two objects, and readily believes that he perceives the prototype of the similitude. It is thus that a man in a dim light sometimes thinks he sees various objects, when he perceives only things which somewhat resemble them.

2. *Incomprehension* occurs where the whole of a subject does not present itself to our view at once, but we discern only a part, and then assume that we have comprehended the whole. We should remember that many facts may often be observed, which tend to prove a certain conclusion, while possibly more may be found on the other side, which are very apt to be overlooked, when they oppugn some favorite opinion. The particular facts overlooked depend greatly on the individual's views and feelings, so that one overlooks the very things which another observes, and the latter overlooks all that the former observes.

A common instance of this fallacy is, where one thing is assumed to be a sure sign of another as uniform antecedent, concomitant, or consequent, when farther observation would show that it is no such thing. Many superstitious opinions originate in this way, such as the belief in omens, prodigies, lucky and unlucky days, places, and persons, dreams, fortune-telling, and the power of charms and incantations, although, in producing these results, the paralogism is generally combined with the sophisms of causation and generalization.

In cases of this kind the instances which favor the conclusion are carefully noted and remembered, while the other, and much more numerous, class, are either wholly overlooked or forgotten. When a quack descants on the supposed cures alleged to have been effected by his nostrum, he says nothing of the much greater number of cases in which no benefit was derived from it, or it even produced serious evil.

Another common case of this paralogism is, assuming that a few individuals of a class which we may have seen or known, are fair specimens of the whole, when there is no proof of this, and the fact may possibly be the reverse. Some travelers describe the character of a foreign nation, when, in truth, they never saw one of them in his ordinary mode of life, except innkeepers and their servants. So people have not unfrequently formed an erroneous opinion of mankind at large, from the general character of their friends and acquaintances.

Another frequent instance is, assuming that the mode of effecting a certain end, with which we are familiar, is the only one, or at least the best, whereas several better ways may be known to others.

Another common case is, where a person thinks he can safely afford to purchase several things, because he can purchase any one of them without inconvenience. He never considers the result of purchasing the whole of them, but while he thinks of some, he overlooks the rest.

Fallacies of comprehension are greatly fostered by intuitional assumptions, which often lead us to think, either that no observation is requisite, or that we have already comprehended enough, when the case may be far otherwise. For we are not disposed to seek for proofs of what we deem self-evident, nor to search for further proof where we are already satisfied, though without any good ground.

§ 4. PARALOGISMS OF SIGNS.—(1) Illusive Sign.—Its various Sources.—Its Character and Effects.—(2) Non-interpretation of Signs.—Frequent Combinations.

Where a phenomenon is presented to our observation which appears, and which we assume, to signify something which, in reality, it does not, we fall into the paralogism of *illusive sign*. It differs from those of comprehension in our being misled solely by drawing a false inference, and the observation not being necessarily either careless or defective, though, in fact, it often is so. When we view a statue or a scarecrow, and take it for a living person, or when we look at the images produced on a screen by the magic lantern, and take them for real pictures, or when we have ringing sounds in our ears, and attribute them to some external source, we apprehend the phenomena as they actually are, and we are misled solely by drawing erroneous inferences. So certain symptoms are often inferred to indicate a particular disease, when, in reality, they do not, because they equally accompany several other disorders. Yet the symptoms may be observed with perfect accuracy, and it may be quite true that they always accompany the disease which they are erroneously believed to indicate.

We readily fall into this fallacy when the organs of apprehension are diseased or defective. When the optic nerves are inflamed, the ideas of objects of sight sometimes acquire unusual vividness; and hence they are mistaken for actual apprehensions. In such cases, the phenomena are attributed to the causes which usually produce similar phenomena, as these most readily occur to our minds.

Prejudices sometimes produce this fallacy, even where there is no organic or external difficulty. Thus, we are apt to infer that one whom we strongly dislike, is a bad man, because he exhibits some indication which leads to that conclusion, though it may really prove no such thing, and there may be positive proof to the contrary. On the other hand, we are very ready to infer that those we like are penitent, moral or intelligent, because they have done something which faintly indicates such an inference, when there may be clear evidence of the reverse.

To this paralogism belongs a host of common errors regarding the condition and character of others. A bashful man is often assumed to lack abilities or knowledge, while one who exhibits assurance is taken to be a man of talents and intelligence. A fine dress and equipage have sometimes procured the credit of wealth, which was assumed to betoken happiness. Such errors are more common, because the thing assumed to be the sign is often caused by wilful imposition. It is often easy to borrow the language of the wise, the generous or the learned, while those addressed never suspect the imposition, and consequently give the impostor credit for what really belongs to another. So the felon often counterfeits signs, to remove suspicion from himself, while he removes the real signs of his guilt. By such means, almost every virtue has been counterfeited, and every vice attempted to be concealed.

This fallacy is of an insidious character, and very apt to escape detection, because the inferences are generally made with such rapidity that we overlook them, and suppose that we apprehend what we only infer. Hence it is very apt to lead us into further errors, because we assume the false inference as an unquestionable comprehension.

2. *Non-interpretation of signs* is the converse of the preceding fallacy, and occurs where, owing to carelessness or prejudices, we overlook the obvious and conclusive signification of a phenomenon presented to our view, and thus form an erroneous opinion of the subject. A common instance of it is, where we overlook the signification of certain actions performed by a person of whom we think favorably, and consequently form a better opinion of him than his real character warrants. Another similar instance is, when we form too low an opinion of one whom we dislike, because we misinterpret his actions.

Paralogisms of signs are often combined with those of comprehension. We obtain either an inaccurate or a partial view of what is presented to our observation, and we misinterpret what we do comprehend. Thus, we are apt to form erroneous opinions regarding the characters of nations, from that of a few persons, with whom we are not well acquainted, and whose conduct we misunderstand. So, if we extended our acquaintance with the man of supposed wealth, we might possibly find that what procured him the credit of wealth had never been paid for: and a further acquaintance with the man who counterfeited wisdom and virtue, might show him to be a silly knave.

A common and serious instance of this compound fallacy occurs in the undue importance so frequently attached to material interests and sensual enjoyments. Our apprehensions frequently show us only the least important side of the subject: but as this is seen at once, we are apt to look no further, while, in order to form a correct estimate, we must view the whole subject. This often requires toil and time, which we are unwilling to bestow, and self-denial, which we are unwilling to practice. Hence men often form very erroneous opinions, and sacrifice a great good for one of comparatively no consequence. In order to form a proper judgement regarding the comparative merits of different things or courses of action, we must obtain a full and fair view of every one of them; and this may be prevented by our prejudices wholly withdrawing the attention from several points, so that we seriously err without ever suspecting the fallacy.

§ 5. PARALOGISMS OF MEMORY.—(1) Imaginary Apprehension.—Its Origin.—(2) False Association.—Its Character and Effects.—(3) Mistaking Ideas.—Its Character, and what it particularly affects. —(4) Forgetfulness.—Distinction.

1. The paralogism of *imaginary apprehension* consists in confounding the remembrance of apprehensions with that of conceptions, or mistaking a phantasm for an idea, whence we think that we apprehended what we only imagined. We fall into this error only when our remembrance of an occurrence is indistinct or defective: for, in other cases, the remembrance of the circumstances under which the conception was formed, will prevent any such mistakes. It occurs chiefly where some favorite conception has been often considered, or originally absorbed the attention to such a degree that the impressions made by other things were faint, and consequently they have been forgotten, while the remembrance of the conception is, for the same reason, unusually vivid. It is owing to this fallacy that we hear some old men relating, with perfect sincerity, that they saw or performed things which are purely imaginary.

2. *False association* occurs where we mistake the time or place of apprehending a certain thing. This is owing chiefly to the idea of one apprehension becoming associated with that of another which was not contemporaneous with it, whence we infer that the two prototypes were simultaneous. Thus, we often believe that we saw an acquaintance at a particular time or place, when, in fact, we saw him at another time or elsewhere. Such mistakes are of frequent occurrence, and sometimes produce serious consequences.

3. *Mistaking ideas* is where the idea of one apprehension is mistaken for that of another, owing to indistinctness or failure of remembrance, whence we believe that we apprehended a thing when we apprehended only something like it. This fallacy closely resembles that of miscomprehension, with which it is frequently combined, owing to our employing ideas as representatives of their prototypes. It affects number and quantity so much that little dependence can be placed on the remembrance

of a person of ordinary memory, regarding such things, except where his attention was particularly directed to them, at the time of apprehension.

4. *Forgetfulness* is where we wholly forget something, and then assume that we never comprehended it. As we are liable to such forgetfulness, the mere fact that we have no recollection of a thing, does not prove that we never comprehended it. Hence the only safe inference is, that we do not remember it, except it be such that we could not possibly forget it, if once comprehended. But this condition is often overlooked, and we think that we never comprehended, when we only forget. This fallacy is frequently combined with those of comprehension, as we are very apt to forget what we never rightly comprehended. Instances of it occur where we undertake to do something in utter forgetfulness of the fact that we have already done it. Another similar instance is, when we believe that we never saw or said something which we really did, because we have no recollection of having done so.

Memory generally retains the errors of original comprehension. Thus, if we mistook the character of a thing when we saw it, without discovering our error, we shall continue to think that we saw what we took it to be, unless extraneous circumstances show that we were mistaken: but neither the error nor the correction proceeds from the Memory; and hence the consideration of them does not concern our present subject.

§ 6. INTRINSIC PARALOGISMS OF TESTIMONY.—Immaterial Distinction.—Intrinsic and Extrinsic Paralogisms of Testimony.—(1) Ambiguous Expression.—Logomachies.—Equivocation.—Ambiguous Interrogation. — (2) Overlooking Conditions. — Results. — (3) Assuming Conditions.—(4) Obscure Expression.—Frequent source of this Fallacy.—Proper mode of dealing with it.—(5) Wrong Expression.—(6) Falsehood.—Its Extent.—Erroneous Definitions.—(7) Suppressing Truth.—Where to be expected.—(8) Misrepresenting Comprehensions.—Its Extent.—(9) Misrepresenting Testimony.—Where frequently employed.—Frequent Combinations.

The question whether a witness believes his own statements or not, is of the utmost importance when we are considering his character: but it is generally of little or no consequence when we are inquiring whether they are true or false. The real sincerity of a witness who believes his false statements, is quite as apt to mislead us as the mock sincerity of the liar; and the belief or disbelief of the witness is frequently no test of truth. Men often hear and relate truth without believing it; and they still more frequently utter falsehood unawares. Consequently we shall not here draw any distinction between conscious and unconscious errors. The character of the fallacy is generally unaffected by the belief or disbelief of the witness.

Paralogisms of testimony are either *intrinsic* or *extrinsic*. The former comprise those where the testimony is fallacious in itself: the latter include those which arise from our dealing with it fallaciously. We shall treat of the former in this, and of the latter, in the ensuing section.

1. Expressions bear two or more different significations, which are not distinguished, constituting the paralogism of *ambiguous expression*. Thus, "testimony is fallacious," may signify *all* or only *some* testimony. *All*, again, may signify either *every one separately* or the *whole together*; and *some* may denote *a very few* or *a great many*. So, "a scriptural custom" may signify simply one which is frequently referred to in the Scriptures as existing, without any expression of approval; or it may signify one which is described as having the divine sanction, which is a very different thing.

Many logomachies arise from overlooking the ambiguity of expressions. Such questions as "who was the greatest man that ever lived?" can be only disputes about words, till the parties have agreed in which of its numerous significations the word *great* is to be understood. In all such cases, there are, in reality, as many different questions as there are different significations of the expressions used.

A common instance of this paralogism is, *equivocation*, or employing expressions which are true in one sense, but not in another, while the witness furnishes no indication which of the two he would have us adopt. Sometimes we cannot ascertain whether the language is figurative or literal, as when commendatory epithets are used, and they may possibly be ironical. This proceeds sometimes from a design to mislead, and at other times, from mere inattention or oversight. The ancient heathen oracles, and the ambiguities purposely used in making contracts, furnish many instances of the former, and hasty compositions, of the latter.

Sometimes the ambiguity assumes the form of an interrogation, which appears to admit of only one answer, but, in reality, admits of several, some true and some false. This form is distinguished as *ambiguous interrogation*. It is frequently employed by witnesses who are unwilling to make a direct misstatement, which they attempt to avoid by an apparently candid appeal to the interrogator, as if the answer sought must be obvious to himself. The fallacy lies in the ambiguity of the question, and is to be avoided by ascertaining which of the several significations is meant.

2. Expressions which are true only on certain conditions, are stated as if they were true absolutely, forming the paralogism of *overlooking conditions*. When it is said that "a man can live without eating on Saturday," this may be true if he did not fast before, but very false, if he has fasted during the previous fortnight. This paralogism is one of the principal means of riveting the numerous errors arising from partial views of a subject. The language employed may directly express nothing but what is strictly true; yet it naturally implies something which is false. An ancient author might say: "the Sun appears in the south at noon, in every part of the known world," and the statement would be strictly true; but the same expression is quite false at the present day.

3. Expressions which are true absolutely, are assumed to be true only on certain conditions, forming the paralogism of *assuming conditions*. A common instance of it is, where a man assumes that some substance, habit, or course of life, will not injure him, on account of his peculiar constitutional character, when, in fact, it inevitably injures every one, just as fire burns, or ice chills.

4. The expressions are so obscure that it is impossible to ascertain their precise meaning, or whether they have any, forming the paralogism of *obscure expression*. A common instance of it is, using figurative language in such a manner that we cannot form any definite notion of its meaning, whence we are liable to misunderstand it. The proper course, in such cases, is, to attach no meaning whatever to the language: for otherwise we shall very probably attribute to it a meaning different from what was intended.

5. *Wrong expression* is where a person employs words

which may be sufficiently clear and precise, but yet convey a meaning different from what he intended. Sometimes improper terms are employed: in other cases, they are unobjectionable, but they are wrongly arranged or constructed. Typographical blunders, and slips of the tongue, furnish numerous instances of both these forms.

6. *Falsehood* is, where a person, either knowingly or otherwise, makes a statement which is substantially, if not wholly, untrue, with the expectation of being believed. Many expressions are false, in their ordinary acceptation, which do not come under this head, because the real meaning is well understood, as in the case of figures of speech; but here the witness speaks with the intention of producing belief; and the difficulty generally lies in the fact that there is no clear indication of the misstatement. It is frequently believed, owing to the apparent veracity and sincerity of the witness, the smallness of the temptation to depart from truth, or the indolence or prejudices of the party addressed. It is also of very common occurrence, because there are so many ways of being mistaken, and so many temptations to misrepresent, even where the witness is well aware of the truth.

A pernicious instance of this paralogism is, false definition. This is very apt to escape detection, owing to the ignorance of the party addressed, and the presumption that he who undertakes to give a formal definition of a thing, understands it well. As definitions affect fundamental points, errors of this kind are very apt to convey radically erroneous views of the whole subject.

7. *Suppressing truth* is where an important part of the testimony, or of what was witnessed, is entirely omitted, or the facts are garbled, and the favorable or unfavorable alone stated. A frequent instance of it is, depicting a few individuals of a class as fair specimens of the whole, when, in reality, they are not. This may not be directly stated; but things are represented so as to lead to that inference. Every large class generally contains persons of very different characters; and it is easy to select a few that differ widely from the great majority, who alone determine the character of the class. By such means very different pictures may be exhibited of the same community, the one as much darker than the reality as the other is brighter. So, by selecting the good and bad

qualities of an individual, two very erroneous exhibitions may be easily given of his character, yet both containing some portion of truth, so that they are readily received as accurate by those who like them.

8. *Misrepresenting comprehensions* occurs where the truth is either exaggerated, extenuated, or mixed with more or less of positive error, in a statement of what the witness personally observed or experienced. This is often done by employing terms either weaker or stronger· than truth requires. Many actions or characters may be viewed in a favorable or an unfavorable light, as several virtues border on corresponding vices; and language affords abundant means of representing it under either aspect, so as to contain some portion of truth.

Hence there is a boundless field for the exercise of this paralogism. Penurious selfishness may be dignified under the name of " a prudent regard to our own interests," while a proper degree of economy may be styled "niggardly parsimony." Indifference to the welfare of others often passes for " good nature," while one who warns us of serious danger, which we wish to overlook, is termed "an officious intermeddler," and a well-timed rebuke passes for "scolding." An action may be called " brave" or " rash"—" prudent" or " cowardly"—" energetic" or " severe"—" liberal" or " extravagant." Real sublimity may be termed " unmeaning rant," while bombast passes for " sublimity." Vapid declamation is termed " true oratory," while genuine eloquence is characterized as " tumid bombast." Ruling a nation with a rod of iron has sometimes been termed " administering the government with a strong hand," while inefficient imbecility was styled " mild and paternal sway."

Instances of this fallacy occur in the justificatory or defensive statements of a party charged with having done something wrong; and its influence may often be readily seen by comparing the narratives which two adverse parties give of the same occurrence.

9. *Misrepresenting testimony* is, where reliable testimony is misrepresented, so as to corrupt it materially. This is done in various ways. Sometimes it is mistranslated; and the error often escapes detection, owing to the original being unknown or inaccessible to the party addressed. Sometimes it is misquoted, and various garbled passages are strung together, with some important

additions or alterations, so as to convey a meaning very different from the original. At other times confident references are made to testimony, as proving a certain proposition which it does not prove, sometimes with the expectation that it will never be consulted by the party addressed; and the fallacy is occasionally masked by a bold statement that the testimony is clear and conclusive, while possibly it may be all the other way.

This paralogism is of frequent occurrence in controversy. An opponent's views are misrepresented, and then follows a refutation of imaginary statements or opinions; or the doctrines of a party are erroneously assumed to be the same as those of some person connected with it. Fallacies of this kind occur so frequently that little confidence can generally be placed in a professed controversialist's representations of an opponent's views, except where he quotes fairly from some unexceptionable authority; and it is unsafe to assume that he does so, without proof.

The two paralogisms of misrepresentation are most frequently couched in spoken or written language: but they are sometimes communicated by various other means, such as inaccurate maps or pictures, or by sounds and tastes or odors, which do not rightly represent what they profess to do, as if a man should present an apple, to give us a notion of the smell of a rose, or imitate the note of the raven, as a correct representation of the cuckoo's.

Intrinsic fallacies of testimony are frequently combined with each other. Ambiguous, obscure or wrong expressions accompany and strengthen falsehood, suppression and misrepresentation, so that it is sometimes very difficult to ascertain their true character. This can be done only by a close and careful application of the criterions of testimony, which frequently requires time and toil, while the adoption of the fallacy requires neither. Hence they have been a very frequent and powerful means of producing and perpetuating error.

§ 7. EXTRINSIC PARALOGISMS OF TESTIMONY.—(1) Adopting a Mean. —Erroneous Assumption.—(2) Counting Witnesses.—Relation of Numbers to Credibility. — (3) Credulity. — Why prevalent. — (4) Scepticism.—Its Origin and Operation.—(5) Overlooking Testimony.—Its Origin.—(6) Indiscrimination.—Why prevalent.—Frequent Combinations.

1. *Adopting a mean* is, where there are contradictory

or inconsistent testimonies, and we assume that the truth lies between the statements of the different parties, while it may be quite otherwise. It is so much easier to apportion the difference between conflicting testimonies than to ascertain their true character, that this fallacy is of frequent occurrence.

Some persons generally assume, in all cases of this kind, where their prejudices are not concerned, that the truth lies somewhere between the assertions of the different parties; and they take it to be nearest to those of the one who speaks with most confidence, whereas it is often beyond the statements of either party, or wholly with one, and that the party who speaks with least confidence; and we can never safely assume that it is intermediate without conclusive proof.

Wherever the statements are not only inconsistent, but contradictory, the truth cannot by possibility lie between. If one says Cæsar was a tyrant, and the other says he was not, one must be wholly right and the other equally wrong. It is also evident that there is never any rational and tenable mean between truth and falsehood. If the creed of the theist is true, that of the atheist must be quite false: and if the peculiar doctrines of the Trinitarian are true, those of the Arian must be as erroneous as the Socinian's.

2. *Counting witnesses* occurs where the testimony is estimated by the number of the witnesses, regardless of their character, the many being thought entitled to much, and the few, to little credit. A frequent instance is, the credit given to common rumor, where, as is usually the case, its origin, and consequently its value, are unknown. The prevalence of this practice is shown by the proverb, "what everybody says must be true"—the predicate meaning simply general report, which is often quite false. Testimonies should be weighed, not counted; and consequently all those which have no weight, ought to go for nothing. One unexceptionable witness is entitled to implicit belief, while one million of worthless witnesses are entitled to none. But as counting is, in this case, a much easier process than weighing, it is often adopted.

3. The paralogism of *credulity* occurs where we receive as satisfactory, testimony which is palpably unreliable, and which we could very readily perceive to be such, by a moment's consideration. This fallacy is very preva-

lent, because it is fostered by several strong prejudices. The "Mississippi System" of Law, in France, and the "South Sea Scheme" of Blount, in England, which caused such wide-spread ruin, in the early part of the eighteenth century, are two notable instances of this kind. Other instances occur in the ready belief given to the hyperbolical exaggerations of partisans.

4. *Scepticism* is, where testimony which might be easily ascertained to be conclusive, is rejected as unsatisfactory. This fallacy is simply the converse of the preceding, and has a similar origin. Hence both are often adopted by the same person, in reference to the same subject. The favorable testimony is believed, though worthless, and the unfavorable is rejected, though conclusive. Instances occur in the frequent rejection, by many, of scientific truths, established by evidence of which they could easily ascertain the conclusiveness.

5. We either wholly overlook, or pass by without any serious attention, accessible testimony which materially affects the point under consideration, and assume that we have properly considered all the testimony, forming the paralogism of *overlooking testimony*. This fallacy naturally flows from the disagreeableness of a long investigation of evidence; and hence it is of frequent occurrence. We have a common instance in the errors into which many historians have been led, by failing to consult important testimonies within their reach.

6. *Indiscrimination* is where we either receive the whole of a testimony as satisfactory, because some parts of it are so, or reject it all as incredible, for a similar reason. The most faithful witness sometimes makes slight mistakes himself, or is misled by others, in matters of little moment, while his testimony may be very correct, in the main. On the other hand, the least credible witness generally relates some portion of the truth, even where his statements are substantially false. Yet, as it is much easier to accept or reject in the mass than to sift out truth from error, this fallacy is very prevalent.

The paralogisms described in this section are frequently combined with those of comprehension and other fallacies of testimony, a combination which often produces unhesitating conviction.

§ 8. PARALOGISMS OF MISINTERPRETATION OF LANGUAGE.—Nature and Origin of this class of Paralogisms.—(1) Misunderstanding Archaisms.—(2) Misinterpreting Technicalities.—(3) Misinterpreting Ambiguities. — (4) Confounding different senses. — (5) Overlooking the Idiom. — (6) Following Etymologies.—(7) Mistaking the Style.— (8) Misplacing the Accent. — (9) Misconstruction.— (10) Mistaking Expressions.—Its Sources and Effects.—(11) Ignorant Interpretation.—Its Origin.—(12) Misconception.—Frequent source of it.—(13) Fallacious Implication.—(14) Mistaking Allusions.—Where frequent. — (15) Fallacious Propriety. —Why frequent.—Effects of these Paralogisms.

This class of fallacies comprises those cases in which we misunderstand or misinterpret statements which may be perfectly fair and correct in themselves. They proceed from the defects of language, ignorance, carelessness, or prejudice. The following enumeration includes the most common:

1. We affix to a term its ordinary signification at the present day, when, in reality, it is used in an antiquated sense, which we may term *misunderstanding archaisms*. It is apt to occur in interpreting ancient laws, or other compositions of a remote period, as where the word "publish" is taken to signify *print and offer for sale*, while it means to *utter* or *proclaim in public*, or where to "prevent" is interpreted to *keep back*, when it means *to go before*.

2. We overlook the nature of the composition, and interpret words in their ordinary sense, when they are used technically, or conversely, which may be termed *misinterpreting technicalities*. By a *gale* a seaman understands a high, strong wind, while, in ordinary language it means only a moderate breeze.

3. *Misinterpreting ambiguities* is where we attach to an ambiguous expression a sense different from what was intended. This we are very apt to do where we know little of the subject or the language, and consequently misunderstand the context. Many instances occur in various interpretations affixed by commentators to ambiguous passages in ancient authors.

4. *Confounding different senses* is where we blend various significations of an expression, and unconsciously understand it now in one sense, and then in another, so that we form a confused conception of its import. The various confused significations attributed to such words as *law*, *idea*, and *nature*, are instances of this fallacy.

5. *Overlooking the idiom* occurs where we attribute to a foreign expression the exact import of the corresponding words in our vernacular, while there are important differences in their significations. This is apt to occur where our knowledge of the idiom is imperfect, and we are not well acquainted with the history, institutions and manners of the people whose language we interpret. *Virtus*, in old Roman authors, does not generally mean *virtue*, but *valor;* *temperantia* meant moderation in desires and pursuits, as well as in eating and drinking, while *humilitas* denoted abjectness or meanness, a very different thing from what we now understand by *humility*.

6. *Following etymologies* is where we assume that a derivative or compound term has the exact signification of its original, when, in fact, they mean things widely different. This fallacy is of frequent occurrence in interpreting dead languages, where too much importance is often attached to etymology, owing to our other means of ascertaining its signification being very scanty. We often *trow* things that are not *true;* we may form *projects* or *speculations* without being *projectors* or *speculators;* and we are not bound to observe a *holiday* as a *holy day*.

7. *Mistaking the style* occurs where that which is figurative is interpreted literally, or conversely. Such errors readily spring from overlooking the nature of the subject, or the differences between the style of one language and another. Some languages employ figures more freely than others; and hardly any two use them precisely alike, in all cases. A common instance of this fallacy is, where figurative idiomatic phrases are taken in their literal acceptation. Such expressions as " a man beside himself," or " out of his mind," puzzle persons ignorant of the English idiom, while the corresponding expressions in their languages may be equally obscure to all those who are not familiar with them.

8. *Misplacing the accent* is, where we attach a wrong meaning to an expression from mistaking the accented word. The proposition " he who sins shall die," points out either the subject or the nature of the retribution, according as we accent the subject or the predicate. So if only the last word of the ninth commandment be accented, it is made to forbid belying only our neighbours.

9. *Misconstruction* is where some word is assumed to qualify the wrong term, the true construction being misunderstood. This fallacy abounds in translations. The expression *Aio te Romanos vincere posse*, may be rendered "I say that thou canst conquer the Romans,"— or "I say that the Romans can conquer thee." So *redibis nunquam peribis*, signifies "thou shalt return; thou shalt never perish"—or "thou shalt never return; thou shalt perish"—according as we assume the pause before or after *nunquam*.

10. *Mistaking expressions* occurs where we mistake the language employed, and assume that something has been uttered essentially different from the reality. We are very liable to adopt this fallacy, where the expression is strange to us, and yet resembles one with which we are familiar. Many popular and typographical errors originate in this way, as—*sparrow grass* for *asparagus*, *animals* for *mammals*, *Candia* for *Cardia*, and *Persians* for *Pierians*. Sometimes the negative particle is overlooked; and thus a meaning is assumed the direct contrary of what was intended. This paralogism has caused various slanders and bitter quarrels.

11. Owing to mere ignorance of the language, we attach to an expression a meaning which it does not bear, forming the fallacy of *ignorant interpretation*. It arises chiefly from mistaking a term for another which it resembles, or adopting the first definition given in a dictionary. Many errors found in translations have originated in this way. In various translations of *Exodus*, Chapter xxxiv., *verse* 7, one part is rendered so as to contradict, not only the whole tenor of Scripture, but the immediate context.

12. *Misconception* is, where we comprehend only a part of the meaning of an expression, and unconsciously miss what is possibly the most important part. This fallacy frequently prevents truth from producing its legitimate effects on the mind, because that which is most effectual continues unknown, while the individual believes he understands the whole subject. The meaning attached to many terms,—such as *patriotism, benevolence, humility, justice, modesty*, and *bravery*—depends greatly on the character of the party addressed, so that four persons may receive as many different impressions from the same term. In order to comprehend the force and sig-

nificance of another's language, we must frequently, not only attend to his circumstances, but also enter into his feelings. It is for this reason that persons of different characters and principles so often misunderstand, and consequently misrepresent, each other.

A frequent instance of this fallacy is, where, in reading History or Biography, we attribute to the language the same signification which it would bear, if used by a contemporary countryman of our own, when its real signification is widely different.

13. *Fallacious implication* is, where we assume that the words imply something which, in reality, they do not. This is apt to occur where we are not well acquainted with the subject or the character of the author. A pernicious instance of it is, where the erroneous language of men, recorded in Holy Writ, is attributed to God, and thus fallaciously inferred to be true.

14. *Mistaking allusions* is, where we assume that the speaker or writer alludes to one thing when, in reality, he alludes to another. This fallacy is of frequent occurrence in interpreting ancient and foreign languages, as we are apt to overlook the differences between ancient times or foreign countries and our own. Commentaries on ancient authors furnish many instances of this paralogism.

15. *Fallacious propriety* is, where we force on the language an erroneous meaning, because we think it must have been the one intended, as, otherwise, the sense conveyed would, in our opinion, be false or improper. The sense intended to be conveyed may be untrue or improper, or we may think so when it is otherwise, because our views of the subject may be erroneous or defective. Instances of this fallacy occur in the false glosses and misinterpretations of Scripture arising from the erroneous assumption that any other interpretation would give a false or improper sense.

As we are generally unwilling to believe that our own views are wrong, and always desirous of finding them tally with those of men whom we respect, this fallacy is of frequent occurrence.

Fallacies of misinterpretation are apt to be very pernicious in their effects, because they frequently refer to subjects of great importance, while their existence is unsuspected by their victims. Not only is language an imperfect weapon wielded by an arm which is ever liable to miss its aim, but its heaviest and most direct blows are often parried by carelessness, ignorance, prejudice, or stolidity.

CHAPTER XVIII.

OF SOPHISMS, OR FALLACIES OF INTERMEDIATE REASONING.

§ 1. SOPHISMS OF CONFUSION.—Nature of this class of Fallacies.— (1) Sophistical Connection.—Relations of Conclusions and Premises.—Influence of Prejudices.—(2) Inferring the Converse.—Why it often escapes detection.—Relations of a Proposition to its Converse.—(3) Altering Propositions.—How sometimes disguised.

This class of fallacies comprises those cases in which we draw inferences not implied in our premises, owing to our mistaking one proposition for another which resembles it, but is yet essential different. The following are the principal kinds:

1. *Sophistical connection* is, where premises are assumed to be true or false, according to our opinion of the conclusion deduced from them; or, conversely, the conclusion is assumed to be true or false, according as it is implied in the premises or not.

False premises are often employed to prove true conclusions, and unobjectionable premises are often employed to prove false conclusions, while false conclusions may be implied in false premises. The premises and the conclusion, again, may be both true, while the latter is not implied in the former: but the only legitimate inference is, that the conclusion is not proved by those premises; for it may be conclusively proved by other unobjectionable premises.

Where the conclusion is implied in the premises, the former stands as high as the latter, but no higher. If these be cognitions, so is the conclusion; and the same relation holds between them when the premises are only probabilities: that of the conclusion is as strong as the latter's, but not stronger. It must be carefully observed, however, that the converse does not hold true. False or doubtful premises do not prove false or doubtful conclusions, nor exclude other proofs that they are true. To warrant us in assuming that a conclusion is false, we must have positive proof that it is so; and the fact that the premises may be false, or that the conclusion may not be implied in them, furnishes no such proof.

This fallacy is one into which we are very readily led by prejudices. We are apt to pay too little attention to our premises, when they establish a favorite conclusion; and when the conclusion is strongly repugnant to our desires, we are inclined to reject the premises which prove it, without giving them a careful consideration.

2. *Inferring the converse* occurs where we infer that a proposition is true, because its converse is true, as when we infer that every equiangular triangle is equilateral, because every equilateral triangle is equiangular. This fallacy often produces conviction, not only on account of the great resemblance which a proposition bears to its converse, but from the fact that the converse is often true, although the case is generally otherwise. Although every horse is a quadruped, yet every quadruped is not a horse. We may, indeed, express a proposition in such a way as to render its converse true, by making its subject and predicate identical: but propositions of this class are only an exception to the general rule; and the cases in which the converse is true accidentally (as in the instance mentioned above) are too few to be taken into account.

A common example of this sophism is, where we infer that one thing is identical with another, because both belong to the same general class. White is a color, and black is a color: therefore black is white. So, every wise man carefully considers the future, and John does so: therefore he is wise. It is assumed, in all such cases, that the converse is true; and hence the fallacy. Although white is a color, every color is not white.

3. *Altering propositions* is where we erroneously assume, in the course of an argument, that something has been already proved, or has appeared to be true, which is employed accordingly, as a sound premise, while it differs materially from that whose place it thus usurps. The previous proposition generally bears some resemblance to the other: but it is less extensive, or it has been proved only with certain qualifications or restrictions, which are afterwards overlooked, or it is proved true only in another sense from that assumed, or it lacks something important which is afterwards added.

This fallacy is greatly fostered by the defects of language; and it is often disguised by being combined with ambiguous expression. Here the ambiguous terms are used in different senses, throughout the argument, while some of these are not true, in the sense in which they would require to be so, to render the reasoning valid. When it is said that "three and two are five, and five is one number: therefore three and two are one number" —the word *are*, in the first proposition, should have the sense of "are identical with," in order to sustain the conclusion, as understood; and, in this sense, the proposition is false. Three and two are no more five than so many copper coins are one silver coin. They are only *equivalent* to five; and, in this sense, the conclusion is proved: three and two *are* equivalent to one number. So it can be proved that "some men are wise," and "some men are fools:" but it does not thence follow that fools are wise, since the expression "some men" denotes totally distinct objects, in the two propositions.

An amusing instance of this combination is found in the old story of Protagoras the Sophist, and his pupil Euathlus. The former taught the latter Rhetoric, on condition that he should be paid for his services, if his pupil were successful in pleading his first case. Euathlus having failed to commence practice, Protagoras sued him; and he pleaded his own case. So this was his first case. Therefore Protagoras was bound to win: for, if the decision were in his favor, Euathlus was bound to pay, in virtue of the judgement; and if it was otherwise, Euathlus gained his case, and, therefore, was bound to pay, in virtue of the contract. But, on the other hand, Euathlus was bound to win: for, if the decision was, that he should pay, he had lost his first case, and, therefore, should pay nothing, by the terms of the contract: but if the decision was otherwise, this left him free.

§ 2. SOPHISMS OF GENERALIZATION.—Nature of these Fallacies.—(1) Sophistical Extension.—Frequent source of it.—With what often combined.—Influence of the Combination.—(2) Sophistical Inclusion.—(3) Sophistical Contraction.—(4) Sophistical Exclusion.—(5) Sophistical Combination.—Why often undetected.—(6) Imaginary Universality.—Why prevalent.—Caution.

This class consists in generalizing further than our actual knowledge warrants, or the converse: and there are six principal kinds of it.

1. *Sophistical extension* is where we assume that things found together in certain cases, are always connected, or that a certain property is common to every one of a class, since it belongs to several of its individuals or tribes. In other words, we mistake an empirical for a scientific generalization. Some men have been deceived by several of their neighbours; so they infer that all mankind are dishonest: many have been found, under certain circumstances, to lie; therefore it is inferred that nobody can be safely believed, under any circumstances. Others find that certain things related by historians are untrue: so they conclude that all history is unworthy of credit. Many things are uncertain, and many falsehoods have been believed as truths: therefore they infer that nothing is certain.

Many fallacies of this kind originate in the fact that a certain property has, not unfrequently, been found to extend beyond the cases first observed; and hence it is erroneously inferred that it extends indefinitely. A negro of Central Africa is apt to infer that all men are black or brown, while some natives of Northern Europe once inferred that all mankind are white. The experience of so many cases leads to a tacit inference that they are the effects of some constant and unvarying cause, which, however, may have no existence.

This fallacy is frequently combined with those of comprehension. Not only is there no real induction, but the facts on which we reason are either inaccurately or partially observed; and a more extended or careful view would speedily explode the generalization. Judicial Astrology is a notable instance of this combination. It assumed that the position of the planets, at the time of a person's birth, determined his character and future destiny, although many men, born at the same time, have, in every age, exhibited the most different characters and fortunes, on which the aspect of the planets had not, in reality, the least influence.

2. In *sophistical inclusion* we assume, without satisfactory proof, that several things which agree in some respects, agree in others also, and thus infer that all belong to a particular class, when, in fact, they may not. Thus, naturalists have sometimes ranked a species with a genus to which it did not belong, on their own principles. So men often attribute to others characters which

they do not possess, because they exhibit certain appearances which accompanied the supposed character in other cases, but does not, in the case in question. In the same way, the opinions of certain persons have been classed with a school to which they do not belong, because they have several things in common; and all diseases have sometimes been classed under a few general heads, so that affections essentially different, and requiring different treatment, were classed together. Other common instances of this sophism are, that things essentially different are quite alike, because they happen to be called by the same name, and that language faithfully expresses the realities of nature, in every instance, when, in fact, it fails to do so, in many important cases.

3. *Sophistical contraction* occurs where an induction is mistaken for an empiricism, and it is assumed that the facts beyond those immediately observed are different, when, in reality, they are as uniform as the course of nature. This fallacy is the converse of sophistical extension: and we have an instance of it in the common remark that a certain kind of diet is very conducive to the health of the speaker, although he admits that it has been found injurious in other cases; and, in fact, there is conclusive proof that it must always be so. Another common instance is, where a man infers that he can, by some means or other, escape the certain and inevitable consequences of a dissipated life, while he is well aware that they have followed in other cases.

4. In *sophistical exclusion* we assume, without satisfactory proof, that things which differ in some respects, differ in others also, and consequently infer that they belong to different classes, when possibly they do not. It is the converse of sophistical inclusion, and leads us to exclude an individual or a species from its proper class. Thus it is often assumed that a man does not possess a certain character, because he does not exhibit peculiarities which accompanied such a character in other cases, but in fact do not, in his case. Another instance is, where it is inferred that things are materially different because they are called by different names, while they may all be modifications of the same thing, and essentially alike. Naturalists fall into this sophism when they form a mere variety into a distinct species, or exclude a species from its proper genus, because it does not ex-

hibit peculiarities which are erroneously believed to characterize the class from which it is excluded.

5. In passing from several special propositions to a general one, which professedly embraces them all, and no more, something material is added or excluded, forming the fallacy of *sophistical combination*. It frequently escapes detection, especially in elaborate arguments, because several of the particular propositions are unexceptionable, and the attention is diverted from the objectionable part, which may form only a small, though possibly an essential, portion of the general proposition. Forensic arguments frequently exemplify this fallacy, by drawing, from the testimonies given, some general conclusion, which involves a material assumption of which no satisfactory proof has been given.

6. *Imaginary universality* is where we assume that a proposition is true or false universally, because it has been proved to be so generally. This sophism is of frequent occurrence, because there is a general prejudice in favor of excluding all exceptions to general laws or rules, while these may exist notwithstanding. Most men are rational and two-handed, but some are not. We can never safely admit the universality of a proposition except where this has been rigidly proved; and this can never be done, beyond the bounds of necessary truth.

§ 3. SOPHISMS OF CAUSATION.—(1) False Cause.—Important Distinction.—(2) False Effect.—Error regarding Experience.—(3) Confounding Cause and Effect.—(4) Hypothetical Causes.—(5) Mistaking the chief Cause.—(6) Mistaking the chief Effect.—(7) Mistaking the ultimate Cause.—(8) Sophistical Explanation.—(9) Sophistical Induction.—(10) Sophistical Proof.—Relation of Proof to Cause.—(11) Sophistical Relation.—Why prevalent.—(12) Excluding Causes.—(13) Excluding Effects.—Why prevalent.—(14) Imaginary Effect.—Caution.—(15) Imaginary Cause.—Frequent Combinations.—By what this class of Sophisms is particularly affected.

This class comprises those sophisms in which we err regarding causes and effects. The following enumeration includes the most common.

1. *False cause* occurs where a mere antecedent or concomitant is inferred to be a cause, while the effect may depend wholly on other agencies. A uniform antecedent or concomitant is a *sign* of the consequent: but, before we are justified in considering it a *cause*, we must know that it is concerned in producing the effect.

Many superstitions are examples of this fallacy. A pagan fails to present the usual offerings to the gods: he is taken sick, and at once attributes this to the supposed anger of the offended deities. Another has used some silly charm, before engaging in an important undertaking: he succeeds, and attributes it to the influence of the charm. Medicine and politics, also, furnish numerous instances of this fallacy. A person takes a medicine, and recovers from his disease: the cure is attributed to the remedy, although possibly it may have only retarded his recovery, which was effected by the healing powers of the system, in spite of the pernicious effect of the medicine. So politicians have sometimes attributed public prosperity or adversity to certain laws, which, in fact, operated wholly the other way.

2. *False effect* is, where a consequent or concomitant is inferred to be the effect of an agency of which it may possibly be quite independent. A man's success, in one case, and his failure, in another, have been frequently deemed the effects of agencies by which they were not, in the least, influenced. The Moon is incessantly changing, and so is the weather: hence changes of the former are speedily followed by changes of the latter, as a matter of course; and yet these have often been believed to be effects of the former, although further observation would show that the two classes of phenomena have not the least connection with each other.

Where a certain change follows the supposed cause, it is sometimes confidently declared that experience proves the point, when all it proves is, that one followed the other, and it may have been its effect no more than night is the effect of day, or winter the effect of summer. The premonitory symptoms of a disease are its uniform antecedents: but, instead of being its cause, they are only early effects of the common cause.

3. We mistake the order of cause and effect, making that the cause which is, in reality, the effect, and conversely, forming the sophism of *confounding cause and effect*. We are very liable to fall into this error where the cause and effect appear simultaneously, or where the effect reacts and strengthens the cause, or where two things mutually produce each other, so that what is, in one case, the effect, is, in another, the cause. Thus loose thinking produces loose acting, while the latter increases

and confirms the former, or even produces it independently. A pernicious instance of this fallacy is, the opinion that severe training renders children hardy, because many who bore it are so. They are not hardy, because they were treated severely; but they bore such treatment unflinchingly, because they were originally very hardy; and they would be hardier than they are, if they had been more rationally treated. The sophism has caused the deaths of many thousands of children, and loss of health and strength to still greater numbers.

4. *Hypothetical causes* occurs where supposed agencies are inferred to be the causes of the effect in question, when possibly they have no existence. This fallacy is frequently exemplified in the motives assigned for the actions of others, where these are not, in reality, known. A man of good principles is apt to assign motives to which the party in question may be an entire stranger, while persons of a different character are inclined to attribute everything to bad and sordid motives, such as usually sway themselves. This fallacy abounds in Scholastic Physics. Water rose in a pump, because nature abhorred a vacuum: descending bodies moved faster and faster, because their motion was violent, and not natural: air occupied the upper regions, because that was its proper place: muscular contraction was effected by the animal spirits—and so forth.

5. *Mistaking the chief cause* occurs where one of several agencies is inferred to be the sole or chief, when it is possibly only a minor cause, having comparatively little influence on the effect. A man whose constitution has been ruined by dissipation, takes a slight cold, and dies. His death is attributed to the cold, although it would have produced no serious effect on him, if his constitution had retained its original vigor. Another complains of bad health, which he attributes chiefly to hard study, labor, or exposure, when, in fact, it has proceeded mainly from luxurious living.

6. *Mistaking the chief effect* is, inferring that a minor or secondary is the chief effect. A broken down debauchee will sometimes complain of a comparatively trifling evil result of his vicious courses, and overlook the fact that he is a total wreck: so a politician sometimes inveighs against a law as very bad, because it produces some evil, while he overlooks the fact that it produces a much greater amount of good.

7. *Mistaking the ultimate cause* is, where we infer that some intermediate is the ultimate cause, and requires no further explanation, when, in fact, the last at which we have arrived must be attributed to some remoter cause. Instances of this fallacy abound in various treatises on Physics. When a phenomenon has been traced to electricity, for example, it is inferred that the ultimate cause has been disclosed, whereas the motions of electricity result from one or more agents beyond itself, as certainly as it is the immediate cause of the phenomenon in question; and, until that has been unfolded, we continue ignorant of the ultimate cause.

8. *Sophistical explanation* occurs where it is inferred that the cause has been clearly proved when only some explanation is given, which possibly refers the phenomenon to some acknowledged law, but which assigns no real cause. "India rubber contracts, when it has been stretched, because it is elastic: the particles of solids stick together, because they are adhesive: and streams flow downward, because all bodies gravitate towards the center." Such expressions only refer a particular phenomenon to a general law, and unfold the cause of nothing. They only state that the phenomenon is owing to what produces something else, on which no light is thrown.

9. *Sophistical induction* is, where inductive truths are mistaken for efficient causes. This sophism differs from the preceding in attributing causal power to a mere law, whereas the other refers us to some real, though undefined, cause, which produces the general phenomenon. Instances of it are found in statements which attribute to the laws of electricity, magnetism, and gravitation, the numerous changes exhibited by their several phenomena, and which result from some force that is unexplained, or even unsuspected. A law is merely a command, or a general truth, and cannot be an efficient cause of anything. Laws and agents are totally different things; and this fallacy consists in confounding them, and attributing to the former what cannot possibly belong to anything but the latter.

10. *Sophistical proof* occurs where we mistake the proof for the cause, and attribute to the former what belongs to the latter. When we have proved that a certain agent operates, which uniformly produces a particu-

lar effect, we have implicitly proved the existence of the latter. But the converse does not hold true: we may prove that a certain effect exists without proving anything regarding its particular cause. The appearance of ice on the waters is a proof that there has been cold weather: but, instead of being the cause, it is an effect. This fallacy is fostered by the habit of employing the same conjunctions to denote causes and proofs.

11. *Sophistical relation* is, inferring that effects resemble their causes, or that effects of a common cause are all alike in kind, and proportional to its intensity; or, conversely, that different effects must proceed from agencies different in kind, which are proportional to the effects. This sophism is very prevalent, because the inferences frequently hold true, and hence the exceptions are overlooked. Yet these are neither few nor unimportant. Thus, most of those agencies which excite pleasant sensations, produce pain when they act with intensity, such as heat, light, and various sounds. So the effects of property on a man's well-being are by no means proportional to its amount. The contrary inference is a very common and pernicious instance of this fallacy. An appeal to experience is generally necessary, in order to ascertain whether the relations inferred by this sophism really exist, and also to ascertain the common origin of the varying effects of the same cause, acting with different degrees of intensity.

12. *Excluding causes* is where we attribute a phenomenon solely to certain agencies, and exclude or overlook others, which operate in its production. The fact that an agency does not by itself produce a given effect, by no means proves that it is not instrumental in its production. Water alone will not decompose organic substances; yet it is an active agent in such decomposition. So the Sun has a very perceptible effect on the tides, and causes the difference between spring and neap tides, although the Moon's influence is much more conspicuous. A common instance of this sophism is, denying that the art of reading is not an important means of moral and intellectual discipline, because its effects may be nugatory, unless aided by other agencies. So it is sometimes inferred that certain bad habits have not operated in producing disease or premature death, because other agencies have operated, and some persons have been long-lived who were addicted to those habits.

13. *Excluding effects* is, where we exclude or overlook certain effects of an agency, and infer that some others are the sole effects. This frequently happens in investigating the results of laws and of morbific agencies. An instance of it occurs in the common opinion that the immediate effects of alcohol on the human frame are its sole effects, and that it is not instrumental in producing chronic disease or permanent insanity.

14. *Imaginary effect* attributes to a certain cause an effect which does not, in reality, exist. Sometimes there are phenomena resembling that which is assumed to exist; but they are materially different. In other instances, the assumption is partially true, yet it differs essentially from the reality. A non-existing phenomenon can have no cause; and consequently any argument ostensibly proving its cause, must be sophistical. Hence, before inquiring into the cause of a supposed or assumed phenomenon, we should ascertain whether it is a fact. Cases are not rare in which much time has been spent in investigating the causes of imaginary effects, which never existed. An instance of the sophism occurs when it is believed that a man recovered, by means of a certain treatment, from a disease which he never had. So a politician sometimes attributes to certain laws or public acts, good or evil which never existed.

15. *Imaginary cause* occurs where an effect is attributed to an agency which never existed, and which evidently can produce no effect. Consequently any proof that it has done so must be fallacious. A physician sometimes attributes the cure of his patient to a medicine which was duly prescribed, but which the latter never took; and a politician has been known to attribute good or evil to acts of the ruling powers which were never performed. This and the preceding fallacy are always combined with some paralogism, which they disguise and render less liable to detection.

All the sophisms of causation are frequently combined with those of comprehension and signs. Facts are either overlooked, misapprehended or misinterpreted; and then we reason sophistically upon a wrong view of the subject. A certain phenomenon, for instance, is assumed to be a uniform antecedent, when careful and continued observation would show that it is only an occasional antecedent; and then it is sophistically inferred to be the cause.

This class of fallacies is particularly affected by ignorance and prejudices, since the truth is not generally obtruded on our Comprehension, but must be learned by a close and cautious view of the subject.

§ 4. SOPHISMS OF PROBABILITY.—Nature of these Fallacies.—(1) Inferring the Probable.—(a) Inferring Hypotheses.—(b) Accumulating Probabilities.—Important Distinction.—(c) Friends' Opinions.—(d) One-sided Arguments.—(e) Harmonizing Conclusions.—Why frequent.—(f) Contingent Connective.—Frequent Combination.—(g) Incomprehensible Connective.—Important Distinction.—(h) Inconclusive Investigation.—(i) Sophistical Leap.—Distinction.—(2) Rejecting the Improbable.—(a) Discordant Opinion.—Its extensive Influence.—(b) Overlooking the Alternative.—Why prevalent.—Its Operation.—(c) Rejecting Theories.—Frequent source of this Sophism.—(d) Severing Probabilities.—(e) Enemies' Opinions.—(f) Mortifying proofs.—Effects of this Sophism.—(g) Imaginary Absurdity.—Cause of its great Influence.—(h) Sophistical Distinction.—Where frequent.—Means of determining whether a Distinction is material.—(3) Varying Probability.—(a) Exaggerating Probability.—Frequent Combination.—(b) Diminishing Probability.—(c) Exaggerating Improbability.—(d) Diminishing Improbability.—Combinations.—Why such Fallacies prevalent.—Two Sources of them.—Why Fallacies of Probability are very prevalent and influential.—Frequent Combinations.

This class of sophisms consists in confounding probability with certainty, or mistaking the character of the probability: and there are three principal kinds of it, each of which comprises several sorts.

1. We infer that a proposition is true, because it appears probable, while it may, in reality, be false. This we term *inferring the probable*, of which the following sorts are common.

(*a.*) *Inferring hypotheses* is where we infer that an hypothesis or supposition is true, because it is rendered probable by various facts, when it may possibly be disprovable by conclusive arguments or evidence. The history of science exhibits many instances of this fallacy, from the crystaline spheres, epicycles and eccentrics of the ancient astronomers, to the habitable Moon and planets of later ages. An hypothesis often appears so beautiful, harmonious, complete, and systematic, that its author or his disciples are ready to adopt it as established, without requiring proof that it is true, although it may possibly be easy to discover facts by which it is completely exploded.

A frequent instance of this modification is, inferring

that a certain operative agency, which uniformly produces the effect in question when it is not counteracted, must have actually produced it, while possibly the ordinary result was prevented by some counteracting agency. Or, conversely, it is inferred that a known effect must have proceeded from a particular agency, when it may possibly have resulted from any of several others. By this means persons have been charged with crimes that were never committed, and the misdeeds of one have been attributed to another, against whom there were suspicious circumstances.

(*b.*) *Accumulating probabilities* occurs where it is inferred that a proposition is true, because it is supported by several dependent probabilities, or probabilities of probabilities. These are confounded with concurrent and independent probabilities, where every one strengthens the preceding, instead of weakening it, as happens with the former class. The implicit credit frequently given to the statements of persons who derived their information from each other, is a common instance of this fallacy. It is not perceived that, in such cases, the more witnesses, the less credibility.

(*c.*) *Friends' opinions* is, where we infer that a proposition is true, because our friends, or persons whom we deem competent witnesses and good judges, state or believe it, when there is no conclusive proof. Many men hardly ever investigate a subject of any difficulty independently: they inquire what such and such men say or think of it, without ever seeking or finding any conclusive proof; and then they adopt those men's opinions as known truths. A common case of this form is, where we receive as conclusive the testimony of a witness of whom we think favorably, while it establishes only a probability.

(*d.*) The sophism of *one-sided arguments* is, where arguments have been adduced which render the proposition in question probable, and, therefore, it is inferred to be true, while other arguments, which militate against it, are overlooked. One of the most common cases is, the erroneous opinions which many hold regarding their own characters, desires, pursuits or expectations. The arguments which favor the bright side are seized with avidity, and adopted as conclusive, while those of a contrary tendency are overlooked, although they may possibly be more cogent than the others. Hence the erroneous views which men have so frequently held regarding the character of themselves, their nation, age, religious denomination or political party; and hence pride and self-sufficiency, coupled with contempt for others, and a fixed determination to persist in present belief and courses, right or wrong.

(*e.*) *Harmonizing conclusions* is, where we infer that a proposition is true, because it appears to harmonize with some real or supposed truth, or it agrees with our own opinions, or its reception as true removes a difficulty. This form has been a fertile source of error, because one truth always harmonizes with another, and we are strongly inclined to adopt as true what tallies with our own views and opinions. Instances of it occur in the numerous cases in which the false statements of historians, travelers or newsmen have been received as true, because they accorded with the opinions of those to whom they were addressed. Another common instance is, where we receive a charge against a person as true or false, according to our opinion regarding his previous character, while the evidence may prove the reverse.

(*f.*) *Contingent connective* is, where a contingent truth is employed as the connective of a syllogism, instead of an intuition. A common instance of this form is, where an induction is employed as a connective, as when it is argued that a certain person must die, since all men are mortal, or that an animal must have such an organization, since every one of the species has it. The fallacy is often combined with that of false extension, the assumed induction being only an empiricism, as when it is inferred that a certain man must be of such a character, since every one of his nation bears it, when, in fact, it is only a prevalent character among them, and by no means universal.

(*g.*) *Incomprehensible connective* occurs where a proposition is employed as a connective, because its contradictory is incomprehensible, whence it is erroneously assumed to be self-evident, when it may possibly be false. Instances of it occur in some mathematical demonstrations and philosophical speculations. Thus, it has sometimes been assumed as self-evident that a vanishing quantity must have some finite value, and that duration cannot be eternal, nor extension infinite. A notable instance is, the ancient puzzle which professed to prove that the swift-footed Achilles could never overtake a turtle, because the latter always advanced a little, while the man was traversing the intervening space. This was assumed to be subdivided indefinitely; and, as we cannot comprehend an infinite number, we are apt to infer that it would take an endless time to traverse an infinite number of small spaces, whereas the time of traversing the intervals would become shorter, exactly as they became smaller, and one minute can be divided and subdivided without limit, as well as one mile.

There are many truths that transcend our comprehension: and, consequently, we can never safely assume that a proposition is necessarily true, because we cannot comprehend its contradictory. This raises only a probability that it is true; but, instead of being necessarily so, there is sometimes conclusive proof that it is false. The incomprehensible and the self-evidently impossible, although often confounded, are widely different things.

(*h.*) *Inconclusive investigation* is, where we have investigated the character of a proposition, and we infer that it is certainly true, while our labors establish no more than a probability, because the criterion of truth has not been properly applied, on one or more points. Instances of this form abound in historical, scientific, and judicial investigations, as well as in common life.

(*i.*) The *sophistical leap* occurs where the proposition has appeared to be very probable, and, therefore, it is inferred to be certain, the wide gulf that separates probabilities from certainties being overleaped. Future evidence or discoveries may disprove the strongest probability, and show that it is wholly false, or at least annulled by an equally strong counter probability, whereas they cannot, in the least, invalidate a certainty, although they may corroborate it. When once established, a certainty continues unchangeable ever afterwards, while probabilities are subject to change from year to year, or even from hour to hour.

2. *Rejecting the improbable* is, where we infer that a proposition is false, because it appears to be improbable, when it may possibly have been proved conclusively. The following are the most common sorts.

(*a.*) *Discordant opinion* is, where we infer that a proposition is false or unproved, because it conflicts with our own opinions or experience. Thus one error often leads to several others, and the rejection of many truths. The fact that there are antipodes was long rejected by the great mass of mankind as an absurdity; and, until recently, accounts of showers of stones having fallen from the sky, were generally classed with the old story of Vulcan having fallen from heaven upon Lemnos. This fallacy has been very prevalent, owing to the general tendency of men to make their individual views and experience the standard of truth, and to reject whatever is incompatible with them, although the former may be erroneous, and the latter very narrow.

(*b.*) *Overlooking the alternative* is, where we infer that a proposition must be false, because there are proofs that it is highly improbable, while, in reality, its rejection involves a much greater improbability, or even an impossibility. This form is of frequent occurrence, because a very high degree of improbability is readily confounded with an impossibility. The infinity of extension and the eternity of duration, appear highly improbable, as they are undoubtedly incomprehensible: yet, if we deny those attributes, we must admit that the former has boundaries, and that the latter had a beginning, two things which are absolutely impossible.

Owing to this fallacy, men often adhere to an opinion which involves much greater difficulties than that which they reject for its improbability. They dwell upon the difficulties of the latter, and overlook the fact that these are not conclusive, and that greater difficulties are involved in the alternative, which they must receive, if the other is rejected. Thus, many historical statements have been rejected as improbable, where the falsity of the testimony would be a much greater improbability, or even an absurdity.

(*c.*) *Rejecting theories* is, where an hypothesis which has been proved to be true, is rejected, because it appears improbable. This is frequently owing to its clashing with some favorite, but groundless, hypothesis of the rejecter: but it often proceeds from his being too ignorant, indolent, or narrow-minded to appreciate the proofs by which the hypothesis is established.

(*d.*) *Severing probabilities* consists in rejecting conclusive circumstantial evidence, because the separate circumstances establish only a probability, which is believed

SEC. 4.] SOPHISMS OF PROBABILITY. 303

to be rebutted by a greater antecedent probability against the conclusion, while the combined force of all the evidences is overlooked. The fallacy is like maintaining that twenty men cannot raise a weight, because none of them separately can do it. A common instance is, where a man charged with a crime, is believed to be innocent, on account of his previous good character, and the evidence against him being wholly circumstantial, although, taken altogether, it may be quite irrefragable. In the same way, another is believed to be guilty, because of his bad reputation, and none of the exculpatory circumstances being conclusive; although all of them taken together may be quite so.

(*e.*) *Enemies' opinions* consists in inferring that a proposition is false, because it is rejected by those of whom we think highly and adopted or testified by others to whom we are opposed or unfriendly, while the proof by which it is sustained has never been properly, if at all, examined, and, for anything that appears to the contrary, it may be quite conclusive. This sophism is the converse of friends' opinions, and operates similarly. A common instance is, where we reject as unsatisfactory the conclusive testimony of a witness whom we dislike, for which reason we think it highly improbable, and therefore not credible.

(*f.*) *Mortifying proofs* is, where an established proposition is rejected, because it is highly distasteful to our wishes or feelings, whence we infer that it must be false. Prejudice makes it appear very improbable, and, therefore, it is inferred that there must be some radical defect in the argument or testimony by which it is sustained. Thus, when it has been proved that a man's conduct, on a particular occasion, has been highly culpable, he immediately appeals to the supposed purity or excellence of his motives, as invalidating the argument, which is not in the least affected by their character. So men are apt to believe that arguments or testimonies which militate against the excellence of their own character, conduct, or position must be fallacious, although they may be quite the reverse.

This sophism fosters selfishness and bad passions, and throws strong obstacles in the way of reforming evil practices, or correcting erroneous opinions. It is the counterpart of the fallacy of one-sided arguments, which it generally accompanies and strengthens.

(*g.*) *Imaginary absurdity* is, where a proposition is inferred to be false, because it is erroneously believed to contradict self-evident or palpable truth. Conclusive proof has often been thus rejected. It is so pleasant and easy to adopt current opinions as true, so difficult, in many cases, to find truth, and so unpleasant to discover we have believed error, that a great portion of mankind have been misled by this fallacy, on some occasion or another. The faithful statements of travelers and historians have been rejected as false, on no better grounds than that they were at variance with the views of those who heard or read them, regarding the criterions of truth.

(*h.*) *Sophistical distinction* is, where we infer that the proposition in question is unproved, because it appears to differ, in some respects, from that which has been proved, while the difference is, in reality, quite immaterial. When the question has been proved in effect, differences in the mode of expression, or some immaterial particular, are evidently of no consequence. This fallacy is sometimes adopted by controversialists, when other arguments have failed.

In determining whether a distinction is material, we must look to the nature of the inquiry, and see how it affects the point in question: for a difference which may be of no consequence in one investigation may be very important in another. Thus, the particular day and hour of witnessing a phenomenon may be of little consequence to a naturalist, while the time of witnessing an act may be of the utmost importance on a criminal trial.

A common instance of this sophism is, where unimpeachable testimony, regarding an immediate comprehension, is rejected, upon the ground that the witness is not a proper judge of the matter, the comprehension being confounded with inferences that may be drawn from it, which are a very different thing. Any one who possesses the use of his faculties, may be quite competent to prove what he apprehended, although it may require a person skilled in the subject to draw the proper inferences.

3. The degree of probability or improbability is materially mistaken, forming the sophism of *varying probability*, of which there are four varieties.

(*a.*) *Exaggerating probability* is, where we infer that

the degree of probability is much greater than what the proof, in reality, establishes. It is frequently combined with the sophistical leap: a low probability is inferred to be very high, and then it is inferred to be a certainty.

(*b.*) *Diminishing probability* is, where we infer that the probability is much less than has been shown; and it is the converse of the preceding fallacy.

(*c.*) *Exaggerating improbability* is, where an improbability is shown, and we infer that it is much greater than the reality. It is the converse of the first.

(*d.*) *Diminishing improbability* is, where an improbability is proved, and we infer that it is much less than the reality, which is the converse of the third.

These fallacies are generally combined with each other, because the same desire which leads men to exaggerate one view of the subject induces them to diminish the contrary: and they are of frequent occurrence, because probability does not generally admit of accurate measurement, and its real character is not easily determined. A common instance of the combination is, where men embark in hazardous undertakings with a confident expectation of success, because there is a slight probability in its favor, which they magnify into a very great probability, while they equally diminish the probabilities of failure. So, favorite opinions, based on slight probabilities, are often held tenaciously, as being extremely probable, while the contrary, and really more probable, opinions are unhesitatingly rejected.

Fallacies of this kind may arise either from mistaking the character of the probability in the first instance, or by drawing inferences from it, and then overlooking the character of the premise.

Sophisms of probability are very prevalent, because they save the pain of suspense and the labor of further investigation, and because we are frequently obliged, in the ordinary business of life, to act upon strong probabilities as if they were certainties. Their influence is often increased by the Imagination forming vivid and agreeable pictures of what we desire to be true, and equally vivid, but repulsive, pictures of what we desire to be false. These erroneous representations excite strong feelings, which withdraw our attention from the weak points of the case, and confine it to what makes in favor of that which we desire to be true, whence the fallacy is frequently unsuspected.

Different kinds of this class of fallacies are very frequently combined with each other. We adopt one proposition the more firmly, because the contrary is deemed so improbable, and conversely, while we frequently exaggerate or diminish the real probabilities. A common instance of this combination is, the erroneous conclusions we form regarding our future lives or the results of our undertakings or designs. We can easily find facts which establish a probability that we shall live long, succeed in our principal undertakings, and secure permanent happiness, as well as an improbability of the reverse. We are apt to overlook the possibly stronger probabilities which lead to a different conclusion, and to overcolor both sides of the picture.

All fallacies of this class are frequently combined with paralogisms of comprehension, signs and testimony, without which they would often be detected by a full and accurate view of the whole case; but when we overlook one side of it, and form an erroneous opinion of the other, that may appear to be very probable which we should readily perceive to be the very reverse, if we viewed the matter aright. Those who purchased lottery tickets often thought there was a great probability that they would draw a high prize, although it was quite clear to any person who carefully considered the matter for a few seconds, that the probability was all the other way, as the high prizes were very rare, and only an exception to the rule. So many satisfy themselves that they will act prudently, live long, resist temptations to vice, and enjoy prosperity. They overlook the fact that, owing to their character and circumstances, the probability is all the other way: and the events turn out accordingly.

CHAPTER XIX.

OF ABERRANCIES, OR FALLACIES OF IRRELEVANCY.

§ 1. ABERRANCIES OF CONFUSION.—Nature of this class of Fallacies.—(1) Irrelevant Illustration.—Legitimate Object, and Abuse, of Illustration.—(2) Indefinite Terms.—Proper course of dealing with this Fallacy.—(3) Irrelevant Analogies.—Requisite to validity of reasoning from Analogy.—Frequent Combination.—(4) Deciding by Character. — Distinction. — Why this Aberrancy prevalent.— Frequent Combination.—(5) Deciding by Consequences.—Truth preferable to Error.—Distinction.—(6) Deciding by Motives.—Why these no Criterion of Truth.—Relation of Motives to Arguments.—Frequent Combinations. — (7) Deciding by Appearances. — Frequent effect of this Aberrancy.—(8) Irrelevant Induction.—Why prevalent in Political Discussions.—(9) Irrelevant Empiricism.—Requisite to establish an Empiricism.—Frequent Combination.—(10) Irrelevant Objection.—Relation of Difficulties to sound Arguments.—Criterions.—Caution.—Proper mode of dealing with Objections.—(11) Irrelevant Modification.—(12) Homonymous Expressions.—Frequent Combination.—(13) Verbal Illusion.—Why prevalent. — Distinction.—(14) Illusive Contradiction.—Frequent Combination.—(15) Confounding Means and End.

IN these fallacies, one conclusion is assumed to be tantamount to another, while they are essentially different, and it is inferred that one has been proved, because the other may have been proved. The following enumeration includes the more common:

1. *Irrelevant illustration* consists in confounding mere illustrations with proofs. The legitimate object of illustrations is, to throw light on a proposition which is proved and directly explained independently. Where all our knowledge of a subject is derived from illustrations, we generally form erroneous conceptions of the thing illustrated, and thus think we know what we only conceive. Where a thing entirely unknown is illustrated by something familiar, but not well understood, we are apt to think that we understand both, when we understand neither; and consequently the mere illustration is mistaken for a proof that it is true, while no legitimate proof has been given.

One of the most common forms of this fallacy consists in drawing a comparison between the thing to be proved and something else, and then assuming that what is known to be true of the latter holds equally true of the former. Thus, communities have been compared to individuals, and it was then assumed that·they all have a period of youth, manhood, old age, and extinction.

2. *Indefinite terms* occurs where obscure, ambiguous or figurative language is employed in such a manner that it is impossible to determine what it means; and, therefore, the conclusion actually established may be very wide of that which ought to be proved, if anything at all has been established. Where an essential part of an argument consists of language whose exact meaning we cannot ascertain, it is impossible to determine whether

any conclusion has been established: for the objectionable part may involve a fallacy; and consequently the whole should go for nothing, so far as proving a conclusion is concerned.

This fallacy abounds in various treatises on mental science. Loose generalities are substituted for definite statements; and, in many instances, gross absurdities are veiled under misty plausibilities.

3. *Irrelevant analogies* is, where an analogous conclusion is proved, which is assumed to establish the question, when, in reality, it does not. In order to render such argument valid, it must distinctly appear that the two cases are essentially alike, both in the conclusion and in all those points on which it depends: for, if they differ in these respects, other resemblances will avail nothing. It is not sufficient that they *may* resemble each other in these respects: it must appear that they actually do so: else the analogy may fail in some essential point. We may err by assuming, without proof, either that things which are alike in some respects, are alike in others also, or that things which differ in some respects, differ in others also.

A common instance of this fallacy is, proving something of one man, and then inferring that this holds true of another, who resembles him in some respects, while there is no proof that the similarity extends so far as it should do, in order to render the reasoning valid. Thus, one sick person's symptoms may resemble those of another, while their diseases may differ so widely that what cured one will only injure the other. So it has been argued that popular education must be injurious, because, if a horse knew enough, he would throw his rider, while it is not shown (and, in fact, it is not true) that the relation of a horse to his owner is the same as that of a people to their rulers.

To this fallacy belongs the practice of applying general maxims to cases essentially different from those to which they properly apply, and thus drawing erroneous inferences regarding the case in question. This form is often combined with the paralogism of misinterpretation, the maxim being misunderstood, as well as misapplied. Other instances of this aberrancy are—assuming that the future will be like the past, and that a thing will never be, because it has been shown that it never was—assuming that a thing must be useful for one purpose, because it has been shown to be useful for another—assuming that a thing must be totally bad, and should never be used, because it is liable to be abused, or to produce evil—proving that a thing ought to be, and then assuming that it actually is, or that it ought not to be or that it would be useless, and then assuming that it is not—that a man will act in a certain way, because it is shown that ordinary prudence requires him to do so—and proving that there is no known reason why a conclusion should not be true, and then inferring that it is true, while there may be unknown proof that it is false.

The aberrancy is frequently combined with the sophism of false cause, by assuming that, in the analogous case, a certain phenomenon resulted from a cause of which, in reality, it was wholly independent, and which is absent in the case under consideration.

4. *Deciding by character* occurs where a conclusion is assumed to be true, because it is shown that it is advocated by good men, or alleged to be a good doctrine, or it is assumed to be false, because it appears that it is maintained by bad men, or it is alleged to be a bad doctrine. Good men have often advocated errors, while bad men have held many opinions which were correct: and truth has frequently been decried as pernicious, while error has been upheld as fostering virtue. Hence the truth or falsity of a proposition is a very different inquiry from that of the character of its advocates, or its alleged nature; and we wander from the point when we turn to discuss them, while the matter before us is, the professed proof that the proposition is true or false.

In many cases the alleged character of the parties and nature of the proposition, are widely different from the real, so that the argument, besides being irrelevant, is otherwise fallacious. Yet, owing to the influence of several strong prejudices, its true character is not even suspected. It is pleasing to indolence to think that it can determine the truth or falsity of a proposition by merely glancing at the character of its advocates, which is assumed to be what is alleged by their friends or enemies: and it is soothing to the feelings to conclude that a doctrine is true, when it flatters avarice, pride or vanity, and false, when it wounds these emotions.

5. *Deciding by consequences* is, where certain good

consequences are held to flow from a doctrine, and therefore it is inferred to be true, or, conversely, it is inferred to be false, because certain evil consequences are alleged to flow from it. A man sometimes finds that, if a certain conclusion is true, he is bound to abandon some favorite practice, or relinquish some lucrative occupation; and, therefore, he is strongly disposed to reject such a doctrine, though it may have been proved quite conclusively, and, in fact, the difficulty which he encounters is in favor of the disagreeable conclusion. That it should militate against a vicious practice or a pernicious pursuit, rather proves it true than otherwise. Yet the victim is very unwilling to believe the bitter truth, although, in reality, the sooner he does so, the better for his own permanent welfare.

Truth is always more favorable to happiness, in the long run, than error, although it may occasion some temporary pangs, or deprive us of some fleeting or fancied good. The world has yet to see the first instance in which a person did not gain much more than he lost by discarding error and adopting truth, while there are innumerable instances in which men clung to errors, under the belief that the contrary doctrines were inimical to their happiness, when, in reality, those errors destroyed their happiness, while the adoption of truth would have secured an opposite result. We should, therefore, address ourselves fearlessly to the proofs, while we are inquiring whether a proposition is true or false. Its bearings on our future condition is a distinct question, which should never be allowed to interfere with the former.

To this fallacy belongs the practice of pronouncing on the wisdom or folly of a certain course, according to the subsequent results. These may depend on circumstances which no human foresight could anticipate, and make a course afterwards seem foolish which previously appeared the most eligible. On the other hand, a very silly scheme sometimes produces favorable results, owing to accidental circumstances, which were not foreseen, and of which there was very little probability.

6. *Deciding by motives* occurs where the conclusion is decided according to the motives which are alleged to actuate the advocate. A man may advocate the cause of truth from bad motives, or that of error from the reverse, and hence motives are no criterion of the conclu-

sion. In examining a man's general character, or the bearings of an act which he has done, the motives which influenced him may form a very pertinent and important subject of inquiry: but the truth or falsity of a proposition which he advocates is a very different matter. A conclusive argument is not a whit invalidated by being urged from bad motives, while a worthless argument is not in the least aided by the good motives of its pleader.

A common instance of this aberrancy is, assuming that an argument must be worthless, because it is inconsistent with some other opinion expressed by the advocate, as if a good argument was refuted by the fact that its maintainer had formerly employed a bad one, or expressed himself differently on the same subject.

This fallacy is often combined with those of deciding by consequences and character. Certain bad consequences are alleged to follow from the conclusion, and it is asserted that the advocate must be a bad man, and therefore he must be influenced by bad motives, and therefore the conclusion must be false, or conversely. Possibly the supposed consequences are wholly imaginary, or the individual may not see that they follow. Many hold opinions that really imply consequences which they neither see nor admit. Doctrines produce their legitimate consequences in the long run, on the majority of those who embrace them: but they do not always do so in the first instance, or in every individual case.

Not unfrequently the aberrancy is combined with sophisms of causation and probability. The individual has, in reality, acted from several different motives, or those alleged are only probable, or perhaps wholly imaginary, so that, in fact, no conclusion whatever is proved.

7. It is shown that there are several indications which go to prove a certain conclusion, and it is inferred that it is absolutely proved, forming the aberrancy of *deciding by appearances*. It often escapes detection, partly owing to the influence of strong prejudices, and partly from the indications establishing a probability in favor of the conclusion.

Instances of this fallacy occur where it is inferred that a certain person is a good man, because it has been shown that he professes to be, and is generally considered such, and that he has performed some good actions, while a wider and closer investigation would prove the reverse.

So it is often inferred that a man is a bad character, because he has been charged with having committed wicked acts, when, in truth, the charge is quite groundless, and the acts were harmless and proper; and, even if they were not, they would not prove the conclusion. Particular actions do not prove general character, any more than the latter prove the former; nor does the performance of certain virtues and abstinence from certain vices, prove a man virtuous.

8. *Irrelevant induction* is, where a thing is proved to hold true of numerous cases, and this is assumed to prove that it holds true of a whole class, when there is no proof to warrant such an inference, which is justified only by facts that logically involve it. Instances of this aberrancy occur even in Mathematics, as where the binomial theorem had been assumed to hold true generally simply because it did so in many cases. It is very common in political discussions, since parties readily assume that what has been true, in several cases, will always hold true, where the assumption helps them in defending a weak position.

9. *Irrelevant empiricism* is where a thing is proved of several individuals, and it is then assumed to apply to others, of which it has not been proved, and of which, possibly, it does not hold true. The satellites of the Earth, Jupiter and Saturn, all revolve round their primaries from west to east; but some of Saturn's revolve in the contrary direction. To establish an empiricism, it is requisite that every individual object embraced in the conclusion, should have been examined, and found to possess the attribute in question.

The two preceding aberrancies are sometimes combined with each other, and with paralogisms of comprehension. While the proposition in question is professed to have been established as an induction, a wider and more accurate observation would show that it is not true even empirically. Phrenology furnishes a good instance of this combination. Not only do its advocates fail to show that the alleged conformity between certain forms of the head and certain mental characteristics hold true generally, but they fail to prove that it has been hitherto found to hold true; and careful observation will readily show that no such conformity exists, and that the cases of nonconformity are much more numerous than those of casual conformity.

10. *Irrelevant objection* occurs where it is shown that the proposition in question is liable to certain objections, or that difficulties attend its reception, whence it is inferred to be false, while, in reality, those difficulties neither disprove the conclusion nor invalidate the arguments by which it is established. The alleged doubts or difficulties are generally founded on the objector's vague or erroneous views of the whole subject, or at least of an essential part of it, and consequently possess no real weight. Difficulties attend every department of human knowledge: but they do not affect conclusive proofs.

Sometimes the objection applies only to a misrepresentation of the proposition in question, made by a party who does not rightly understand it, and not to the real question. It is not uncommon for persons to defend doctrines which they do not rightly understand; and, therefore, we should first ascertain what a proposition really is, before we regard objections, which may possibly apply only to what it is erroneously represented to be.

A common instance of this aberrancy is, where a proposition is argued to be false, because it is alleged to be inconsistent with some known fact, or to lead necessarily to some erroneous inference, when, in reality, such allegations are quite irrelevant. Sometimes they are false, and, in other cases, although true, they are, in reality, quite consistent with the proposition in question.

A sound argument can no more be inconsistent with any truth than one truth can be incompatible with another. If, therefore, the objection conclusively proves that the proposition in question cannot possibly be true, it is sound, but otherwise not. It avails nothing that it establishes a strong probability against it, if this is rebutted by conclusive evidence, or even by a stronger probability on the other side. If the proof in favor of the proposition is conclusive, every objection must evidently be futile, and should go for nothing.

In no case should we reject as invalid a proof which appears to be quite the reverse, till we have given or obtained a demonstration of its fallaciousness. Sometimes we may think this is attainable, when a serious attempt to find it would show us that the objection is worthless, and the proof irrefragable. It frequently happens that the proof exhibits something which is objectionable, but which does not affect its substantial validity, as where something false is assumed in an argument which does not affect its soundness, and may, in fact, be a mere illustration. Yet the objector is apt to fasten on the flaw, and hold it forth as conclusive. We should, therefore, observe the relation of the objectionable part to the whole proof, and ascertain whether the objection is not irrelevant, even admitting that its primary premises are quite true.

11. *Irrelevant modification* consists in proving a proposition which is a modification of the one in question, but yet materially different. A common instance of it is, where a conclusion is proved conditionally, when it ought to have been proved absolutely. Thus jurists have sometimes professedly shown that a particular form of government was the best for a particular nation, and then assumed they had proved it is the best for every nation. So it is sometimes assumed that a certain regimen or medical treatment is best for everybody, because it has been shown to be the best for persons of a particular constitution; and many lives have been thus lost. Another instance is, showing that a certain law, institution, or custom had a rational origin, and then assuming this as proving that it is good universally, while, in fact, it may be very bad when circumstances have greatly changed.

12. *Homonymous expressions* consists in proving a proposition which sounds very like the one in question, and may be readily mistaken for it, while it is essentially different. It is favored by the fact that many words and expressions are very similar in sound, but yet materially different in signification. A man may do many foolish things without being a fool, and many good things without being good. This aberrancy is often combined with the preceding; and the combination occurs not unfrequently in public addresses. Thus, a speaker proves that everybody within a certain narrow circle thinks or does so and so; and this is assumed as proving that all mankind do so.

13. *Verbal illusion* occurs where the name of a thing, or a definition of it, is given, and this is assumed to be tantamount to communicating or acquiring a knowledge of its nature. We are very ready to adopt this fallacy, because the name is apt to be confounded with the char-

acter, and it is so much easier to master the former, or read a definition, than to acquire a real knowledge of the latter. Thus, many have thought that they had acquired a good knowledge of Botany, when it went no farther than to enable them to tell the name of a genus. So teachers have often thought that they communicated to their pupils a knowledge of the subject, when they only taught them words whose real import they never understood. Words can convey no real instruction unless their import is properly understood and remembered, which is a very different thing from merely hearing or reading and repeating them.

14. *Illusive contradiction* consists in confounding the contrary of a proposition with its contradictory, and it is inferred that it is true, because its contrary is shown to be false. Thus, it is often assumed that a thing is bad, because it has been proved that it is not good, or that a certain line of conduct is commendable, because the contrary course is reprehensible, as if the sky must be white, because it is not black. So it is often assumed that a man is austere, because he is not gay, or penurious, because he is not extravagant, or rash, because he is not timid, and so forth.

The fatalists' argument is a notable example of this aberrancy. "A thing will either happen or not," and this is confounded with "happening if I act thus, and not happening if I act otherwise." When a man's house is on fire, the real question is, whether his efforts will affect the result; and this is not, in the least met, by saying that it is destined either to burn or not to burn, which is only saying that it either will burn or it will not, a self-evident truth, indeed, but quite irrelevant.

Another instance of this fallacy is, confounding "believing" with "not disbelieving." As we must either believe or not believe, it is assumed that we believe everything which we do not formally disbelieve, whereas we have no real belief or disbelief in a thing of which we are ignorant. Confidence in a man's veracity is a very different thing from believing his opinions, which we cannot do till we know what they are.

This aberrancy is frequently combined with the sophism of false connection. Because a conclusion does not follow from the premises, it is inferred, not only that it is false, but that the contrary is true. Thus, a controversialist often assumes that his own position is proved, because he has detected a flaw in some reasoning employed to prove the contrary conclusion. The compound fallacy is of this sort: "this reasoning fails to prove that crows are black; therefore they are white." To warrant us in receiving a proposition as true, it is not sufficient to disprove certain arguments employed to prove the contrary doctrine: we must have positive and conclusive proof that it is true. For the contrary proposition may be provable by other arguments, or both propositions may be false.

15. *Confounding means and end* consists in proving that the end or object to be effected by certain means, is good, and then assuming that the latter are good. A proper end may be sought by wrong means, of which numerous cases are found both in History and common life. Consequently the character of the means must be ascertained, on other grounds than the goodness of the end. In many instances, the object is really bad, and only supposed to be good: but the fallacy is not the less complete, even where it is really good. Fraud is unjustifiable when it is employed to propagate truth, as well as when it is used to sustain error.

§ 2. ABERRANCIES OF APPEALS TO AUTHORITY.—Nature of this Class.—(1) Universal Belief.—Why not conclusive Proof.—(2) General Belief.—(3) Conflicting Opinions.—Its Foundations.—(4) Modern Opinions.—Why not Proof.—(5) Sages' Opinions.—Distinction.— Sources of Error.— Frequent Combination.— (6) Many Arguments.—Distinction.—Various Forms.—(7) Pretended Refutation.—Proper mode of dealing with Refutations.—(8) Irrelevant Admission.—Its Characteristic.—Frequent Effect of it.—With what often combined.—Why Aberrancies of this class are common.—How to be avoided.—Important Distinctions.

In this class of aberrancies, it is shown by others, or is found by ourselves, that certain persons have believed the proposition in question, which is considered tantamount to proving that it is true, while, for anything that appears to the contrary, it may be false. The following are the most common kinds.

1. *Universal belief* occurs where a conclusion is either assumed to be true, because it is found that all mankind have believed it, or it is assumed that it must be false, because it appears that all mankind have disbelieved it. There may be various sources of illusion, common to all

mankind; and hence universal belief is by no means a conclusive proof that a proposition is true. All mankind, for many ages, believed things self-evidently or demonstrably false, and rejected as false things self-evidently or demonstrably true. We uniformly believe that our dreams are real, till we awake, when we discover the contrary. So it was universally believed, for many ages, that the Sun, Moon, and stars revolve daily round the Earth, while the doctrine that the Sun is at rest, and the Earth in motion, was rejected as an absurdity.

2. *General belief* is, where we infer that a proposition is true, because it appears that the great majority of mankind have believed it, in all ages. If universal belief does not prove a proposition true, much less will the belief of a majority only, especially when we consider how frequently and readily opinions are adopted without any proper investigation. It has been a generally received opinion that seeds germinate more quickly during the crescent Moon than when it is in the wane; yet a very slight investigation shows that it is quite groundless.

3. *Conflicting opinions* consists in inferring that a proposition is doubtful, because it appears that different opinions are held regarding it, when its truth may have been conclusively established. This aberrancy proceeds on the absurdities that any man's views are as likely to be correct as any other's, and that we cannot be certain a conclusion is true, as long as there is anybody so ignorant, indolent, stupid or prejudiced as to reject it. Where we have unquestionable proof that it is true, the opinions of such persons to the contrary are entitled to no weight.

4. It is inferred that a proposition is true or false because it appears to be deliberately held or rejected by the men of the present day. This we call the aberrancy of *modern opinions*. No age is exempt from error; and it has sometimes happened that old opinions have been re-adopted, on good grounds, after having been long rejected. The astronomical opinions of Aristarchus of Samos were rejected by all astronomers, for nearly two thousand years; yet they have been demonstrated to be correct.

5. *Sages' opinions* occurs where it appears that wise or distinguished men held a certain opinion, and, therefore, it is inferred to be correct, although there may be

positive proof to the contrary. History furnishes innumerable instances in which such men, who were considered guides, held conflicting views, or opinions which were afterwards demonstrated to be quite erroneous. We must distinguish between testimony, regarding matters which can be certainly known, and mere opinions, which are generally based only on probabilities. An unimpeachable witness may be safely believed, when he relates something within his knowledge: but this does not, in the least, warrant us in adopting his mere opinions as ascertained truths, especially where there may be accessible proofs that they are quite erroneous.

Distinguished men are sometimes as much under the influence of prejudices as any others: and a person of ordinary abilities, who examines a subject carefully and impartially, will often arrive at truth, where men of greater talents err, through inattention, haste or prejudice. A conceit of their own abilities has often injured such persons, in the pursuit of truth, more than their intellectual superiority benefited them. In all the most important investigations, the great requisites are, attentive, unprejudiced and persevering examination, with a fixed desire to discover truth. For the want of these, neither a great reputation nor brilliant talents will, in the least, compensate.

This aberrancy is often combined with those of miscomprehension and misinterpretation of language. A person's real views are mistaken or misinterpreted; and then his authority is adduced to support opinions which he never held. Thus, the usages of our ancestors are often quoted to support or oppose some proposed change. As our ancestors' circumstances were very different from ours, their actual usages form no criterion of what they would have done, in our circumstances. It is not likely that our pagan forefathers would discard revealed religion, if placed in our circumstances. Time produces many changes, so that what is eligible in one generation may be the reverse in another. So it is frequently maintained that the experience of competent judges has already decided the question, when the thing really experienced is essentially different. It was once believed the experience of intelligent surgeons proved that scalding oil is useful in dressing wounds, whereas it is now well known that it proves the reverse.

6. It appears that the point in question is sustained by various arguments, whence it is inferred that it is duly proved, forming the aberrancy of *many arguments*. No number of arguments can prove a conclusion, as long as every one of them is invalid; and many such have been employed to prove false conclusions, of which the Ptolemaic Astronomy and the Aristotelian philosophy furnish various instances. We should remember that, as in the case of witnesses, it is the weight of arguments that avails, and not their number: while one conclusive argument proves a proposition, beyond any reasonable doubt, a thousand fallacious arguments establish nothing. Yet this fallacy has often produced conviction, because its victims, while possibly distrustful of the particular argument under consideration, relied on the combined force of the others, and so on, in an endless round.

7. *Pretended refutation* is, where an argument and the conclusion based on it, are rejected, because they have been professedly refuted, when, in reality, the refutation is fallacious. It is often added that the conclusion in question is an exploded opinion, which nobody but very ignorant or silly persons now hold. In many instances, the argument is grossly misrepresented, and nothing is refuted but what was never held, or something essentially different from the matter in question. Sometimes the refutation is directed against the fallacious arguments of an ignorant or incompetent advocate, and never touches the real proof. At other times, the main parts of the argument are overlooked, while a few immaterial errors are detected; and this is assumed to be a complete refutation. Consequently we should examine the argument, and compare it carefully with the professed refutation, before we receive the latter as of any weight. Error never appears more clearly in its true character, than when it has had a fair hearing and a searching examination.

8. It appears that a certain conclusion follows from premises admitted by a party who denies it, whence it is inferred that it must be true, forming the aberrancy of *irrelevant admission*. It is perfectly fair to argue that a man is bound to admit a conclusion necessarily implied in premises which he admits or believes: the fallacy lies in assuming that the conclusion has been established absolutely, while the admitted premises may be false. If a

man admits that the main object of public punishments is, the reformation of the offender, he admits, by necessary implication, that capital punishments are wrong; but this admission is false, the reformation of the offender being only a secondary object in public punishments, although it is often the sole object of private chastisement.

This fallacy has sometimes confirmed controversialists in their errors. The opposite party made admissions which proved their tenets; and they never saw or suspected that those admissions were unwarrantable. The aberrancy is often combined with the paralogism of misrepresenting testimony. A man's admission or expressed belief is misrepresented; and then it is assumed that the consequences necessarily implied in the misrepresentation are established absolutely and conclusively.

Fallacies of appeals to authority are very prevalent, on account of the greater ease and pleasure with which a conclusion can generally be settled by such appeals, instead of being decided by a proper investigation of proofs, while numerous strong prejudices often intervene, and rivet the error. The proper course is, to ascertain whether there is conclusive proof of the truth of the proposition in question. If there be, it must stand, although high authority reject it: if there be not, it has no good title to be classed with cognitions, and it may possibly be false, no matter who believe the contrary.

Preponderating authority may be very properly made our guide where certainty is unattainable, and we can arrive only at probabilities: but mere human belief can never prove any proposition, because it is never exempt from error. We should not, therefore, confound the question whether the proposition under consideration is true, with the very different inquiry what others have thought of it, as is generally done by those who are misled by this class of fallacies. To establish or refute a proposition by legitimate proof, is a very different thing from showing that men have believed it to be true or false.

§ 3. ABERRANCIES OF APPEALS TO DESIRES.—Nature of this Class.—(1) Inferring the Agreeable.—Its operation.—Distinction.—(2) Rejecting the Disagreeable.—Frequent Practices.—Sneers and Ridicule.—Distinction.—Combination.—How these Fallacies are fostered.—Their two-fold Origin.—Their Influence and general Character.—How combined.—How they may be avoided.

SEC. 3.] APPEALS TO DESIRES.

Here the proposition in question excites strong emotions or sensations, whence it is inferred that it is true or false, while its real character is not properly, if at all, investigated. Of this class there are two kinds.

1. A conclusion becomes very agreeable, and it is thence inferred that it is true, forming the aberrancy of *inferring the agreeable*. Here the facts or arguments that militate against the conclusion are seldom considered with any degree of attention: they are sometimes dismissed with a sneer or a sarcasm; and sometimes they are not noticed at all. In many cases the proposition in question is merely characterized by eulogistic epithets, and its adherents are called by corresponding terms, whence it is assumed to be true. There is no error to which the strongest terms of commendation cannot be easily applied, while there is no truth to which the most opprobrious epithets cannot be applied, with equal facility.

A common and pernicious instance of this fallacy is, inferring that the course which promises present enjoyment and immunity from present pain, is the best, when the case may be far otherwise; and thus many have sacrificed their future all for a fleeting present gratification. The notion of enduring present pain, or foregoing present pleasure, is so disagreeable that they assume the future is of less importance; and the excitement withdraws the attention from a careful consideration of it.

Another common instance is, where it appears that we shall gain wealth, ease, distinction, the approbation of friends, or the patronage of the powerful, if we believe the proposition in question, and it is assumed that, therefore, it is true. Although such inducements tend to produce conscious hypocrisy, rather than real belief, yet it has frequently caused conviction, the prejudices excited having confined the attention to one aspect of the subject, and concealed its true bearings. But the advantages to be derived from believing a proposition are a totally different thing from its truth or falsity, with which they ought never to be confounded.

2. Something appears which renders a conclusion disagreeable, whence it is inferred to be false, forming the aberrancy of *rejecting the disagreeable*. It is simply the converse of the preceding fallacy; and the same remarks apply to both, by merely reversing the terms. Some-

times the proposition in question is held up to ridicule, and called by offensive or contemptuous epithets, and then it is assumed to have been proved false, a course which has often imposed on the unthinking, who seem to believe that an argument can be refuted by a simple exclamation of scorn or ridicule.

The strength of the language in which the conclusion is decried, generally increases as the evidence of its truth becomes clearer. When a man is too prejudiced to yield to testimony or arguments, which prove that his conduct or opinions are wrong, he frequently stifles his convictions or suspicions by a free use of ridicule or abusive language. But, for the reason already mentioned, these have no logical force; and they are generally employed by those who can adduce no valid proof in support of their own opinions, or against their opponents'.

A frequent instance of this aberrancy is, where it appears that a person will undergo serious loss or suffering, if he disbelieves a certain proposition, or adopts the contrary, and, therefore, it is inferred that it is false. This result has frequently been produced seemingly, and sometimes really, by means of imprisonment, corporal chastisement, torture, loss of situation, or disapprobation of relatives and friends.

This aberrancy is generally combined with the preceding, and the combination has frequently imposed upon the weak-minded, the ignorant, and the unprincipled. The advantages held out, on the one hand, and the disadvantages either threatened or actually inflicted, on the other, have produced, in countless millions, belief in dogmas which they might easily have ascertained to be totally false.

Fallacies of appeals to desires are greatly fostered by the perfect ease with which they can be used. To produce facts or substantial arguments for or against a proposition, requires time and labor, while it is always very easy to call it by laudatory and pleasing epithets, on the one hand, or by offensive and contemptuous terms, on the other. The former indirectly imply that it is so well established otherwise, or so perfectly evident, that more formal proof is not required, or, conversely, that it is so absurd, or so conclusively refuted already, that no elaborate argument is now required to disprove it, while, for anything that appears to the contrary, the truth may be the very reverse.

These aberrancies are very apt to escape detection in public addresses, because the strong feelings excited by mutual sympathy between all present, both speaker and hearers, concentrate the attention on certain points, and withdraw it from others, so that the worthlessness of the argument is overlooked, although, in many cases, the slightest analysis of it would readily show that it proves nothing at all.

Such fallacies are not only employed by others, to mislead us, but we are very liable to fall into them, in our own original investigations, since our minds tend toward the agreeable, and against the disagreeable, without any prompting from others: and indeed this native tendency is requisite, in order to effect extraneous imposition by such means, which are rarely effectual, except where we are prepared to adopt them.

Fallacious appeals to desires have been one of the principal means of propagating and riveting error, in every age and country. They are so numerous that it would be tedious to describe the various species in detail; and they are all so similar in their operation, and run so much into each other, that such a description would be of little use. The most common and pernicious are those which relate to religious, ethical and political matters; and the two kinds are generally combined, the one side being depicted as very attractive, in order to secure an easy belief, and the other represented as quite repulsive, in order to deter scrutiny.

These aberrancies, however, are generally combined with fallacies of comprehension, testimony, probability, or appeals to authority, without which their influence would be comparatively small, since all know that wishes are not proofs. Sometimes the facts are first misrepresented; then a probability is established on this foundation; then this is attempted to be fortified by fallacious appeals to authority; and, finally, friendly and hostile passions are excited, to secure the reception of the error. In other cases, some of these processes are omitted, or the order of arrangement is reversed, the opposing views being rendered odious or contemptible in the first instance, by being grossly misrepresented.

As in the case of the simple aberrancy, we may mislead ourselves by such combinations, as well as be misled by others. Thus, we often form a very erroneous opinion regarding the character, condition and position of ourselves, our friends, denomination, or country. As it is pleasant to think well of ourselves, and unpleasant to think otherwise, we look at the bright side of our own characters and at the dark side of others, while we exaggerate both; and thus we easily find probabilities in favor of our own superior excellence, and improbabilities against our being in the wrong, while, on the other hand, we find, with equal facility, probabilities that others are in the wrong, where they differ from us, and improbabilities that they are right. Our opponents or neighbours, by reversing the proceeding, and looking at their own bright side and our dark side, readily arrive at an opposite conclusion.

Such combinations have fostered selfishness, pride, national or sectarian animosities, and an obstinate persistence in wrong courses. Thus, we often flatter ourselves that we would act much better than others have done, if we were placed in their circumstances, because we mistake the nature of these, and form erroneous judgements regarding the mode in which we have acted, when placed in similar circumstances. If we corrected these errors, we might find that our superiority is wholly imaginary.

Combinations like the preceding are sometimes aided by the production or presence of apprehensible objects. The production of a bloody knife has sometimes led to the condemnation of the accused, where there was no satisfactory proof of his guilt: and the sight or smell of liquor has often convinced a reformed drunkard that a glass would do him good, a few minutes after he was rightly of the contrary opinion. Sometimes the same effect is produced by broad allusions, which strongly affect the Memory or Imagination of the party addressed.

Fallacies of this kind are to be avoided by laying aside desires, and addressing ourselves attentively to the proofs, either for or against the conclusion. We must disregard both adulation and vituperation (which are generally used only by the advocates of error), and examine the proof with due care and attention. When we have ascertained where truth lies, then, and not before, it is proper to give way to the feelings which it inspires, and to designate doctrines and opinions by what we know to be their true character.

CHAPTER XX.
TABLE OF FALLACIES.

I. PARALOGISMS.

1. *Paralogisms of Intuition*
   - (1.) Intuitional assumption.
   - (2.) Intuitional rejection.

2. *Assuming what is attempted to be proved*
   - (1.) Assuming the question.
   - (2.) Reasoning in a circle.

3. *Paralogisms of Comprehension*
   - (1.) Miscomprehension.
   - (2.) Incomprehension.

4. *Paralogisms of Signs*
   - (1.) Illusive sign.
   - (2.) Non-interpretation of signs.

5. *Paralogisms of Memory*
   - (1.) Imaginary apprehension.
   - (2.) False association.
   - (3.) Mistaking ideas.
   - (4.) Forgetfulness.

6. *Intrinsic paralogisms of Testimony*
   - (1.) Ambiguous expression.
   - (2.) Overlooking conditions.
   - (3.) Assuming conditions.
   - (4.) Obscure expression.
   - (5.) Wrong expression.
   - (6.) Falsehood.
   - (7.) Suppressing truth.
   - (8.) Misrepresenting comprehensions.
   - (9.) Misrepresenting testimony.

7. *Extrinsic paralogisms of Testimony*
   - (1.) Adopting a mean.
   - (2.) Counting witnesses
   - (3.) Credulity.
   - (4.) Scepticism.
   - (5.) Overlooking testimony.
   - (6.) Indiscrimination.

8. *Misinterpreting Language*
   - (1.) Misunderstanding archaisms.
   - (2.) Misinterpreting technicalities.
   - (3.) Misinterpreting ambiguities.
   - (4.) Confounding different senses.
   - (5.) Overlooking the idiom.
   - (6.) Following etymologies.
   - (7.) Mistaking the style.
   - (8.) Misplacing the accent.
   - (9.) Misconstruction.
   - (10.) Mistaking expressions.
   - (11.) Ignorant interpretation.
   - (12.) Misconception.
   - (13.) Fallacious implication.
   - (14.) Mistaking allusions.
   - (15.) Fallacious propriety.

II. SOPHISMS.

1. *Sophisms of Confusion*
   - (1.) Sophistical connection.
   - (2.) Inferring the converse.
   - (3.) Altering propositions.

2. *Sophisms of Generalization*
   - (1.) Sophistical extension.
   - (2.) Sophistical inclusion.
   - (3.) Sophistical contraction.
   - (4.) Sophistical exclusion.
   - (5.) Sophistical combination.
   - (6.) Imaginary universality.

3. *Sophisms of Causation*
   - (1.) False cause.
   - (2.) False effect.
   - (3.) Confounding cause and effect.
   - (4.) Hypothetical causes.
   - (5.) Mistaking the chief cause.
   - (6.) Mistaking the chief effect.
   - (7.) Mistaking the ultimate cause.
   - (8.) Sophistical explanation.
   - (9.) Sophistical induction.
   - (10.) Sophistical proof.
   - (11.) Sophistical relation.
   - (12.) Excluding causes.
   - (13.) Excluding effects.
   - (14.) Imaginary effect.
   - (15.) Imaginary cause.

4. *Sophisms of Probability*
   - (1.) Inferring the probable, including
     - (a.) Inferring hypotheses,
     - (b.) Accumulating probabilities,
     - (c.) Friends' opinions,
     - (d.) One-sided arguments,
     - (e.) Harmonizing conclusions,
     - (f.) Contingent connective,
     - (g.) Incomprehensible connective,
     - (h.) Inconclusive investigation,
     - (i.) Sophistical leap.
   - (2.) Rejecting the Improbable, including
     - (a.) Discordant opinion,
     - (b.) Overlooking the alternative,
     - (c.) Rejecting theories,
     - (d.) Severing probabilities,
     - (e.) Enemies' opinions,
     - (f.) Mortifying proofs,
     - (g.) Imaginary absurdity,
     - (h.) Sophistical distinction.
   - (3.) Varying probability, including
     - (a.) Exaggerating probability,
     - (b.) Diminishing probability,
     - (c.) Exaggerating improbability,
     - (d.) Diminishing improbability.

III. ABERRANCIES.

1. *Aberrancies of Confusion*
    - (1.) Irrelevant illustration.
    - (2.) Indefinite terms.
    - (3.) Irrelevant analogies.
    - (4.) Deciding by character.
    - (5.) Deciding by consequences.
    - (6.) Deciding by motives.
    - (7.) Deciding by appearances.
    - (8.) Irrelevant induction.
    - (9.) Irrelevant empiricism.
    - (10.) Irrelevant objection.
    - (11.) Irrelevant modification.
    - (12.) Homonymous expressions.
    - (13.) Verbal illusion.
    - (14.) Illusive contradiction.
    - (15.) Confounding means and end.

2. *Aberrancies of Appeals to Authority*
    - (1.) Universal belief.
    - (2.) General belief.
    - (3.) Conflicting opinions.
    - (4.) Modern opinions.
    - (5.) Sages' opinions.
    - (6.) Many arguments.
    - (7.) Pretended refutation.
    - (8.) Irrelevant admission.

3. *Aberrancies of Appeals to Desires*
    - (1.) Inferring the agreeable.
    - (2.) Rejecting the disagreeable.

PART IV.
A SPECIAL SURVEY OF THE PRINCIPAL BRANCHES
OF KNOWLEDGE.

CHAPTER XXI.

CLASSIFICATION OF KNOWLEDGE, ACCORDING TO ITS SUBJECTS.

§ 1. SCIENTIFIC KNOWLEDGE.—Knowledge General or Particular.—Definition of Science.—Requisites to a Science.—What determines its Boundaries.—When new Sciences may be formed.—Three Classes of Sciences.—I. Subjects of Mathematics.—Analysis and Geometry.—Subdivisions of each.—Distinction.—II. Subjects of the Physical Sciences.—Their relation to Mathematics.—Their Divisions.—(1) Subjects and Subdivisions of the Mechanical Sciences.—(2) Of the Ethereal.—(3) Of the Organical.—(4) Of the Geographical Sciences.—III. Subjects of the Mental Sciences.—(1) Logic.—(2) Psychology.—(3) Theology.—(4) Morality.—(5) Jurisprudence.—Its Subdivisions.—Distinction.

In regard to its nature, all knowledge is either *general* or *particular*. The former comprises—(1) all cognitions which hold true of a whole class, as "the three angles of a triangle are equal to two right angles"—"the lion is carnivorous"—(2) those which express a fact that does not materially vary for ages, as "the Earth is about 96 millions of miles from the Sun"—"the Nile flows northward into the Mediterranean"—and (3) those which, although they may be essentially particular, affect a whole class, as "Adam was the progenitor of all mankind." The latter class includes—(1) cognitions expressing things which materially change, from age to age, as "the population of London is about three millions"—and (2) those which express particular occurrences, as "Napoleon Bonaparte died in 1821." The former class may be called *scientific*, as general cognitions form the main and essential part of every science, and other facts are employed only for the sake of proof and illustration, or to guide future researches regarding points not yet ascertained.

A *science* is, a systematic body of general truths, relating to an important subject. To render a branch of knowledge a science, it must possess the following characteristics.

1. Its propositions must be general: for a series of particular facts cannot evidently form a science.

2. These propositions must be so numerous as to form a body of knowledge: for a few propositions can no more form a science than a few sticks and stones can form a house.

3. The truths must be real cognitions, and not merely believed, or fallaciously argued, to be such. True science always consists of things which are known by immediate discernment, satisfactory testimony, or conclusive inferences from unobjectionable premises.

4. The truths must be arranged according to some principles of classification, so that a person ignorant of the subject can master the whole, as it is laid down, provided he possesses ordinary faculties, and the requisite preparatory knowledge. The best materials, thrown together without order or connection, are no more a science than a heap of building materials is a house.

5. The subject must be important, either intrinsically or for its bearings on other subjects: else it would be unworthy of notice.

6. The cognitions must be distinguishable from those of other sciences: else there would be no occasion to class them separately.

7. The truths must all relate to one general subject, or to closely kindred subjects: otherwise the materials would be incongruous, whence would result confusion, obscurity, and error.

The boundaries of a science are determined partly by the nature of the subject, and partly by the objects which it aims at effecting. It should embrace every important cognition strictly belonging to its subject, and having no close relation to any other. But, in some cases, cognitions are related, in nearly equal degrees, to two or three sciences: and here we should be guided by the rule that they ought to be placed as the interests of study require. They should be so arranged as to render the attainment and retention of knowledge as easy as possible.

Inductions which do not properly belong to any existing science, and are too few in number to constitute a separate one, should be classed with that to which they bear the closest affinity: and when a sufficient number of them has been established, they should be formed into a new science. So, when some subdivision of a science, which is of a distinct nature, has become extensive, it should be classed as a separate science. But this course is not desirable simply because the subject may have become very extensive: for it would separate from each other things which are all parts of one closely connected whole.

The sciences consist of three classes, the *mathematical*, the *physical* and the *mental*, the subjects and divisions of which are as follows.

I. The mathematical sciences treat of the relations and properties of abstract quantity. As this consists of number and magnitude, Mathematics consist of two corresponding parts, *Analysis* and *Geometry*, the former of which treats of numbers, and the latter, of magnitudes.

Analysis employs various symbols or signs, some of which express quantities—as 1, 2, 3, a, b, x, y, z—and others express the relations of quantities, or the operations to be performed on them—as $+, -, \times, =, <, \sqrt{}$. It is of two kinds, *Special*, where the symbols have all a special signification,—as 1, 2, 3—and *General*, where these have a general signification—as a, b, x, y. The former is commonly termed *Arithmetic*, and the latter, *Algebra*.

Algebra is subdivided into the *Elementary* and the *Higher*. The former employs only *constants*, or symbols and functions which have but one value throughout the processes or operations in which they appear: the latter employs *variables*, or symbols and functions which vary in value in the same expression or operation, while the different values are frequently indefinitely, or, as it is often expressed, infinitely small.

Geometry treats of the four kinds of magnitude, lines, angles, surfaces and solids, the mathematical signification of which differs from the physical. The former includes only the various forms of pure extension, and excludes all conception of material substances. Thus, a physical line has always some breadth and thickness, but a mathematical line has none. Geometrical magnitudes are represented to our apprehension, however, by physical symbols, which greatly assist us in studying the science.

Geometry consists of two parts—*Synthetical* or *Pure*, which treats of its objects directly, without the aid of Analysis—and *Analytical* or *Algebraic*, in which the propositions are investigated by means of Analysis. The latter represents the magnitudes by numerical quantities, while the former represents them directly.

Analytical Geometry is subdivided into *Determinate*,

which discusses problems that admit only of a limited number of solutions, and *Indeterminate*, which treats of problems that admit of an indefinite number of solutions. In the former, the symbols which represent the unknown quantities have only one or a few values, while, in the latter, they vary indefinitely, and consequently it enables us to discuss the general properties of geometrical quantities.

II. The physical sciences treat of physical or material nature, including whatever is directly cognizable by our senses. They sometimes employ mathematical propositions very extensively, in deducing inferences from fundamental principles. Yet they always differ essentially from Mathematics, in treating of material beings, and not of abstract quantity. They may be divided into the three following classes. (1) The *inorganical*, which treat of inorganic nature exclusively. (2) The *organical*, which treat only of organic nature, or of plants and animals. (3) The *geographical*, which treat of both. The first class comprises the *mechanical*, which treat of ponderable matter, and the *ethereal*, which treat of imponderable matter. *Ponderable* matter is that which gravitates, or possesses weight: *imponderable* matter is that which does not gravitate, or possesses no weight.

1. The mechanical sciences treat chiefly of the mechanical properties of matter, the forces dependent on these properties, and the motions or equilibrium which they produce. *Mechanical* properties are those which are directly cognizable by our senses, and at the same time tend to produce motion or rest in the bodies in which they inhere, such as weight, rigidity, elasticity, fluidity, roughness and smoothness. The following are the sciences belonging to this subdivision.

(*a.*) *Mechanic*, which treats of solid bodies, or those whose parts firmly cohere. It consists of three parts—*Static*, which treats of solid bodies in a state of equilibrium—*Dynamic*, which treats of the motions of such bodies, and the forces by which they are produced—and *Mechanism*, which treats of the mechanical properties of solids, the communication and distribution of motion, and the principles of machinery and engineering, irrespectively of the moving forces.(19)

(*b.*) *Hydric*, which treats of liquids, or those bodies whose parts do not cohere, or only very slightly, but yet are not repelled, and move with little mutual friction. It consists of *Hydrostatic*, which treats of liquids in a state of equilibrium—*Hydrodynamic*, which treats of liquids in motion—*Hydromechanism*, which discusses the mechanical properties of liquids, the means of raising, conducting and confining them, and the principles and construction of water works.

(*c.*) *Pneumatic*, which treats of gaseous bodies, or those whose parts, instead of cohering, are mutually repelled from each other.

(*d.*) *Acoustic*, the science of sound and hearing. As air is the ordinary medium of sound, this science is closely connected with the preceding.

(*e.*) *Astronomy*, the science of the heavenly bodies. It is subdivided into *Practical*, which treats of the use of astronomical instruments, and the apparent magnitudes, positions, aspects and motions of the heavenly bodies — *Descriptive*, which discusses their real conditions, motions, sizes, and distances — and *Dynamical*, which investigates the nature and effects of the forces that control their motions, and thence deduces their future positions, so that these can be accurately laid down in tables. Descriptive Astronomy is subdivided into *Heliography*, *Selenography*, *Planetography*, *Cometography* and *Asterography* or *Sidereal Astronomy*, which treat respectively of the Sun, Moon, planets, comets, and fixed stars.

2. The ethereal sciences treat of the properties of the imponderable agents, all of whose phenomena appear to depend on different kinds of undulations, or small waves, propagated through an invisible and very subtile medium termed *ether*, whence I have designated them as above. They consist of the following sciences.

(*a.*) *Optic*, the science of light and vision. It is subdivided into *Physical*, which treats of the nature of light and vision,—and *Mathematical*, which discusses the production of images, and the consequences deducible from the general laws of reflection, refraction and polarization.

(*b.*) *Thermotic*, the science of heat. Like the preceding, it may be subdivided into the *Physical* and the *Mathematical*.

(*c.*) *Electric*, the science of electricity. It may be subdivided into—*Electrostatic*, which treats of electricity in equilibrium—*Electrodynamic*, which treats of electricity in motion—and *Electromechanism*, which investigates the nature and force of the electric fluid, and the peculiar principles of machinery worked by electric forces. Electrodynamic includes *Galvanism*, which treats of galvanic or ordinary electric currents, and *Magnetism*, which treats of magnetic electricity

3. The organical sciences are the following.

(*a.*) *Botany*, the science of plants, or inanimate organisms. It consists of *Phytology*, which treats of the structure and development of plants in general—and *Descriptive* Botany, which describes the peculiar properties of the various subdivisions, classified according to their structure.

(*b.*) *Zoology*, or *Natural History*, which describes the various kinds of animals. It consists of as many subdivisions as there are classes of animals, such as *Mammalogy* (history of mammals) — *Ornithology* (history of birds)—*Herpetology* (history of reptiles)—*Ichthyology* (history of fishes)—*Malacology* (history of mollusks)—*Entomology* (history of insects)—*Helminthology* (history of worms) &c.

(*c.*) *Anatomy*, which describes methodically the various organs of animals, and the specific functions of each. It is subdivided into *General*, which treats of the organs of animals generally—*Comparative*, which discusses the analogous or corresponding parts of the various classes —and *Human*, which gives a description of all the organs of the human frame.

(*d.*) *Physiology*, the science of the phenomena of animated or living beings. Like Anatomy, it is subdivided into *General*, which discusses the structure and functions of animal organisms generally— *Comparative*, which treats of such as are peculiar to the various classes of animals—and *Human* which is confined to man alone, and includes *Hygiene*, the doctrine of the laws of health, and *Ethnology*, which investigates the origin and physical characteristics of the various races of men.

(*e.*) *Pathology*, which treats of the phenomena and laws of morbific and curative agencies. It comprises *Nosology*, which treats of the general nature and phases and the classification of diseases—*Ætiology*, the doctrine of the causes of diseases, including *Toxicology*, which treats of the action of poisons—*Symptomatology* or *Semeiology*, including *Anatomical Pathology*, which treats of the characteristic symptoms of diseases, and of the indications of their future course and results—and *Therapeutic*, which discusses the remedies and proper treatment of diseases, including the nature, operations and results of medicines and other curative agencies. These branches are properly blended with each other, in many instances; and they are all closely connected with the arts of Surgery and Medicine.

4. The geographical sciences treat of the Earth, and such of its natural phenomena as do not belong to any of the preceding sciences. They consist of Geography, Chemistry, Mineralogy and Geology.

(*a.*) *Geography* gives a general description of the Earth, and the scientific phenomena presented by its several parts, including a general account of their animal and vegetable productions. It may be subdivided into *General* or *Mathematical*, which treats of the form, size, motions and density of the Earth, and the means of determining the positions and distances of places on its surface, and delineating them on a sphere or plane— *Geognosy*, which discusses the natural phenomena presented by the land—*Hydrology*, which treats of the oceans, seas, and streams—and *Meteorology*, which gives an account of the general properties of the atmosphere and its phenomena, including aerial, aqueous, and luminous meteors, and the extensive and interesting subjects of winds and climates.

(*b.*) *Chemistry* investigates the nature of the simple substances of which all ponderable objects are composed, and describes both these elements and the compounds formed from them, including the laws of their composition and decomposition. It is subdivided into *Inorganic*, which treats of the simple elements, and such compounds as are found in inorganic substances—and *Organic*, which treats of such compounds as occur only in organic bodies.

(*c.*) *Geology* treats of the rocks or mineral masses that compose the Earth, and of the organic remains contained in them. It may be subdivided into *Petrology*, which describes the various rocks, and investigates their origin —and *Palæontology*, which describes the organic remains found in the rocks, and investigates the structure of the original organisms, and the circumstances in which they existed.

(*d.*) *Mineralogy* discusses the chemical composition, the mechanical properties, the crystaline form, and the situation of solid minerals and crystals. It may be subdivided into *Mineralography*, which treats of the chemical composition and other properties of minerals, apart from their crystaline form—and *Crystalography*, which treats of the crystaline form of minerals.

III. The mental sciences are those which chiefly regard mind, or things imperceptible to the senses, and treat of other matters only as connected with their main subjects. They are included in the following enumeration.

1. *Logic*, which has been already defined.

2. *Psychology*, the science of the human mind. It unfolds the nature of the mental faculties, both intellectual and emotional, and discusses everything regarding them which does not fall within the province of Logic.

3. *Theology* treats of the existence and attributes of the Deity, and the relation in which man stands to him, as an intelligent and immortal being. It consists of *Natural* Theology, which investigates the evidence regarding God and the future destination of man afforded by the works of nature—and *Biblical* Theology, which pursues the same subjects, under the additional light derived from the Sacred Scriptures.

4. *Morality* or *Ethic* inquires into the nature and sanctions of duty in general, and investigates the principles which determine particular duties.

5. *Jurisprudence* is the science of juridical law. It consists of two parts—*Public* Jurisprudence, which investigates the foundation and sanctions of government, and discusses the structure and peculiarities of its various leading forms—and *Private* Jurisprudence, which investigates the subjects and principles of private law. Each part may be subdivided into National and International. *National Public* Jurisprudence treats of the internal government of a nation. It exhibits the functions of the various departments of government, and the proper modes of conducting public affairs. *International Public* Jurisprudence discusses the intercourse between different states, or between individuals and foreign governments. *National Private* Jurisprudence unfolds the juridical rights and duties of private members of a state, in relation to each other. *International Private* Jurisprudence shows the mutual rights and obligations of

private subjects of different governments, in their intercourse with one another.

The science of jurisprudence is to be distinguished from an exposition of the laws of a particular state, two things which differ as much as Hygiene and an account of the mode of living of some particular community. The former is as unchangeable as the nature of man: the latter changes from age to age, or even from year to year, so that it is not a science at all, but merely an art, to enable a man to expound or administer the existing laws of the state.

§ 2. MIXED KNOWLEDGE.—What is meant by this term.—Subdivisions.—Art.—(1) Philology.—Dead and Living Languages.—(2) Ethnography.—Archæology.—(3) Technology.—Three principal kinds of Arts.—Distinctions.

By *mixed knowledge* are understood those branches of which general propositions and particular statements form essential parts. It may be subdivided into *Philology*, which treats of words or language—*Ethnography*, which describes states, communities, and towns—and *Technography*, which discusses the modes of operating in the various arts. An *art* is, a body of rules for effecting some known end, with such directions and explanations as may be requisite for their due application.

1. Philology may be divided into three main parts: (1) *Special* Philology or *Grammar* unfolds the elements and structure of some particular language, and lays down rules for obtaining an adequate knowledge and command of it. It consists of two parts—(*a*) that which treats of *dead* languages, or those which have ceased to be spoken by any community—and (*b*) that which treats of *living* languages, or such as are used by communities, as their ordinary speech. (2) *Comparative* Philology treats of the agreements and diversities exhibited by one or more groups of kindred languages. (3) *General* Philology discusses the structure of language in general, and the conditions requisite in order to its fulfilling the objects of language.

2. Ethnography comprises descriptions of the following subjects. (1) The political divisions and institutions of the various states, including the nature of their governments and laws. (2) Their races of men, populations, and resources. (3) Their religious tenets, and moral

condition. (4) Their social institutions, manners, and customs. (5) Their language, science, and literature. (6) Their arts, manufactures, and commerce. (7) Their cities, towns, and remarkable edifices. (20)

The condition of all these subjects generally varies, more or less, from age to age; and the description may apply either to the present or the past, in which case it is frequently termed *Archæology* or *Antiquities*.

3. Technology comprises the three following classes of arts.

(1) The *mechanical*, which aim at effecting some change in material elements, to minister to the necessities or the comfort and convenience of mankind, such as *Tillage* (including *Agriculture, Horticulture,* and *Arboriculture*), *Pasturage*, or the art of managing flocks and herds, *Metallurgy, Spinning, Weaving,* and *Architecture*.

(2) The *intellectual*, or those whose immediate object is, to solve problems, although some material change is often sought, as a further end. Such are *Government*, the art of executing the public laws—*Statesmanship*, the art of obviating difficulties in the administration of these laws, and improving them where bad or defective—*Diplomacy*, the art of conducting negotiations with foreign governments—*Law*, the art of expounding and administering the private laws—*Rhetoric*, the art of persuasion and the communication of truth—*Education*, the art of training and instructing the young—and *Navigation*, the art of directing the course and finding the position of a ship at sea.

(3) The *emotional*, or such as are designed chiefly to excite agreeable feelings. Of this kind are *Poetry, Music,* and the imitative arts, including *Painting, Photography,* and *Sculpture*. They differ from Rhetoric in making pleasing feelings their chief end, whereas the former uses these only as means towards its main object of producing conviction and action. They also differ from the sciences in making instruction and mental discipline only secondary ends, while the sciences reverse the case, and make these their chief objects.

§ 3. PARTICULAR KNOWLEDGE.—Its most important Subjects.—(1) History.—Its chief divisions.—(2) Chronology.—(3) Biography.—Its principal Divisions.

Exclusive of those which relate to science or art, particular facts of general interest or importance, belong chiefly to History, Chronology, or Biography.

1. *History* is, a narrative or continuous account of past events, regarding communities or classes of mankind, including a view of their more immediate causes and effects. It consists of a great variety of parts, according to the country, time, or subjects embraced. That of the same age and country comprises the following (1) *Ecclesiastical*, or that of religious affairs and morals. (2) That of science, literature, and language. (3) That of legislation and law. (4) That of political and military transactions. (5) That of domestic life and manners. (6) That of the arts, manufactures, and commerce.

2. *Chronology* is, an investigation and exposition of the dates of historical events, for the purpose of their being duly arranged in the order of their occurrence.

3. *Biography* is, an account of the lives and characters of remarkable persons, the circumstances which directly influenced them, and the effects which they immediately produced. It bears much the same relation to individuals that History does to states or classes of men. It comprises the following divisions. (1) *Religious and moral*, containing the lives of persons distinguished for their piety and benevolence. (2) *Scientific and literary*, including the lives of men of science, scholars, and authors. (3) *Professional and Artistic*, containing the lives of persons distinguished for great improvements or skill in the professions and arts. (4) *Political and military*, embracing the lives of celebrated rulers, statesmen, and warriors. (5) *Miscellaneous*, including the lives of persons distinguished in several respects, or for something peculiar or remarkable connected with them, exclusive of any personal excellence or achievement. These various parts may be subdivided according to time and place.

§ 4. TABLE OF THE PRINCIPAL BRANCHES OF KNOWLEDGE.

I. THE SCIENCES.

1. *Mathematics.*

(1.) Analysis
- Arithmetic.
- Algebra
  - Elementary.
  - Higher.

(2.) Geometry
- Synthetical.
- Analytical
  - Determinate.
  - Indeterminate.

2. *The Physical Sciences.*

(1.) Mechanical
- (a.) Mechanic — Static. Dynamic. Mechanism.
- (b.) Hydric — Hydrostatic. Hydrodynamic. Hydromechanism.
- (c.) Pneumatic.
- (d.) Acoustic.
- (e.) Astronomy — Practical. Descriptive. Dynamical.

(2.) Ethereal
- (a.) Optic.
- (b.) Thermotic.
- (c.) Electric — Electrostatic. Electrodynamic, including Galvanism and Magnetism. Electromechanism.

(3.) Organical
- (a.) Botany — Phytology. Descriptive Botany.
- (b.) Zoology — Mammalogy. Ornithology. Herpetology. Ichthyology. Malacology. Entomology. Helminthology, &c.
- (c.) Anatomy — General. Comparative. Human.
- (d.) Physiology — General. Comparative. Human, including Hygiene and Ethnology.
- (e.) Pathology — Nosology. Ætiology, including Toxicology. Symptomatology, including Anatomical Pathology. Therapeutic.

(4.) Geographical
- (a.) Geography — General. Geognosy. Hydrology. Meteorology.
- (b.) Chemistry — Inorganic. Organic.
- (c.) Mineralogy — Mineralography. Crystalography.
- (d.) Geology — Petrology. Palæontology.

3. *Mental Sciences.*

(1.) Logic.

(2.) Psychology.
(3.) Theology { Natural. / Biblical.
(4.) Morality or Ethical Science.
(5.) Jurisprudence { (a.) Public { National. / International. } / (b.) Private { National. / International. } }

II. MIXED KNOWLEDGE.

1. *Philology.*
    (1.) Special { Dead Languages. / Living Languages. }
    (2.) Comparative.
    (3.) General.

2. *Ethnography.*
    (1.) Political Divisions and Institutions.
    (2.) Races of Men, Population, and Resources.
    (3.) Religion and Morals.
    (4.) Social Institutions, Manners, and Customs.
    (5.) Language, Science, and Literature.
    (6.) Arts, Manufactures, and Commerce.
    (7.) Cities, Towns, and Remarkable Edifices.

3. *Technology.*
    (1.) Mechanical Arts.
    (2.) Intellectual Arts.
    (3.) Emotional Arts.

III. PARTICULAR KNOWLEDGE.

1. *History.*
    (1.) Ecclesiastical.
    (2.) Scientific and Literary.
    (3.) Legal.
    (4.) Political and Military.
    (5.) Social.
    (6.) Artistic and Commercial.

2. *Chronology.*

3. *Biography.*
    (1.) Religious and Moral.
    (2.) Scientific and Literary.
    (3.) Professional and Artistic.
    (4.) Political and Military.
    (5.) Miscellaneous.

CHAPTER XXII.

OF MATHEMATICS.

§ 1. PECULIARITIES OF MATHEMATICS.—General Nature and Subjects of Mathematics.—Errors regarding them.—Distinction between a Theorem and a Problem.—Mathematical Definitions.—How the essential nature of Mathematical Quantities is known.—Axioms.—Mathematical Reasoning.—Characteristic of Analysis.—Its relation to Geometry.—General Principle which connects the two.—Unit of Measure.

MATHEMATICS adopt, as primary premises, the existence and some self-evident properties of certain abstract quantities, the essential peculiarities of which are either accurately defined or known by Intuition; and they deduce from these premises, by means of ordinary reasoning, and independently of experience, a long series of connected inferences and conclusions, which express the properties and relations of those quantities, and constitute the body of these sciences. Hence Mathematics are based, in no degree, on experience or testimony, and they are totally independent of every other department of knowledge, while their truths are all necessary, and never contingent, so that they are universally true, independently of time and place.

Some have considered mathematical truths only hypothetical, upon the alleged ground that mathematical quantities do not exist in nature. But the existence of such quantities is as self-evident as that of time and space. Although the most slender wire has some breadth, yet, there is a mathematical line running through it which has none. Such quantities not only exist, but they are wholly independent of the physical objects which may have suggested to us their nature, or in connection with which we frequently consider them.

Another error regarding Mathematics is, that they are based chiefly on observation. It is easily seen that they are quite independent of observation, which shows only what is, at a particular time and place, while these sciences are confined to propositions which are necessarily true, in all times and places.

The distinction between a *theorem*, or something laid down to be proved, and a *problem*, or something proposed to be done, is not fundamental, but only formal, since the thing to be done is effected by means of some theorem, which is proved, though not formally stated. The problems are, in fact, corollaries, or easy inferences from the theorems; and every problem might be stated as a theorem, and the mode of forming the figure, or calculating the quantity, appended as a corollary.

Mathematical definitions are generally suggested by obvious properties of material objects with which we are familiar: but we abstract from the definition something which is present in the object; and we frame it so that we can reason from it with rigid accuracy, irrespectively either of the additional peculiarities or the variations found in physical objects. Such definitions are not only very precise and intelligible, but they give us a knowledge of the essential peculiarity of the thing defined.

The essential nature of the thing defined is known intuitively, by simply considering the definition, although physical symbols facilitate an understanding of the definition. Indeed, in various instances, we know the nature of the quantity without any definition. Thus we know the nature of a straight line and a plane rectilineal angle, as soon as we see two straight lines crossing each other, while all the definitions of them that can be given are only verbal, and help us to understand the nature of the quantities as little as definitions of red or blue.

Mathematical reasoning does not differ from any other reasoning. The intuitive principles employed are termed *axioms:* but they differ, in no respect, from other principles of reasoning, and they are employed in the same way. Several of them, indeed, are formally stated; but this makes no difference in the reasoning, which is independent of any such statements. Not only are these not made, in many treatises on Mathematics; but those which do give them, employ many which they do not state or formally refer to.

The characteristic of Analysis is, that it immediately regards number only; and it indicates magnitudes solely because these are expressible by numbers. The processes employed in Analytical Geometry are not more applicable to Geometry than to other subjects that fulfil the same numerical conditions. Geometrical properties are ascertained by means of Analysis, because they bear certain obvious or ascertainable relations to the numbers expressing the magnitudes of the various parts of its figures. Synthesis is connected with Analysis by the general principle that *the properties and relations of magnitudes correspond to those of the symbols which rightly represent them.* Thus, if one line is three inches long, and another five, the lines are to each other in that ratio.

The letters employed in Analysis do not properly represent quantity, but only the numbers that are assumed to measure them. A unit of measure is always assumed, in Analytical Geometry: otherwise Algebra would be as useless as it is in ordinary arithmetical calculation. But no unit need be expressed, because the results are unaffected by the particular one which is supposed to be employed.

§ 2. USES OF MATHEMATICS.—(1) They form an excellent Mental Discipline.—(2) They are extensively employed, in many Arts and Sciences.—Applications of the various Parts.

1. Mathematics are an excellent means of initiating the mind into habits of close and continuous reasoning. The study is comparatively easy; and it is, in a great measure, free from the prejudices and illusions which accompany several other branches of knowledge, while these sciences abound with long and rigorous chains of reasoning, which must be attentively examined, before any real progress can be made in the study.

2. These sciences are of great utility on account of their numerous applications in the arts and sciences, many of which are wholly dependent on them: and their assistance is more or less requisite, in order to obtain an extensive and accurate knowledge of most of the physical sciences. Their applications in the arts are so frequent and familiar that it can hardly escape the knowledge of anybody; and, even in History, the dates of many occurrences have been settled, and some anachronisms detected, by means of these sciences, where the problems could be solved in no other way.

Ordinary numerical problems can be solved by means of Arithmetic: but those which require us to operate with a quantity, before its value is determined, or to express general properties, can be solved only by means of the concise and general symbols of Algebra. For we cannot, by any other means, either remember the various parts of the process, so as to perform the requisite operations aright, or discuss satisfactorily the relations of the several quantities.

All problems relating to magnitudes require the aid of Geometry, although that of Analysis also is generally more or less requisite.(21)

§ 3. STUDY OF MATHEMATICS.—Most important Points, at the Commencement.—Principal things to be guarded against.—Superior Methods.—Aids of Generalization.—Selection of Propositions.—Working Problems.—Positive and Negative Results.—Signification of isolated Negative Quantities.—Source of Difficulty.—Extended Significations.—Distinction.—Imaginary Quantities.—Advantages of Analytical Geometry.—Uses of the Synthetical.—Means of extending and improving Mathematics.—Effects of the exclusive Study of Mathematics.—How obviated.

The most important point, in commencing the study of any branch of Mathematics, is, to obtain precise and accurate views of the fundamental principles and the import of the symbols, in order to which the nature of the things must be considered, apart from the definitions.

The principal defects against which the student should be on his guard, are, vague or inaccurate definitions and sophistical reasoning. Although mathematical demonstrations profess to establish the conclusions beyond the possibility of any doubt or uncertainty, yet they sometimes fail to do so, more especially in the higher departments, in which they occasionally establish only a probability. Consequently the student should ascertain how far the demonstration extends, and distinguish what is proved from what is not.

Wherever there is a choice, the learner should prefer the most general methods, as they are the most comprehensive and powerful. This will both save time, and enable him to master difficulties which are not easily surmounted without such aids. He should, therefore, obtain a good knowledge of the Higher Analysis, which presents no serious peculiar difficulty, and which is very superior to the comparatively feeble and prolix methods of the older mathematicians.(22)

Attention to the symmetry and regularity of expressions is an important means of discovering the more general laws: and it may be said with truth that a proper series of symbols, and a due arrangement of terms, are equally advantageous to the student and the original investigator.

Those propositions which are either necessary links in the chain of demonstration, or valuable for their applications, are amply sufficient for the purposes both of mental discipline and practice; and the student should beware of spending much time on the endless list of curious problems of no application. Indeed particular problems and examples should occupy little time, and be employed merely as elucidating theorems, which is generally their only real use. The practical problems which occur in the sciences and arts based on Mathematics form the best exercise, either for mental discipline or the attainment of readiness and skill in calculation. Working problems mechanically, by rules whose real character is not known, stultifies the mind almost as much as repeating by rote demonstrations which are not understood.

In Arithmetic, every expression and result is viewed as positive: but, in Algebra, the case is frequently otherwise, because the precise nature of the problems or of the quantities sought may not be clearly understood at the outset, or the problem admits of several symmetrical solutions.

We can always understand the precise signification of an isolated negative quantity, such as $-a$, by remembering the self-evident truth that $+a-a$ (or $a-a$)$=0$: and, therefore, if we know what a or $+a$ signifies, we need have no difficulty in understanding what $-a$ means: for we must interpret it so that, when we prefix $+a$, we are brought to 0, zero, or the starting point, which is well known, and through which we pass, in going from $+$ to $-$ or the reverse.

The learner is apt to be puzzled by assuming that $-a$ denotes a quantity *less than nothing*, whereas it denotes the same amount as $+a$, but taken reversely. Thus, if $+a$ mean such a distance from a certain line measured to the *right*, $-a$ means that distance measured from the same line to the *left*: if the former mean so much measured *upward*, the latter means so much measured *downward*, and so on.

In Algebra $+$ often means simply that the quantity to which it is prefixed is measured in a certain direction, and $-$, that it is measured in a contrary direction. The former sign is prefixed to the quantities which are deemed additive, positive or increasing, being those which were first considered; and the latter is prefixed to those which are deemed subtractive, negative or decreasing, being those which come into view in examining the less obvious aspects of the proposition. But, so far is this from indicating absolute addition or subtraction, that the signs might frequently change places, without any inconvenience. Thus, in Analytical Geometry, distances to the right of the vertical co-ordinate are indicated by $+$, and those to the left by $-$, evidently because we write and read towards the right: but had the science originated with those Asiatics who write and read the other way, they would probably have reversed this use of the signs; and this would be naturally as proper as our method.

The student must not confound *abstract* with *concrete* numbers, nor attempt to apply to the former what properly belongs only to the latter. All abstract numbers are essentially positive: and, therefore, to speak of multiplying one abstract number $-a$ by another abstract number $-b$, is, to heap one absurdity upon another, because abstract numbers less than nothing cannot exist. But when we come to concrete numbers, or those which denote quantities measured in a certain way, or particular kinds of quantity, the case is greatly altered; and we must then apply the properties of abstract numbers only so far as they hold good.

In dealing with concrete numbers, we are not bound to stop at the zero or starting point, because quantity exists equally on both sides of it; and we may proceed on one side as well as the other, only indicating on which side the quantities lie, by prefixing $+$ to the one, and $-$ to the other. In abstract numbers, on the other hand, we cannot pass zero: for, when we reach that, our quantities wholly vanish. Concrete numbers are employed upon certain assumptions made at the outset, the nature of which must be marked and attended to, if we would avoid confusion and error, whereas no such assumptions can be legitimately made in regard to abstract numbers. In order to render demonstrations and processes relating to concrete numbers perspicuous and valid, they must conform to the restrictions imposed by the particular assumptions made.

Even those expressions which might appear to defy

every attempt to assign them a clear and definite signification, become quite intelligible, when their origin and nature are clearly understood. Such are, the square roots of negative quantities, which are termed imaginary or impossible, and which may all be reduced to the form $a\sqrt{-1}$. These arise from some inconsistency or impossibility, involved in the problem whence they originate; and the result shows the nature of the absurdity, and how the problem must be modified, in order to remove it. A consideration of the origin and real import of such expressions shows that they may be employed in operations like others, that they are to be interpreted according to the nature of the problem in which they occur, and that, when they disappear from an equation, the result is not, in the least, vitiated by their having entered into the operation.

Analytical Geometry possesses over the Synthetical the advantages of greater generality and conciseness, and of furnishing means of testing hypotheses and evolving consequences with much greater facility. Hence it is usually preferable even in those cases where the synthetical is applicable, while, in many cases, the latter is quite useless. Yet it furnishes the only means of establishing the principal propositions of Elementary Geometry; and it affords more concise and elegant demonstrations of particular propositions than the other.

As Mathematics are independent of observation, experiment, or testimony, they can be extended and improved chiefly by means of indirect discovery and invention, the only exception being, where a discovery is made accidentally, while performing a process. The higher departments admit of indefinite extension: yet, as the field is already very wide, it is desirable that what is known should be generalized, abridged and elucidated, before we are required to proceed much farther. It sometimes happens, however, that new discoveries totally supersede more tedious and feeble methods previously in use: and we cannot have too many discoveries of that kind; for they abbreviate the sciences, while they render them more powerful in their application. Testing hypotheses analytically forms one of the principal instruments of progress in this direction.

The exclusive study of Mathematics naturally tends to produce credulity, scepticism, one-sided views, and a habit of regarding mere expressions, without paying sufficient attention to what they denote. Moreover, as Mathematics exercise no influence on the feelings or morals, those who study nothing else are, so far as these are concerned, on a level with those who study nothing. But such tendencies are completely obviated by studying the organical and mental sciences. The mechanical and ethereal sciences hardly furnish a sufficient antidote, on account of their close resemblance to Mathematics.

CHAPTER XXIII.

OF THE PHYSICAL SCIENCES.

§ 1. OF THE PHYSICAL SCIENCES IN GENERAL.—Differences between Physics and Mathematics.—Of what the former consist, and on what based.—Directions for Study.—General Uses of Physics.

THIS class of sciences differs essentially from Mathematics in being based on physical realities, instead of abstract quantities and definitions. Their fundamental principles are learned chiefly from experience and testimony; and consequently they extend only to the present system of nature, and their truths, although general, are only contingent, without possessing, to any great extent, the universality of Mathematics. They consist mostly of inductions regarding material objects, and inferences from these inductions. Observations or experiments are generally required to establish the primary inductions; and in many cases, these require to be numerous, and made with much care and skill.

In studying these sciences, we should test the proofs on which the professed primary inductions rest, as some of those stated in several books are false. Besides guarding against vague and erroneous definitions and sophistical reasoning, as in Mathematics, the student must further beware of undue assumptions and fallacies of testimony. The definitions also require a more careful consideration than in Mathematics, as they may fail to express the essential peculiarities of the things defined, or to convey a correct and adequate notion of them, because they are much more complex and difficult to understand than mathematical quantities.

The physical sciences give us accurate, although inadequate, views of nature and its Eternal Ruler. Hence a knowledge of them tends to banish superstition, and to strengthen the foundations of true religion. They also improve the faculty of observation, and teach us to look carefully to our premises, as well as to our inferences, so that they supply the most striking defect of Mathematics, as an instrument of intellectual discipline. They also furnish many truths which are employed in the mental sciences, while their applications in the arts are innumerable.

§ 2. OF THE MECHANICAL SCIENCES.—How these are to be studied.—(1) Mechanic.—Friction, and Strength of Materials.—(2) Hydric.—Inaccurate Definition.—Cohesion and Friction of Liquids.—(3) Pneumatic.—(4) Acoustic.—(5) Astronomy.—Means of ascertaining its Conclusions.—Difficulties.—Law of Gravitation.—Distinction.—Tables.—Eclipses.—Masses.

This class resembles Mathematics so much that most of what was said regarding these, is equally applicable here. But there is occasion for more attention to the fundamental principles, which are much more liable to be fallacious than those of Mathematics, while there is less danger of our time being thrown away on futile investigations. Yet it is possible to miss the most interesting and important parts of the subject, by dwelling too long on others. There are various curious problems in Mechanic, for example, that might not unjustly be classed with the magic squares and Diophantine Analysis of the old mathematicians, and which it would be very improper to study, to the exclusion of Astronomy.

1. Mechanic is based partly on intuitions, and partly on simple observation and experiment, but more on the last than on the second. The force of terrestrial gravity is accurately determined by means of Atwood's falling machine; the velocity of projectiles can be approximately measured by the ballistic pendulum; and the laws of equilibrium are established or confirmed by numerous experiments with weights, levers, pulleys, screws, inclined planes, &c. (23)

Most of the manual arts are, more or less, dependent on Mechanic, which also forms the chief foundation of the other mechanical sciences, and is of frequent application in the organical and geographical sciences.

A very important part of this science is, that which

treats of friction and the strength of materials, things which can be properly determined only by numerous careful experiments. Many lives have been lost, and much property has been destroyed, owing to ignorance or erroneous views of these subjects.

2. Hydric adopts the conclusions of the preceding science, and derives its other primary premises chiefly from observation and experiment. The specific gravity of liquids can be accurately ascertained by means of the hydrometer; the laws of their motions are deduced from observations on streams, and experiments with tubes, artificial canals, and vessels from which water is made to issue through an orifice. Reasoning, apart from experience, is of comparatively little avail in this science: yet, by combining the results of experiments and mathematical principles, much may be learned that is of great use in the construction of pumps, fire-engines, and waterworks, including canals, and also in determining the best forms of ships and their moving apparatus, as well as the best methods of loading and working them.

A liquid has been frequently defined "a substance which communicates pressure equally in every direction." But this definition is inaccurate: for it is applicable only to a fluid whose particles are totally destitute of cohesion, which is not the case with any known liquid, nor especially with water, the principal liquid of which we have any knowledge. The drops pendent from the fingers, after being dipped in it, prove that its particles cohere; and as these are larger as the water becomes colder and denser, it appears that the force of cohesion varies with the temperature.

The fact that all liquids possess friction, has been sometimes overlooked, although this property exerts an extensive influence on their motions. Owing to this, and the preceding erroneous definition, the actual motions of liquids are widely different from what the theories of several writers on this subject indicate: and the influence of cohesion and friction must be accurately ascertained before Hydric can approach perfection.

3. The principles of Pneumatic are based on Hydric, and experiments made with the thermometer, barometer, condenser, pressure-gauge, and eudiometer. It consists mostly of primary facts, and contains few long chains of reasoning. The subject of cohesion cannot cause any

difficulty here, as the particles of all gaseous bodies or aeriform fluids are strongly repelled: but this renders the influence of friction very extensive; and the extreme rapidity with which their density changes, with slight variations of temperature, increases the difficulty of determining their movements, otherwise than by direct observation or experiment.

The great importance of air, as a prime necessary of life, and its extensive agency in many of the most interesting phenomena of physical nature, render this science as important as it is beautiful. To it chiefly belong the properties of steam, and the principles of the steam-engine.

4. Acoustic is based chiefly on Pneumatic, the well-known phenomena of hearing, and experiments made on sonorous substances and the transmission of sounds through various bodies. It is less extensive than the preceding science: yet the importance of hearing, the singular phenomena of vibrating strings and surfaces, and its applications to determine the proper forms and arrangement of rooms for public speaking, render it both interesting and curious. It also derives an extraneous importance from the light which it throws on the ethereal sciences. To it also belong the theory of Music and the vibrations of sounding bodies, which reveal several remarkable peculiarities of solid bodies, not discoverable otherwise.

5. The immensity and grandeur of its subjects vindicate for Astronomy the conspicuous place which it has always held among the physical sciences: for it treats, not only of worlds, but of countless systems of worlds, at distances of which we can form no adequate conception. The astounding magnitude of the Sun, the all-dissolving heat and intolerable glare of light on its surface, its huge swift-rolling waves of fluid fire, and its vast ever-varying Tartarean shades, form most sublime subjects of contemplation. Nor is this science devoid of practical applications. Without its aid, it is impossible to navigate the ocean in safety, to ascertain the positions of places on the Earth, to determine the hour of the day, or the proper time for cultivating the fields, or even to obtain a permanent standard of weights and measures.

This extensive science is based on numerous observations, made with instruments constructed with greatest attainable accuracy, and used with the utmost care, skill and dexterity. Inferences are then drawn from the observations, by means of Mathematics, Dynamic and Optic, a competent knowledge of which is indispensable to the successful study of the heavens.

By means of a transit instrument, and a clock which shows sidereal or star time, the astronomer finds the meridian altitude and time of culmination of such bodies as occupy the same apparent places for any considerable period; and this determines their apparent positions on the celestial sphere. The altitude and azimuth instrument, or an equatorial, enables him to ascertain the apparent position of an object which is not on the meridian.

The telescopes and micrometers which form a part of those instruments, enable the observer to measure very accurately the apparent size of such bodies as exhibit any disc or visible surface, and also to ascertain their appearances. By observing from day to day, or from hour to hour, the places of those objects which change their relative positions, their apparent paths, as well as their varying phases, are exactly determined, whence their real motions are ascertained by forming hypotheses, and testing them by the proper criterions. It is thus found that the phenomena accord only with the supposition that the planets, including the Earth, have a diurnal motion on their axes, and another around the Sun, while the satellites revolve round their respective primaries—that the Sun also revolves on its axis and slowly round a very distant center—and that many, if not all, the fixed stars have a similar motion in space, although they are too remote to ascertain whether they have any diurnal motion.

The distances of the heavenly bodies may be ascertained by measuring the differences in their apparent positions when viewed from two points of the Earth's surface or orbit, whose distances from each other are known, the mean radius of the Earth or of its orbit being adopted as the unit of measure. The process is the same as that by which we ascertain the distance of a terrestrial object, when we take the angles which it forms with a line of known length, measured from the two extremities of the line. In each case, we have a side and the three angles of a triangle, to find the other sides, which is very easily done, from the well-known theorem that the sides are proportional to the sines of the opposite angles. The only difficulty lies in making the measurements with sufficient accuracy.

When once we know the Earth's distance from the Sun, that of any other planet from the Sun is readily ascertained from the law that the squares of the times of their revolutions round the Sun, are proportional to the cubes of their mean distances, a law which is a necessary consequence of those of motion and gravitation.

The distance and apparent or angular magnitude of a heavenly body being known, it requires only the solution of the simplest problem in Trigonometry to determine its real diameter: for Radius is to the sine of half the angle of apparent magnitude, as the distance is to the semidiameter. But this method is inapplicable to the fixed stars, because they show no discs; and their magnitudes can only be conjectured from comparing the amount of light which they give with that which the Sun would impart at the same distance, the brilliancy of a body varying inversely as the square of its distance.

Even this loose method fails in the case of most of the fixed stars, because they are so remote that their distances cannot be ascertained even approximately, as the diameter of the Earth's orbit bears no measurable proportion to their distances. It is also to be observed that no astronomical quantity admits of being measured with as much accuracy as terrestrial magnitudes: a close approximation is all that can generally be attained; and, not unfrequently, we can obtain only a loose approximation, as in the case of the distances of fixed stars.

After ascertaining the distances, magnitudes, and real motions, of the members of the solar system, the laws of the forces that control these motions are ascertained by comparing them with those which determine the motions of a common projectile, and employing the ordinary laws of motion, and the principles of Mathematics, in deducing the consequences necessarily implied in the phenomena. This process was first applied to the Moon, then extended to the Earth's motions around the Sun, and afterwards to the other planets.

Thus was established the law of gravitation, which accounts for all the motions of those bodies, and enables the astronomer to determine the perturbations of the planets, or the irregularities in their movements caused by their mutual influence on each other, so that their future positions can be accurately ascertained, and laid down in tables. It is observable, however, that the law only expresses the fact of the existence of a certain force, and the modes in which its influence varies: it gives no explanation whatever of the origin or real nature of the force.

When once accurate tables have been constructed, the future places and phases of the bodies can be found with comparatively little labor, as we are furnished with all the principal elements of the calculation, ready for use. Thus, if we wish to ascertain the number and characteristics of the eclipses that will occur next year, we first ascertain the positions of the Sun and Moon at its commencement. Then from the motions of these bodies in their orbits, as given by the tables, we ascertain how often the Moon will pass through the Earth's shadow, and the Moon's shadow strike the Earth, till the end of the year. The former will be the number of lunar, and the latter, of solar, eclipses.

The ordinary rules of Trigonometry enable us to determine the times and peculiarities of the various eclipses, from knowing the relative positions of the Earth and the Moon, when the phenomena occur. Those peculiarities depend on the distances of the Earth from the Sun and Moon, and the latter's position in its orbit. Thus, in calculating a solar eclipse, the size of the Moon's shadow, at the Earth's distance, is wholly determined by their respective distances from the Sun; and the particular parts of the Earth which the shadow will cross, can be ascertained from the position of the Moon's node, at the time of conjunction, while the duration of the eclipse is readily found from knowing the velocity of the Moon in its orbit.

Knowing the Earth's distance from the Sun and Moon, and the latter's course and velocity in its orbit, the laws of motion and gravitation enable us to determine the proportion which the Earth's mass or weight bears to the Sun's, the latter being adopted as the unit of measure. The problem may be otherwise solved by comparing the influence of the Earth on a falling body with that of the Sun on the Earth. In the case of those planets which are accompanied by satellites, their masses are ascertained by comparing their influence on these with that of the Sun, either on themselves or on their satellites. The masses of those planets which have no satellite, are ascertained approximately, by observing the perturbing influence which they produce on some other body, whether planet or comet. The masses of the satellites are ascertained from the perturbations in their motions produced either by the Sun or by each other. That of the Moon may be farther ascertained from its effects in changing the Earth's axis, and its comparative influence on the tides.(24)

§ 3. OF THE ETHEREAL SCIENCES.—General Character of these Sciences.—Their special Use as a Study.—(1) Optic.—Origin and probable nature of Light.—(2) Thermotic.—Ultimate Source of Heat.—(3) Electric.—Origin of Electricity.—Electrostatic.—Galvanism. — Magnetism. — Electromechanism.—Probable Nature of Heat and Electricity.—Connection of these Sciences.

These sciences resemble the mechanical in deducing numerous inferences from the primary facts, by means of Mathematics: but they differ from those, as well as from all the other physical sciences, in their subjects being of a more subtile nature, and less apprehensible by our senses. Consequently they are based much more on experiment than on simple observation; and great care is requisite in establishing the primary inductions. Owing to their resemblance and close connection, they extensively aid and illustrate each other: and the true source and nature of their phenomena require to be understood, in order to surmount some of the difficulties which they present to the investigator.

The study of these sciences is an important means of enabling us to form accurate views of nature; and a knowledge of them is indispensable to right conceptions of the general structure and laws of the material creation. An exclusive attention to the mathematical and mechanical sciences tends to make us estimate everything by its mass, firmness or momentum. But a knowledge of the ethereal sciences reveals to us that the forces which immediately control nature reside in those things which are least perceptible to our senses, and apparently the most feeble and inefficacious. We are thus led to understand how invisible and impalpable mind is the ultimate source of all power.

1. A knowledge of Optic is requisite in order to the proper construction of optical instruments, while it is an

important means of enabling us to preserve or improve the sight. The wide range of vision, and its necessity to our very existence, consequently render this science of the utmost importance, while the properties of light and color are such as to render it one of the most beautiful and interesting of all the physical sciences. Moreover, by transmitting polarized light through transparent bodies, we can ascertain peculiarities in their structure which can be discovered by no other means.

The composition of white light is ascertained by simply passing it through a common glass prism; and other fundamental principles are established with the aid of reflecting, refracting, polarizing, and discolorable substances, of various forms and kinds. The intensity or amount of light is measured approximately, by means of a photometer; and its velocity is ascertained both from astronomical phenomena and from direct experiment.

Light appears to consist of undulations, or small waves, generated by a rapid, vibratory motion of the luminous body's atoms, and propagated through ether, in every direction. The theory that it consists of solid, unconnected particles, which fly from the luminous center, is attended with various insuperable difficulties.

Not only do the rays of light sometimes interfere, so as to destroy each other, but it is impossible that the Sun, for example, could have been throwing off such particles, for any length of time, without being dissipated through the boundless void. It is no answer to this objection to say that the particles are extremely small: for they must possess some bulk; and the smaller their size, the greater must be their number, since they exist simultaneously, at every point. Nor could such rays move in every direction, through every point, without interfering with each other to a much greater extent than they really do.

It is also impossible that the solar rays could fill every point, as they receded from the Sun, owing to the rapidly increasing surfaces of the spheres traversed, unless we adopt the absurd hypothesis that they regularly divided and subdivided as they advanced. Nor could such particles permeate a great thickness of a hard and dense substance, like glass, as light actually does. It is also incredible that a black substance should receive an indefinite number of such particles, without furnishing the least indication that any such thing ever penetrated it, either by an increase of weight or otherwise. Moreover the tremendous momentum of such bodies, moving with a velocity of nearly two hundred thousand miles in a second, would have destroyed our sight, at the very first impulse.

With regard to what are termed the chemical, or discoloring rays of solar light, I am aware that some consider them totally different from luminous rays: but there are some indications that they are only luminous rays, too small to affect our vision, under ordinary circumstances.

2. Thermotic is a most important science, on account of the controlling influence of heat throughout the organic creation. The changes of the seasons strikingly exhibit this influence on vegetation; and it is equally extensive, though perhaps less obvious, in relation to living beings. Thus, if the temperature of warm-blooded animals is reduced a little below the natural standard, they speedily die, and a like result follows when their temperature is raised a little above the normal heat. A variation from the proper temperature of the different parts of the body is also a fruitful cause of disease and death. The extensive applications of heat in the arts also invest the subject of its production with great interest, and render a knowledge of its sources and laws of great importance.

The most interesting questions in this science are, those which regard the origin, distribution, and influence of heat. Its immediate sources are various: but they all seem to be ultimately referable, like light, to a vibratory motion of the atoms of the body whence it proceeds. This may be produced by chemical or electric action, and by friction, or sudden and violent compression. Heat always tends to dissipate rapidly through space; and the circumstances which accelerate or retard its escape deserve a careful consideration. Myriads have lost their lives owing to their ignorance on this subject.

Several of the laws of heat are learned by simple observation; and others are ascertained by means of the thermometer, pyrometer, calorimeter, and various reflecting, refracting, radiating, conducting, convecting and polarizing substances.

3. The origin of electricity appears to be referable to the same ultimate cause as that of heat. It is generally

produced by friction or chemical action: but it is equally caused by heat. Its laws are mostly established by means of experiments, in which a great variety of instruments is used, including the common electric machine, the galvanic battery, the electroscope, electrometer, electrophorus, condenser, proof plane, and galvanometer, the magnet, and electromagnetic apparatus, of various kinds.

Electrostatic claims attention on account of the powerful agency of statical electricity, both in nature and in the arts. A knowledge of it may be said to have divested lightning of its terrors, and, from being a direful enemy, converted it into a useful servant. It not only explains the origin of lightning and thunder, but points out the mode in which we may guard against danger from the former, and how statical electricity may be usefully employed in the arts.

Electrogalvanism, whose phenomena are produced by continuous currents of electricity, is of much importance on account of its extensive application, both in the arts and in many scientific experiments. In chemical analysis, for example, it affords one of the most powerful means of decomposing substances. It also furnishes the means of accurately measuring very small portions of time, and determining the precise instant of an occurrence, by means of the chronoscope and the electric clock, with their proper appendages.

The phenomena of Electromagnetism arise wholly from electric currents: and terrestrial magnetism appears to consist of currents caused by the varying action of the Sun, during the diurnal revolution, whence its phenomena exhibit the same changes which mark the weather and the seasons.

The value of the magnet in Navigation renders this branch of Electric one of great importance, while it derives additional interest from its application in the treatment of many diseases.

Electromechanism is an important branch, on account of its various applications in the arts. The invention of the electro-magnetic telegraph forms an era in the history of human intercourse; and the application of electric currents as a source of motive power becomes invested with great interest when we consider that metals must sometime supersede coal, as a source of motive power.

Q

As heat is produced, transmitted, reflected, refracted, and polarized, precisely like light, and travels with nearly the same velocity, we may infer that it, also, consists of undulations of ether: and the fact that we can obtain an indefinite amount of heat from a small body, by means of friction, leads to the same conclusion.

Again, the close connection and marked resemblances between heat and electricity, indicate that the latter consists of ethereal undulations, as well as the former.

Hence we come to the conclusion that the imponderable agents are all essentially alike, and that their different phenomena arise from different kinds of undulations. This conclusion is strengthened by the close resemblance which the various phenomena exhibit, and the fact that one frequently produces another. The imponderability of all of them leads to the same conclusion; for, as the weight is proportional to the mass, in all ponderable bodies, it follows that we cannot add solid matter indefinitely to a body without appreciably increasing its weight. The conclusion is further corroborated by the fact that the agents all follow the same law of intensity, which is uniformly as the inverse square of the distance.(25)

§ 4. OF THE ORGANICAL SCIENCES.—Characteristics and Foundations of this Class.—Directions for Study.—Three important Principles.— Monsters.—(1) Botany.— (2) Zoology. — Caution.— (3) Anatomy.—(4) Physiology.—Hygiene.—Ethnology.—(5) Pathology.—Proper Bases of the Medical Arts.—Two guiding Principles.—Distinction.—Foundations of Pathology.

These sciences are distinguished from the preceding by the general absence of long chains of reasoning, as they consist mostly of facts learned directly from observations or experiments, which are frequently aided by the microscope, electric action, chemical re-agents, and so forth. A few observations, made anywhere, may serve for the foundations of most of the inorganical sciences: and when these foundations are once well laid, the rest consists chiefly of simple deductions. But those now under consideration sometimes require that the whole world should be ransacked, in order to furnish their materials, as every country offers something peculiar; and long deductions here give place to extensive observations, and a careful examination of testimonies.

In studying these sciences, we should observe for ourselves, as extensively as our circumstances will permit. This will often enable us to obtain more accurate and lively views of the things discussed than mere descriptions afford, and possibly to correct the errors or extend the statements made in books. We must be constantly on our guard against erroneous testimonies and false theories, which have been very prevalent in this department of knowledge: and we should distinguish the ascertained facts from the plausible, but erroneous hypotheses with which they are sometimes blended. Mathematics are much less applicable in these sciences than in the inorganical, so that they may be studied to advantage by persons who possess only a very limited knowledge of Mathematics.

The principles of classification come frequently into requisition, and a proper application of them is generally a matter of great importance, in all these sciences. They are, therefore, well adapted for investigation by those who dislike long chains of reasoning, and delight in observing and classifying, while they furnish a field which the united labors of many generations will not exhaust.

In examining the structure of minute parts, much aid is derived from the microscope, which enlarges the power of vision to such an extent that it may be said to reveal new worlds: and although these are as diminutive as those unfolded by the telescope are extensive, yet they possess over the latter the great advantage of being completely within our reach.

In all the organical sciences, the following inductive principles guide and assist investigation.

(*a.*) *Every organ performs one or more functions.* Those of some organs, such as the eye and the ear, are easily ascertained; and although those of some parts are not so readily discovered, they may be found out by continued and accurate investigation. Some organs, like the human tongue, perform different functions; and, therefore, when we discover one, we must not infer that it is the sole function.

(*b.*) *One part of an organic being harmonizes with every other part; and the whole fabric is adapted to some particular mode of existence.* Hence an inspection of one part, as a tooth or a foot, may enable us to ascertain the form of another, and the being's general structure and mode of existence.

(c.) *The corresponding organs of similar organic beings perform similar functions.* Hence when those of one organ are known, those of the corresponding organ may be ascertained with little difficulty. The similarity of functions is generally proportional to that of the beings compared: and where this is very close, the functions are usually identical.

These principles do not, of course, apply to monsters or malformations: but the existence of such beings is so rare as not to detract materially either from their value or their general accuracy. Such anomalies are extremely few, compared with the wide extent of organic nature; and even these arise from some violation of organic laws.

1. Besides the interest which it derives from the immense variety, beauty and magnificence of the vegetable kingdom, Botany claims attention on account of its bearings on the necessaries and conveniences of life. Not only our food, but also our clothing, is all derived from vegetation, either directly or indirectly. So is a great number of the most valuable medicines; and the various uses of timber are well known. By describing the qualities and peculiarities of the various kinds of plants, Botany assists us in determining their character, the best modes of cultivating such as are useful or ornamental, and where any required species may be procured.

The facts of this science are learned from observing the plants in their native regions, hot-houses or botanical gardens, or from examining preserved specimens, and from inspecting their minute structure with the microscope, while the component elements are ascertained by chemical analysis.(26)

2. The boundaries between Botany and Zoology are not very obvious, as some animals are so low in the scale that they cannot easily be distinguished from plants; but every being possessing thought is an animal; and this criterion is easily applied, as thought is always accompanied with the power of voluntary motion.

Zoology describes the various kinds of animals, classified according to their organization, including their appearance, general structure, food, instincts, habits, and localities: and a knowledge of it is indispensable to right views of the economy of nature. To the interests of Botany, it superadds that which is derived from sympathy with living beings possessing several faculties in common with ourselves, although they are limited to a comparatively narrow range. It facilitates a discovery of the characteristics of a hitherto unknown species, and enables us to determine the proper mode of treating the lower animals, many of which have often been destroyed through ignorance, when they ought to have been carefully preserved.

The materials of this very extensive science are collected from direct personal observation, the reports of travelers, and the descriptions of naturalists, who lived in the native regions of the animals described, or had access to some fair specimens, such as are sometimes found in zoological gardens. But the proper criterions of testimony should be carefully applied to such descriptions, as several of them are blended with fables or misrepresentations.

3. Anatomy furnishes a striking illustration of the consummate skill with which everything throughout the organic creation has been formed; and it also demands attention on account of its connection with Physiology, Surgery and Medicine. It is based on careful dissection, aided by the microscope, mercurial injections, diluted acids or alkalies, &c. Human Anatomy exhibits a correct view of the most complex and finished work of God that our eyes can behold. General and Comparative Anatomy shows the immense variety which prevails in animal structures, and the numerous modifications found in the same organ, adapting it to various circumstances. It also enables us to classify animals according to their real affinities, which often differ widely from apparent resemblances.

4. Physiology is of the utmost importance on account of its connection with health, the laws of which seldom receive due attention from those who are ignorant of this science. Mere precepts have comparatively little influence on such persons, as they readily evade what they are very unwilling to believe. But when they understand the nature of their physical organization, and the serious injuries which it suffers from various agents and practices deemed harmless by the ignorant, prejudice can no longer resist the inference, and the rules of health soon change their habits for the better. Reckless exposure to deleterious agencies, and habitual indulgence

in health-destroying practices, gradually disappear, till they vanish: and the delusion of supposing that the effects of such conduct can be obviated or speedily removed, is seen in its true light. This science also shows the extreme minuteness and perfect finish of the elementary parts of every tissue. It discloses fibres compared with which the most slender filament that we can see, is a cable, and globules of which it would require millions to form the size of a pea.

The truths of Physiology are established by means of simple observations, and experiments of various kinds, in which much use is made of the microscope. It also derives extensive aid from Chemistry, Botany, Zoology and Anatomy, with the last of which it is very closely connected.

Hygiene, or that part of Human Physiology which discusses the laws of health, establishes these on the structure and functions of the various parts, and the knowledge derived from observation or experiment regarding the modes in which they are affected by particular agents or practices. Health and disease depend on causes which act as uniformly as any agent in the inanimate creation; and, consequently, wherever men fulfil the conditions on which it depends, they will enjoy health, while they will be the victims of disease and premature death as long as they violate these conditions. A knowledge of this part of the science is a most important means, not only of preserving health, but also of curing disease, which can seldom be permanently and thoroughly removed, unless we know the conditions on which health depends.

Ethnology, which discusses the origin and extent of the physical peculiarities that distinguish the various races of men, forms another interesting branch of Human Physiology. Detailed accounts of these peculiarities, as they are exhibited in different countries, properly belongs to Ethnography.

5. The leading principles of Pathology are highly useful in securing obedience to the laws of health, by disclosing the various lamentable consequences which inevitably result from their continued violation. This science also forms a proper sequel to Human Anatomy and Physiology, in a course of medical study. Chemistry and the organical sciences form the only rational and scientific bases of the healing-arts, which are as apt to in-

jure as to benefit while they are merely empirical, and which improve in exact proportion to the advancement in those sciences.

The pathologist is guided by the induction that *morbid action obeys fixed laws, like healthy functions:* and, consequently, he can ascertain the causes of such actions, and the modes in which they may be prevented or stopped, and their bad effects counteracted and removed. He is also aided by the induction that *the reactions of the organism, caused by morbific agencies, tend to produce some beneficial end:* and it is an important object to ascertain what that end is, and how it can be best secured. He should distinguish the sanitary action which occurs in diseased parts from the injurious effects of poisonous or deleterious agencies, which are of a very different character, although they are very apt to be confounded with the former, because they spring from the same causes, and appear simultaneously. While the latter should always be stopped and counteracted with as little delay as circumstances, and a due regard to other hygienic and pathological laws, will permit, the former ought generally to be fostered and aided by all proper means. For their general tendency is, to rectify some derangement or disorder in the organism.

This important and extensive science is mostly established like Anatomy and Physiology: but it has a wider range, and derives more aid from the inorganical and geographical sciences, because disease is more varied in its origin, phases and progress than healthy action. It is beset with difficulties, whenever we pass beyond the immediate phenomena; and the theories with which it abounds, furnish a wide field of investigation.

§ 5. OF THE GEOGRAPHICAL SCIENCES.—General Character of this Class.—Directions for Study.—(1) Geography.—(a) General Geography.—(b) Geognosy.—(c) Hydrology.—(d) Meteorology.—(2) Chemistry.—(3) Mineralogy.—(4) Geology.

These sciences treat both of inorganic and organic nature; and consequently they resemble partly the mechanical and partly the organical. Although their subjects are less vast than those of Astronomy, yet they are not only much more within the reach of our observation, but also more on a level with our faculties; and consequently their phenomena are apt to affect us more powerfully than the vaster scenes of that science. They consist of an immense mass of primary facts, based on observations made in different regions of the globe, and of numerous inferences from these facts. Hence, in studying them, we should guard against fallacies both of testimony and reasoning. As personal observation enables us to test only a very small portion of them, the criterions of testimony should be applied with great care and freedom. In other respects, the remarks already made regarding the study of the other physical sciences, mostly apply to them. As a whole, they offer a very extensive field for future discovery.

1. The immense variety and importance of its details, and their numerous applications, render Geography a very interesting subject of study; and there is hardly any class of persons to whom a knowledge of it is not beneficial.

(*a.*) General Geography forms the basis of all geographical knowledge, and is of the greatest use in the important arts of Surveying and Navigation. It is founded wholly on Mathematics and Astronomy. The exact positions of places on the Earth's surface are determined by finding their latitudes and longitudes, which may be done in various ways.

As the elevation of the celestial pole, or immovable point in the heavens, is necessarily equal to the latitude of the place of observation, this may be found by taking the greatest and least altitude of a circumpolar star, or one which never sets: then half the amount of these gives the latitude, because the greatest elevation is as much higher as the least is lower than the pole. Another simple means of finding latitudes is, by taking the meridian altitude of the sun or a fixed star. As the zenith distance of the equinoctial is equal to the latitude, and the declinations, or distances from that line, are given in tables, a single meridian observation determines the latitude.

The longitude of a place is found by ascertaining the difference in time between it and the first meridian, or that from which longitude is reckoned, and then converting this into degrees, from the proportion of 15 degrees to an hour, every parallel revolving through 360 degrees in 24 hours. One of the most simple modes of finding the difference of time is, to mark the hour on a chronometer which shows the time at the first meridian, when it is noon at the place of observation. This is known by the Sun's being then on the meridian, or at the highest. A more exact, but less simple and easy method is, to find the time at the first meridian from the Moon's distance from some fixed star, with the aid of tables, which give the distances of that luminary from many of the fixed stars, at short intervals. The time of the place of observation may be found from one or two altitudes of a known fixed star, with the aid of tables. This method possesses the great advantage of being independent of time-keepers. Another simple and very accurate method is, to determine the difference of time by means of telegraphic signals; and the use of the electromagnetic telegraph renders this method applicable to all places connected by it.

When the positions of the principal points have been thus ascertained, those of others are found by means of geodetical surveys, conducted on the principles of spherical trigonometry, while minor details are found by means of plane trigonometry and constructions on paper. Maps, globes, and charts are then constructed, by applying the principles of perspective, and using the requisite mathematical instruments.

That the Earth is either a perfect sphere or a spheroid, is evident from the form of its shadow on the Moon, during lunar eclipses, and from the distance of the visible horizon at sea being always nearly proportional to the elevation of the observer's eye. Its magnitude may be easily determined approximately, by measuring the angular depression of the farthest visible part of the ocean from a point of known altitude above the sea level, as we have thus the three angles and a side of a triangle to find the other sides.

A more accurate method of finding the form and dimensions of the Earth, is, to measure an arc of the meridian and of the parallels of latitude, in different parts of the world. In order to this, the difference of latitude and longitude between two places is accurately ascertained; and then the distances between the two meridians and parallels which pass through them are carefully measured: for it is easily seen that these distances bear the same proportion to the circumference of the Earth that the angular distances bear to 360 degrees, or an entire angular circumference. It is thus found that the Earth is a prolate spheroid, whose respective diameters are about 7899 and 7925 miles: in other words, its form and magnitude are those of a solid generated by the revolution of an ellipse about its conjugate or shorter diameter, this being of the former, and the transverse, or longer, diameter of the latter length.

The absolute mass, or weight, of the Earth is ascertained, approximately, in various ways. One is, by determining the influence of a mountain, in deflecting a pendulum from the perpendicular: but as the weight of the mountain and the distance of its center of gravity cannot be ascertained with any great degree of accuracy, this method gives only a rough approximation. A more accurate method is, to determine the comparative force of gravity, at different distances from the Earth's center, in the same vertical line, from the differences in the vibrations of a seconds pendulum, at the two points, whence the absolute gravity of the Earth is easily ascertained.

A third method is, to compare the torsion produced by two large balls of lead on two very small ones, fixed on the ends of a very light rod, suspended by a very slender wire, and brought quite close to the large balls. The weight requisite to produce the same amount of torsion is easily ascertained; and thus the proportion which the Earth's weight bears to the balls can be ascertained, to a great degree of accuracy. The extreme minuteness of the balls is compensated by their close proximity to such a degree as to render their influence measurable, the gravitating force being less as the square of the distance of the body's center of gravity becomes greater. By such means it is found that the Earth is about $5\frac{1}{2}$ times as heavy as a globe of water of its own size, while the Sun's specific gravity is only about $1\frac{1}{3}$.

(*b*.) Geognosy lies at the foundation of several of the physical sciences, while it is indispensable to commerce and intercourse between remote nations. A knowledge of it may also enable the unfortunate and distressed to remove to a country where they can materially better their condition. It is also of great use in the study of Ethnography and History, which cannot be rightly understood without its aid. It is founded on the observations of naturalists, travelers, and mariners, made in the various quarters of the world.

(c.) Hydrology is of the utmost use in Navigation, and furnishes means of extending our knowledge of Meteorology and Geology. Like the preceding subdivision, it is based chiefly on observation: but the causes of the tides are investigated by means of Dynamic and Astronomy. These furnish a striking proof of the Earth's connection with other members of the solar system.

(d.) Meteorology is of great importance in Hygiene, Therapeutic, Tillage, and Navigation. A knowledge of the atmospheric currents, combined with that of oceanic currents and the tides, enables the mariner to navigate the waters with a degree of speed and safety otherwise unattainable, while the electromagnetic telegraph renders the knowledge thus acquired of the utmost practical utility, by warning vessels of the coming storm, before it arrives. So, an acquaintance with the laws of earthquakes furnishes various means of preventing their calamitous effects, by giving intimations of their future action. The materials of this extensive subdivision are derived mostly from direct observations, made with the aid of various instruments, including the anemometer, which determines the force and direction of the winds, the hygrometer, the thermometer, the barometer, the pluviameter or rain-gauge, the tide-gauge, and the seismometer, which measures the force of earthquake shocks.

2. Chemistry affords an insight into the recondite structure and operations of nature, by disclosing the properties of its primary elements, and the laws of their combination and separation. It also bears an important relation to several other sciences, and to many of the arts. It enables the Physiologist to determine the composition of secretions and tissues; it unfolds to the Geologist the composition of rocks; and it gives the Botanist similar knowlege regarding plants. It is likewise of much importance in establishing proper rules of dietetics; and it contributes to the improvement of tillage, by analysing soils, and showing what elements they possess or lack, and how they require to be cultivated, in order to yield abundant produce.

The phenomena of Chemistry should be distinguished from their hypothetical explanations: the character of the latter does not affect the truths of the former, although their causes should continue as unknown as the cause of the yellow color of gold, or the sweet taste of sugar.

This science is based chiefly on experiments, performed by means of heat, electricity, light, and the mutual action of substances on each other. These experiments frequently require much ingenuity and manual dexterity, and a careful application of the most exact and reliable methods of determining quantity, with very delicate and accurate instruments. Many substances are of a complex structure; and it is often no easy task to determine whether a compound, obtained from analysis, was not formed during the process, by a play of the affinities. Another difficulty arises from the extremely small amount of some constituent element, which yet greatly affects the character of the compound.

3. Mineralogy is interesting on account of its applications in mining, metallurgy, and the construction of optical instruments. It also exhibits the beauty and regularity of inorganic nature, in a peculiar manner, as every compound exhibits a characteristic geometrical form. This enables the mineralogist to determine the character of a mineral very easily, where a person ignorant of the science would be readily misled or deceived. He is in no danger of paying the price of a diamond for a piece of quartz crystal, of comparatively no value.

The composition of minerals is determined by chemical analysis; and the exact form of crystals may be ascertained by the goniometer, while Geology and Geography aid us in ascertaining where the various substances may be found.

4. From shapeless masses of rocks and dirt, and the broken, and frequently petrified, organic remains imbedded in them, Geology unfolds the history of the Earth, and the numerous wonders which it exhibits. The origin of seas and oceans, of the lofty mountains and the lowly vales, with their endless diversities, forms a subject in which all intelligent minds must feel much interest; and this is increased when we discover the agencies by which they were produced, and the organisms by which the Earth was formerly tenanted.

This science is also very important in an economical point of view. It teaches us where to look for metallic ores, coal, building and earthenware materials, subterraneous supplies of water, mineral manures, the precious metals, and gems. Much money and labor have been lost in searching and digging for such things, where

Geology would inform us that they were not to be found, while they have lain hid, for many ages, in favorable situations, where it would have readily disclosed their existence. It also assists us in determining the character and capabilities of soils, while the palæontological portion throws light on several difficulties in the organical sciences.

The position, structure, and size of rocks, and of the organic remains found in them, are ascertained from inpecting the face of the country in different regions, and the sections laid bare by flowing water, subterraneous agencies, or works of art. The investigation is much facilitated by the edges of the rocks having been very frequently exposed to view, along the surface of the ground, because they were tilted up from their first position, either by the parts beneath giving way or by an upheaving force having been applied from below. The dip and strike of rocks are most easily measured with the compass and the clinometer, which is a simple modification of the quadrant of altitude. Then, by measuring the horizontal thickness, at right angles to the strike, the real thickness is found by simply solving a right-angled triangle. The nature of the forces which produced the phenomena, and of the organic remains found in the rocks, is ascertained by calling in the aid of Mathematics, and the other physical sciences, as may be requisite.

There is thus furnished a wide field for inferences, regarding the circumstances under which those phenomena originated, and the actual condition of the Earth, during the various periods. The supposition that the Creator employed an endless series of miracles, in order to entrap us into false inferences, is manifestly absurd ; and, therefore, geological phenomena are to be attributed to those agencies which are known to produce similar appearances at present, so far as these are adequate. When they are not, recourse must be had to the usual modes of ascertaining causes, in similar cases.(27)

CHAPTER XXIV.

OF THE MENTAL SCIENCES.

§ 1. OF THE MENTAL SCIENCES IN GENERAL.—General Character of this Class.—Causes of their comparatively slow Progress.—Their principal Foundations, and Importance.—Why frequently undervalued.—Directions for Study.

THE mental sciences exhibit a wide range of primary facts, consisting partly of necessary and partly of contingent truths, while they abound with inferences. Many of their propositions are known by direct discernment, while others are established by an analysis of thought and motives, a process which is never employed in other sciences. Owing to the extreme rapidity of thought, to our performing several mental operations simultaneously, and to one thought's readily leading to another, this analysis is generally attended with difficulty; and it is frequently necessary to have recourse to indirect means of effecting our object. Mere external observation is of little avail, in surmounting the chief difficulties. If to this we add that, owing to the manifest bearings of the points in question on our future condition, strong prejudices frequently interfere, we can easily understand why these sciences have made such slow progress, compared with some others.

The principal foundations of the mental sciences are, immediate discernments, the conclusions of the physical sciences, and inductions from History or Biography. They are of the utmost importance, since they influence our condition much more extensively than any other division of knowledge, and their principles are requisite to regulate our daily thoughts and conduct. They also greatly influence the condition of all other branches of knowledge. For the mind must receive that thorough discipline and those external advantages which only the proper study and general diffusion of these sciences can give, before it is in a condition to enlarge and improve to the utmost the other fields of human investigation, or apply them to their true objects.

The various departments of knowledge are so connected that an improvement in one favors a corresponding advance in others; and this is especially the case with that now under consideration, on account of its extensive influence in improving the mind, removing restraints on proper investigation, and furnishing the various other requisites to success. It teaches mankind the true value of knowledge, and at the same time disposes them to furnish the means for its successful cultivation, and points out how this object is to be secured. They also inure us to those habits of close attention to mental phenomena and recondite, but important, distinctions, which are requisite to a proper solution of many practical problems of the utmost moment.

These sciences also teach us the difficulties attending many of their applications, and thus obviate the dogmatism into which those are incessantly falling who are ignorant of these difficulties. A man totally ignorant of a mechanical art is apt to speak hesitatingly and distrustfully regarding it, whereas those who are equally ignorant of the mental sciences, are apt to speak very confidently regarding points which require for their solution an extensive knowledge and careful application of their principles; and thus they often utter gross absurdities, without ever suspecting the error.

The mental sciences treat of subjects which do not immediately affect the senses like those of Physics, and they are founded, in a great measure, on truths or phenomena familiar to all, whence we are apt to think that we know them sufficiently without study, or that they are of little consequence. Yet they require diligent study, in order to be well understood, while our views of their subjects mold our whole character, and consequently determine our general conduct through life, and our whole future condition, so that, if these views should be radically erroneous, correct and extensive knowledge on other subjects will only render a person more powerful for evil, both to himself and to others.

In studying the mental sciences, we should beware of adopting alleged or supposed, in place of real, intuitions or inductions, and guard particularly against fallacious reasoning, masked by ambiguous or obscure language. For this purpose, great attention should be given to the principal terms, which cannot frequently be rightly understood without closely considering our own thoughts, and comparing them with the definitions, because the subjects are not palpable to our senses, and therefore obscurity and vagueness of style are more apt to occur, and escape notice, than in other sciences.

The student should not only test the accuracy of the definitions, the credibility of the evidences, and the soundness of the reasoning, but he should carefully examine whether statements alleged to be truths of consciousness are so in reality. This is rendered the more requisite by the fact that direct discernment is very frequently appealed to in every part of these sciences, while the statements cannot be rightly tested by any other means, as external observation or the testimony of others is of no avail.

There is here little danger of misspending time in studying truths of no practical importance, as very few of the cognitions are of that kind: but we must particularly guard against adopting mere hypotheses or opinions as cognitions, because the former frequently usurp the place of the latter. We should also fix all the leading truths of these sciences distinctly and permanently in the memory, as otherwise they will be generally misapplied, overlooked or forgotten, and thus fail to secure their principal objects.

§ 2. OF LOGIC AND PSYCHOLOGY.—Boundary between these Sciences.—Importance of the latter.—How distinguished from Physiology.—Why not appreciated by many.—Modes in which its Conclusions are established.—Directions for Study.

Logic is confined to the intellect, and it treats of their products and objects rather than of the faculties. Hence it excludes some of the most interesting departments of Psychology, while it discusses various subjects quite foreign to that science. The one treats of knowledge, and the other, of the mind and its faculties and susceptibilities.(28)

Psychology is a most important science, because a knowledge of the powers, affections and laws of the mind is requisite to our forming accurate conceptions of our present duties and future destination, to determine what we ought to do, in many momentous cases, and to use our faculties aright, in our ordinary conduct. It discusses the physical organs only so far as they regard mental manifestations, while Physiology, on the other hand, views the latter only as they influence the former.

As several of the truths of Psychology are made known to us by immediate consciousness, many fail to see the advantages of studying it systematically; and when they do so, they are apt to fall into the error of thinking that they could not fail to discern, whenever it might be required, every truth which they see to be self-evident as soon as it is clearly and accurately set before them, although History and Biography show the reverse. A proper study of this science is requisite, in order to understand the laws and processes of the mind, and to avoid the grave errors into which we are very liable to fall when we adopt as truths conclusions formed without any careful consideration of the subject. The maxim that first impressions are not reliable, holds true of mental, as well as of physical, objects.

Psychology appeals to our individual consciousness, at every step, except where it introduces other unexceptionable evidence: and it possesses the great advantage of being based chiefly on immediate discernments, from which inferences are deduced as in all other sciences.

In studying this important science, we must beware of obscure language and dogmatic statements requiring proof, which is not given. Particular attention should be paid to the precise signification of terms, as there is no subject on which we are more liable to be misled by considering mere expressions, and overlooking what they denote.

§ 3. OF THEOLOGY.—Sublimity and Importance of this Science.—How the chief Conclusions of Natural Theology are established.—Its Uses and Study.—Biblical Theology.—Its Evidences.—Inquiry regarding its Records, and their Purity.—Its Relation to the other Division.—Observations on Study.—Scripture Interpretation.—Influence of Prejudices.—Two Extremes.—Proper Course.

In the sublimity of its subject, Theology occupies a much higher position than any other science. Although the subjects of Astronomy are so vast as to transcend the powers of adequate conception, yet they are only masses of inanimate matter, inane as the empty void. Theology, on the contrary, investigates the existence and character of the Eternal Ruler, who formed and governs all: and the light which they throw on his attributes, constitutes the greatest value of Astronomy, and all the other physical sciences.

The importance of Theology is not inferior to its sublimity. A knowledge of God's character forms the foundation of all true religion, which is necessary to save mankind from the horrors of superstition and practical atheism, and our views of which determine our condition through eternity, compared with which the longest life shrinks to nothing. Religion affords consolation when nothing else can; and it is the only thing that furnishes enjoyments which will never fade nor cloy. It also leads to the performance of many important duties which no other motive will secure, so that it is as indispensable to steady good conduct as it is to permanent happiness.

Our views of the Most High sway the whole circle of our thoughts and actions. For, when a man learns and bears in mind the true character of the Eternal, his own is gradually improved and elevated, by being assimilated to the object of his constant affection, admiration and reverence, while he who continues ignorant and regardless of this subject, generally becomes the slave of appetite, debasing desires or malignant and selfish feelings, and proceeds from bad to worse. Hence the existence and character of God, and the relation in which we stand to him, as accountable, guilty and immortal beings, are matters to which every one should devote a very careful, diligent and impartial examination, and on which we should particularly beware of assuming that anything is either true or false, without conclusive proof.

The existence of the Deity forms the first object of attention, in Natural Theology. It is proved by the phenomena of nature, both animate and inanimate, since every supposition which attempts to account for these, without a forming and presiding Intelligence, involves demonstrable impossibilities. The character of God is proved by the endless displays of incomprehensible skill, power and benevolence, which are manifested throughout the whole creation.

Natural Theology strengthens the foundations of religion and morality, while it furnishes much exalted enjoyment, by giving us correct views of the works of the Creator, and thus enabling us to participate in his joys, to the extent of our comparatively limited capacities. A knowledge of its principles also furnishes important aid in rightly interpreting the language of Revelation, which assumes these principles as known, or at least knowable, independently of its own teachings.

The elements of this division may first be learned from books, after which we may continually enlarge our knowledge of it, by observing nature, both rational and irrational, with the aid of the principles previously acquired.

In Biblical Theology, the evidences of Revelation form the first subject of consideration. They lie in the nature of the truths revealed, and in historical narratives which are tested by the ordinary criterions. We are then to inquire what are the genuine records of Revelation, in order that we may distinguish them from spurious compositions, which falsely claim to form a part of them. We should also examine the purity of the text of those records, by means of the rules regarding written testimony, and also ascertain the precise limits of Revelation.

After disposing of those preliminary matters, we are to examine what Revelation declares. It is, of course, consistent with the conclusions of Natural Theology: but it goes much farther, and teaches us various things on which Natural Theology sheds no clear light.

The doctrines of Revelation exert a peculiar elevating and purifying influence on those who really understand and believe them, owing both to the impressiveness of direct communications from the Supreme Being, and their thoroughly reformatory and exalting character. They fail to produce such effects only upon those whose reception of them is merely nominal, or who modify and corrupt them, to suit their evil wishes. The teachings of Natural Theology are not sufficiently authoritative and powerful to produce, by themselves, those changes in the conduct and feelings of mankind which are essential to their permanent welfare.

As the whole of revealed religion is contained in the Sacred Scriptures, they should form the principal subject of our study of this division of the science. The characteristic benevolence of the Most High is shown in their being well adapted to the comprehension of every class of sane minds: and although some parts are obscure to the unlearned, these form but a small portion of the whole. All the principal doctrines are expressed and illustrated so clearly and variously, that no diligent student, who is desirous of knowing the truth, need misunderstand them.

Much assistance may be derived from various other works, if judiciously used: but, whenever we have recourse to them, we should beware of fallacies of testimony, misinterpretation of language, and appeals to authority. We should never forget that Scripture is its own best interpreter, and that many works which profess to unfold the doctrines of Revelation, teach most pernicious errors. These we may avoid by a careful and continued study of the Scriptures, which contain the only infallible authority on this subject, and on a right understanding and belief of whose principal doctrines depends our eternal all. We should also bear in mind that one fundamental error may mislead us fatally regarding the whole subject, because the essential doctrines are so connected and dependent that we cannot have right views of some if we are radically mistaken regarding others.

The Bible is interpreted by the ordinary rules of interpretation, since we are evidently addressed in the ordinary language of mankind, which we can best understand, and neither in that of mathematicians nor of enthusiasts. The style is often figurative or poetical: but the figures are such as occur in all impassioned discourse, and are employed with great uniformity. Consequently the language is generally more intelligible than if it had been more literal, because, in the long lapse of time, the latter would have changed more, while its variations would have been less obvious. The occult senses, professed to be found in the Scriptures by the Cabalists and other mystics, are only the dreams of imbecile understandings.

Throughout our study of this most important subject, we require to guard carefully against the influence of prejudices, as there is none on which they are more apt to mislead us. These are the main causes of the great diversity of religious belief that has hitherto prevailed among mankind, and not the inherent difficulties of the subject, which is by no means the most abstruse or difficult branch of human investigation.

On the one hand, multitudes have adopted superstitious and demoralizing doctrines, because they form part of the parental creed, and offer an easy and grateful road to future bliss. On the other hand, many have been led by such doctrines, and the corresponding practices of their votaries, to discard all revealed religion as a delusion or imposture, because they do not care to search into the real state of matters, or to see themselves as they actually are. Both classes are alike in desiring to shun the labor of proper investigation, and the sacrifices which a knowledge of the truth might impose. Hence, the former believe without any satisfactory evidence, and the latter reject conclusive proofs without an impartial examination.

Nature and Revelation are evidently the only reliable sources of religious knowledge: and, therefore, whatever is inconsistent with their teachings, or not legitimately proved from either, should be rejected. But, before rejecting any doctrine, we ought to know that it belongs to this class. On the other hand, whatever is proved by either, should be firmly and unhesitatingly believed.

§ 4. OF MORALITY, OR ETHICAL SCIENCE.—Subjects and Foundations of Morality.—What it aims at effecting.—Its Advantages.—Directions for Study.

This science first inquires into the nature and foundations of duty in general, and afterwards discusses the various kinds of duty specially. It derives its principal primary premises from Psychology, Theology, and Physics, although it also employs various truths derived from other sources.

To determine beforehand the duty of a person, in every combination of circumstances that can arise, is impracticable, since these are endlessly diversified. But we may establish principles, which apply to all cases, and discuss their chief applications, so that the precise line of conduct which duty requires, may be distinctly laid down, in all important cases of frequent occurrence. When once we clearly understand both a principle and its leading applications, we can generally, without difficulty or inconvenience, determine its other applications, since they are always very similar to those already known.

The belief that the mind comprehends the nature of right and wrong, as the eye perceives the distinctions of color, is refuted by the diversities of opinion which have prevailed among mankind on many moral questions. Were that belief correct, there could be no more difference of opinion as to whether a certain course is right, than there is about the color of the sky or the taste of sugar. On the contrary, what is right and wrong cannot frequently be determined without a proper application of moral principles, which are established only by a close investigation of several subjects by no means free from difficulty.

The Holy Scriptures are designed to aid, and not to supersede, the use of our faculties; and, therefore, they leave many important questions to be answered by our own investigations. Thus, we are commanded to do to others as we would have them do to us: but we are not told whether our actual or our reasonable wishes are the test. We can solve such questions readily without the aid of Revelation; and it is evidently no part of the plan of Providence to encourage indolence, and foster mental imbecility, by solving for us problems which we can easily solve for ourselves, by a proper application of our faculties.

Morality discusses the whole subject of duty, once for all, and thus guards us against the various dangers incident to sitting down to solve moral problems, when we ought to be acting. A man who has never considered what his duties are, until he is placed amidst the circumstances in which he is required to act, is very apt to convince himself that what he is strongly tempted to do, is right, whereas, if he had previously examined the subject, while he was free from any such excitement, he would see it in its true light, and the settled view, thus obtained, would enable him to resist the temptation, and to act as duty dictated. Unprincipled or immoral conduct is caused, in no small degree, by failing to consider what duty requires, till the time for action comes, in consequence of which the passion or appetite of the moment controls the conduct.

The influence of prejudices is never more powerful than when we examine what duty requires, with reference to objects of strong desire, while these are placed in full view. For we inevitably see distinctly the sacrifice that must be made, in the event of our conclusions being hostile to the gratification of the desire. Hence the great advantage of ascertaining what our duties are, before such desires have been thus excited, as their influence is thus rendered comparatively feeble.

In studying Morality, the learner should first obtain clear and accurate conceptions of the nature and sanctions of duty in general, and afterwards investigate the fixed and certain relations which the various classes of duties bear to his present and future condition. In order that this science should produce its legitimate effects, it is not enough to read or hear that we should do this and not do that: we must clearly see the sure and inevitable consequences of so doing, and then fix them distinctly and permanently in the memory.

§ 5. OF JURISPRUDENCE. — Nature, Foundations, and Uses, of this Science.—How related to Legislation.—Practical Application.— Importance of discussing Principles in the abstract.—Study.

Jurisprudence resembles Morality in its processes: but it considers man only as the subject of political government; and, therefore, it is much narrower in its range. From premises obtained chiefly from Psychology, Morality and History, it deduces the laws best adapted to man, in a normal state, or that condition in which the community are enlightened, and attentive to everything that seriously concerns them, whether present or future.

A knowledge of this science is requisite in order to understand the true functions, influence, and powers, of government, regarding which many have formed very erroneous views. On the one hand, calamities have been attributed to misgovernment, which were wholly owing to other causes, while things have been expected from a good government which the best cannot confer. Misgovernment is itself an effect of evils which previously existed; and these may continue to exist under any government, as their removal requires more potent agencies than human laws. On the other hand, the sufferings produced by bad laws, or the abuses of power, have often been wickedly laid to the charge of Providence.

It is necessary to understand the elements of Jurisprudence, in order to form correct opinions regarding the nature of true liberty and individual rights: and a citizen cannot secure his own rights, or perform his duty to his country, unless he knows what the former are, and what measures the public welfare requires.

Jurisprudence in the mental sciences resembles Hygiene in the physical. The Hygienist deduces the laws of health from a knowledge of the human organization and the nature of the agents that affect it: his conclusions are not in the least invalidated by the fact that many are ignorant of those laws, that several are too much under the influence of bad habits to obey them perfectly, even after they have been clearly set before them, and that others have contracted diseases which render medical treatment desirable. So the jurist deduces the principles of his science from a knowledge of the mental characteristics of man, and the agencies by which they are affected. Many, or possibly most, nations may be too ignorant, prejudiced, or oppressed, to adopt those principles to any great extent. Yet much is effected by clearly establishing them: for when a people who possess any control over their laws and government, see clearly what duty and their true interests demand, they will take some steps in the right direction, and these will prepare the way for more, until at length their institutions will become conformed to the principles of Jurisprudence, although such a consummation may require many generations.

The duties of the legislator differ from those of the jurist. It is the duty of the former to propose only such changes as the nation are likely to sustain and carry out in practice: for, if he were to go farther, the innovations would be only a dead letter. Every step in the direction, however, is a gain, provided it be real, and there be no subsequent retrogradation. The jurist, on the other hand, aims at establishing scientific principles, applicable in every age and country, independently of any particular changes or circumstances.

When a law is found to be defective, an application of the previously established principles of this science to the circumstances of the case, indicates the change which ought to be made: and, without such guidance, changes would be as likely to deteriorate as to improve: for they would either be made at haphazard or under the influence of strong prejudices, excited by discussing juridical principles for the first time, in connection with the contemplated changes, and the numerous real or supposed interests which they involved. But when those principles are first established and understood, without any reference to party questions or passing events, the changes which ought to be made, may be ascertained without any extraordinary difficulty, since all parties then stand on the ground of established principles, so that prejudices are allayed, and the means furnished for rightly determining the matter in question.

Before commencing the study of this science, the learner should master the principles of Morality, without a knowledge of which Jurisprudence cannot be studied with success. He should then acquire correct views of the foundations, legitimate functions, and proper limits of government and law: and he should always distinguish the existing laws from the principles of this science, with which they have often been confounded. He should also distinguish Jurisprudence from Morality: for, although the two are closely connected, they are by no means identical. Our duty to others often requires from us much more than human laws can prudently demand, and much less enforce, while these laws take no cognizance of the still more extensive and important class of duties with which others have no direct concern.(29)

CHAPTER XXV.

OF MIXED KNOWLEDGE.

§ 1. OF PHILOLOGY.—Two modes of learning a Language.—Ultimate Sources of Knowledge regarding dead Languages.—Grammars and Dictionaries.—Usages.—Authorities.—Branches subsidiary to Philology.—Etymologies.—Requisites to knowing a Language.—How best attained.—Various Significations of Words.—Common Error.—Conversation and Composition.—Comparative Philology.—Its leading Principles and Results.—Affinities of Languages.—General Philology.—Uses of Philology.

A LANGUAGE may be learned either by simply observing spoken usage, or by means of special instructions from others, in a language which we already understand. The elements of the vernacular are necessarily learned by the former method, in the manner already pointed out; and the youth can afterwards learn, from direct inquiries, the exact significations of the more abstruse, rare or vague terms; or he may consult grammars, dictionaries, and scientific treatises, for that purpose.

The ultimate sources of knowledge regarding a dead language are chiefly the following: (1) Living Languages which resemble the dead. (2) Translations made into some living or known language, by persons who learned the dead from spoken usage. (3) Grammars or dictionaries written by such persons, in some known lan-

R

guage. (4) Historical compositions which relate events known to us through some other source. (5) A comparison of works of ancient art with the descriptions or allusions of authors. These sources frequently make known the significations of all the more common words, inflections and constructions, after which reading good authors will farther extend our knowledge, since we can now learn the exact significations of many words and phrases from the context. The learner might now compose a grammar and dictionary of the language, and thus greatly facilitate and abridge the labors of succeeding students.

After having obtained a good general knowledge of the language, by either or both of those methods, we can test the correctness of grammars and dictionaries, by an appeal to the best usage, or the practice of those who spoke, or still speak, the language in its utmost purity. Hence the greater value of dictionaries which quote passages from reputable authors, since they enable us, in a great measure, to test directly the accuracy of their own definitions.

In consulting such dictionaries, we should observe the age and character of the authors quoted. Some may have written at a time when the language was rude and uncultivated, and others, after it had become corrupted and debased. Loose thinking and acting generally produce loose speaking: and hence a decline of morals and science is followed by corresponding deteriorations of language. Some authors, again, are above, and others below, the average of their age: and, therefore, we must consider their particular characters, as well as their age, in order to determine their value as authorities. The lexicographer should point out all the common significations of terms, whether the usage be good or bad: but he ought to indicate the nature of the usage, with respect both to age and character, especially where his work is designed to aid learners in composition: and those dictionaries in which this is properly done, are generally very superior to others, on all important points.

Among different reputable usages, the best is that which best answers the ends of speech: but that of persons who know the language well, and use it carefully, is to be deemed good. In cases of doubt or difficulty, recourse should be had to unexceptionable authority, and the definitions of the dictionary tested by it, as far as may be requisite. Wherever circumstances excite any suspicion of misquotation, we should refer to the original passage, and the part quoted should be compared with the context.

We should distinguish between treatises written by philologists who conversed long and freely with those by whom the language was correctly spoken, and authors whose knowledge of it is derived wholly from books. The former are themselves authorities: the latter are not; and the credit due to them depends wholly on the character of the books which they have read, and the use which they have made of them. Hence their statements are not entitled to be implicitly received, without proof, in any case of doubt or difficulty. Philological treatises written by such persons sometimes contain various errors. Thus, we are told, in Greek Grammars, that, besides the active and the passive, there is a middle voice, representing only actions which the subject of the verb does to himself. But when we come to examine the real usages of the language, we find that every one of the so called middle forms is either active or passive in signification, and represents actions done to others, as well as to the subject. Consequently the middle voice is wholly imaginary.

The acquisition of a language is facilitated by an acquaintance with the ethnography and history of the people who originally spoke it, without which many idiomatic terms and expressions cannot be rightly understood.

The etymology of a derivative generally indicates its signification, since most derivatives closely follow certain rules which prevail in the language, so that a knowledge of the proper rule of derivation, and of the signification of the root, teaches us the signification of the derivative, without reference to any definition. Thus, if we know what *good* and *hunt* signify, we do not require to be told what *goodness* and *hunter* denote. There are numerous exceptions, however, chiefly in nouns, although some occur in other parts of speech: and, consequently, etymology is by no means an infallible guide.

To know a language, we must be acquainted with the signification of its words and phrases, and with their inflections, constructions and collocations. The last three are best acquired, in the first instance, from a good grammar, as any other course would require the labor of many years. After acquiring the simple elements of grammar, we should learn the exact significations of all the common words, except those numerous derivatives regarding whose import we cannot hesitate or err. These are best learned from a classified vocabulary, accompanied with short and easy sentences, containing the words, in various forms. We should then master the rules of composition and derivation, after which we shall know the signification of most of the common words, as soon as we hear or see them. Words of rare occurrence need not occupy our attention till we meet with them in discourse.

In ascertaining the significations of such words as occur in various senses, we should first find the single primary signification which every such word bore, and then learn the secondary or derived significations, and trace the steps by which they arose from the former. This will both aid remembrance, and prevent us from confounding the senses. Very few words have more than three or four significations really different. But an expression which bears only one meaning in its own language may require to be rendered by several in another, owing to differences of idiom; and hence lexicographers sometimes render it in all these ways, without adverting to the fact that some of the significations were purely figurative, and that the language of translation made distinctions which were not made in the original. Hence a multitude of imaginary significations are found in old dictionaries, while the primary is frequently omitted, or not distinguished from others. The best philological works of the present day are distinguished by marked improvements in these respects.

After thus acquiring the elements of the language, the learner should carefully peruse good authors who wrote in it, just as he would read works composed in his vernacular: and, while thus occupied, he should acquaint himself with those parts of the grammar which he had not previously studied. If he were to commence reading regular compositions sooner, his progress would be very slow and tedious, as he would require to be turning incessantly to his dictionary or grammar; and he would frequently fail to learn the true significations of a word, after looking it up many times, because his mind would be distracted by the various renderings given, his attending chiefly to the passage before him, and long intervals elapsing between the references.

If the learner require to write and speak the language, he should use it extensively in composition and conversation: but when he requires only to read it fluently, he need not spend much time in this way.

After acquiring a knowledge of several languages, we may compare them, and note their resemblances and diversities.

Two languages may be found to be similar in their words, but different in their structure, like Latin and French. This indicates that both have a common origin, and that one has undergone great changes, either from having been adopted by a foreign people, or from extensive intercourse or intermixture with another race. The language superseded will be indicated by the structure or various words and idioms which it has communicated to its successor.

Where several languages are very similar, both in words and structure, it is to be inferred that they are the kindred offspring of a common parent, from which they sprung at a comparatively recent period in the annals of mankind. No very definite line of demarcation separates mere dialects from closely cognate languages. The best distinction appears to be, that whenever the parties who speak them, understand each other without an interpreter, they are to be held only different dialects: otherwise they are to be deemed different languages.

If various languages exhibit a close resemblance in structure, but great diversity in their words, it is to be inferred that they had a common origin, and that they have been little affected by foreign intercourse, but that those who speak them have had little intercourse with each other. A good instance of this kind is furnished by the aboriginal languages of America, which all exhibit the same polysyllabic and polysynthetic structure, amid great diversities in their vocabularies.

Where languages greatly differ, both in words and in structure, yet exhibit points of resemblance which cannot be attributed either to chance or intercourse, it is to be inferred that they have had a common origin, but that they have either been altered by foreign intercourse, or that those who speak them have been long separated, or that the original language, and consequently those who spoke it, were quite rude.

The following are some of the conclusions deducible from the application of the preceding principles.

(1) All the languages of mankind have gradually sprung, by natural means, from a common origin, which has long disappeared.(30)

(2) They are divided into great families, the members of which exhibit much closer affinities with each other than with those of any second family, whence it is to be inferred that the various families originated from a common source, subsequent to the general dispersion of mankind.

(3) The families are often subdivided into groups, the members of which are marked by peculiarly close affinities. Some of these groups contain many, and some, only a few languages.

(4) Some exhibit every appearance of having gradually descended from the primitive source, by the changes naturally incident to human speech, without having ever been much affected by foreign elements or intercourse, while others have become so extensively blended with extraneous words, and changed in their structure, by foreign influences, as to have passed rapidly into new and widely different languages. The former may be termed *original*, and the latter, *composite* languages. In the case of the latter, the elements of the original language continue to form the basis or framework of the resulting speech, although the new materials may form the larger portion of the whole.

In order to ascertain the affinities of a language, it is unnecessary to learn all its anomalies and words. Its general structure and inflections, and the nature of the primitive words, determine its whole character.

After acquiring a knowledge of several languages belonging to the most dissimilar families, we can ascertain the principles and structure of language in general. Several works on this subject present erroneous views, as their authors were acquainted only with a few languages of similar structure. Yet good works on this subject are of much use, even to those who understand only the vernacular. They not only tend to improve style, but also facilitate the detection of several fallacies arising from language. Although philologists take good usage as their guide, yet they react powerfully on that of future generations: and their labors do much, not only to fix, but also to simplify, language, and to check, or even remove, defects and anomalies.

The knowledge of a foreign language enables us to consult works composed in it, and to hold oral or written intercourse with those who employ it. The language of a nation is also, in many instances, the only evidence of its origin; and it exhibits various peculiarities regarding its circumstances and character, which cannot frequently be learned from any other source. Comparative Philology removes various difficulties and obscurities attending the study of dead languages. It also furnishes means of determining the origin, migrations and affinities of nations, and thus frequently supplies defects in History.

§ 2. OF ETHNOGRAPHY.—Foundations of Ethnography.—Sources of Errors.—How these may be detected.—Uses of Ethnography.

Ethnography consists chiefly of primary facts, derived from public records, the narratives of travelers, local histories, works of art, and direct personal observations. Owing to the immense extent of the subject, the latter can form only a very small part of it. The facts are also liable to change, more or less, from age to age, or even from year to year: and hence the accounts of the past may not apply to the present. Various errors have also arisen from the national or individual prejudices of those from whose works a great portion of ethnographical treatises is derived. Hence many of them teem with errors. But these may generally be detected by comparing the statements of various writers with each other, or by having recourse to original and reliable authorities. Those who write from their personal knowledge are generally more reliable than such as derived their information from others: but there are great differences, in these respects; and we may generally trust more to the care and veracity of an author than to his means of acquiring knowledge.

A competent knowledge of Ethnography is requisite to a proper understanding of numerous statements and allusions found in the works of authors who treat of times or countries other than our own. It is also indispensable to the proper study of History, while it tends to remove the narrow-mindedness which is apt to accompany an acquaintance with only one state of society. It enables us to compare and trace the effects of different laws, opinions, institutions and manners. Thus we can form a proper estimate of our own condition, while we receive suggestions for its improvement. By the same means, also, we can easily avoid the dangerous error of assuming that our own morals, manners and institutions are perfect standards of excellence, when, in reality, they may be far otherwise. We also learn the wants of others, so that we can take proper steps to supply them. Besides these advantages, Ethnography answers several of the same purposes as Geography, to which it is closely allied.

§ 3. OF TECHNOLOGY.—How an Art is distinguished from a Science.—Foundations of Art.—Its Relations to Science.—Requisites to Proficiency.—Superiority of Arts based on Inductions.—Theory and Practice.—Manual Dexterity.—Source of Improvements.—Importance and Relations of Rhetoric.—Education.—How related to Rhetoric.—Its Importance.

An Art is distinguished from a Science chiefly by its consisting of rules and directions for effecting some object, and a general absence of discussions regarding simple knowledge. Art considers how a certain thing is to be done, while science shows what is. Yet, as the proper mode of accomplishing a certain end, depends on a knowledge of the subject, and this must often be discussed in a treatise on the art, the boundaries of Art and Science are not always well defined. Many sciences include what might properly be termed arts. Thus, Geometry includes the art of drawing geometrical figures, and Astronomy, that of making astronomical observations. Such arts are properly classed with the sciences to which they refer, when they form necessary parts of them.

Art borrows the knowledge which it requires from whatever sources afford it; and it often gives some degree of connection to a great many scientific elements, which have not been incorporated into any regular science. As new applications of the sciences may be made, without any increase of their number, the arts generally multiply much faster than the sciences. Several arts may be based on a single science; and, on the other hand, a single art often derives materials from several sciences.

The dependence of Art and Science on each other is often mutual. The art of the optician is necessary to unfold the mysteries and wonders of the physical creation, whether in the Sun or in a mite, while the optician is dependent on the sciences of Geometry and Optic. The value of the air-pump is felt in several sciences; and the art of calculation is required in a still greater number. On the other hand, Art is based, more or less, on Science; and it is generally impracticable to master any division of the former without acquiring some knowledge of the sciences on which it is dependent. This and careful practice are the principal requisites to proficiency in an art.

While we act only on empirical knowledge, improvements must be accidental, and one advance does not prepare the way for another, as all empirical arts are necessarily chained down to one uniform course, and anything beyond the usual routine is out of the question. On the other hand, if we are guided by inductions, our processes are more certain in their results, and more susceptible of improvement, because we are in possession of the general law. Thus, the mere fact that a certain medicine has been found beneficial in certain cases, does not inform us what are the circumstances on which its efficacy depends, or what are the principles of its operation: and consequently we should be apt to use it where its effects would be positively detrimental. So, if a shipbuilder finds that a vessel of one form sails faster than another, of a different form, this knowledge does not enable him to make any improvements, while a knowledge of Mechanic and Hydric would unfold the principle on which the superior sailing depended, and thus enable him to build vessels of a still better form.

A thing cannot be good in theory and bad in practice: for this is only saying, in other words, that rules based on a wide and correct knowledge of the subject, are inferior to those founded on narrow and inaccurate views. We should not confound what is really with what is only apparently good. A theory may be very plausible, and yet quite erroneous, just as an argument may be very specious, and yet quite fallacious: but a theory which is bad in practice, is bad altogether; and, therefore, whatever is really good in theory must be equally so in practice, if it can be put in practice. The empiric thinks that he alone is guided by experience, when, in truth, he is only guided by a narrow experience, while those whom he decries are guided by scientific generalizations. He also falls, in many instances, into the further blunder of mistaking his own erroneous inferences for the voice of experience. This teaches us simply what we have experienced: and, in order to apply it safely to the future, we must reason on the subject, or, in other words, have recourse to theory.

Although a certain degree of practice and manual dexterity often requires to be combined with theoretical knowledge, in order to render a person an adept in the art, yet the former, without the latter, is generally of little avail. Thus, a man ignorant of Anatomy and Physiology cannot be a good surgeon, however long he may have practiced the art: for, in many cases, he cannot know what operation ought to be performed, or how it can best be effected, while, in other cases, he will operate where no operation is required. Even in the most purely mechanical arts, the hand is greatly aided by an intelligent head: and the value of scientific knowledge, in such arts, is shown by the fact that many important improvements have been made in them by men of science, who were not operative artisans at all, after the empirics had long employed the old method, without ever dreaming of a better.(31)

Among the intellectual arts, Rhetoric has long held a conspicuous place. The Greeks applied it mostly to public speaking, : but it is equally applicable to all didactic discourse. It is based chiefly on Psychology, Logic and Philology; and it discusses all discourse which is designed to produce conviction, whether this be the sole immediate object or not. It differs essentially from the emotional arts, in regarding emotion only as a means of securing conviction, and influencing the conduct of those addressed. Grammar treats of style only so far as literal accuracy is concerned, while Rhetoric considers also those qualities of style which affect the judgement and the feelings. As its principles are drawn from other departments, and it is chiefly occupied with the means of effecting a particular end, Rhetoric is properly classed with the intellectual arts, although some have ranked it as a science.

A person's future condition often depends on his being

convinced of a certain truth, which, if believed, will powerfully affect both his feelings and his conduct. But he rejects it, owing to the advocate's failing to present it in a proper manner. Hence the importance of knowing the best means of communicating truth, and leading those addressed to perform the requirements of duty.

The art of Education has much in common with Rhetoric, since children possess the same powers and susceptibilities as adults, although in an immature degree. Hence the same rules are applicable, when modified to suit the particular condition of those addressed. But here the field is much wider, since the object is, not merely to instruct, but also to train and develop the faculties in a proper manner. The comparatively volatile and negligent character of the young, also, renders it necessary to employ measures which are not requisite in the case of adults, in order to secure proper attention, without which little real progress can be made either in training or instructing.

The subject of education cannot be overestimated, as it forms the moral character, and exerts a very extensive influence on the physical and intellectual. It properly includes, however, the instructions and examples of all with whom the young have intercourse, whether professed teachers, parents or associates.

A knowledge of this subject is useful to parents and guardians, as well as to ordinary teachers, on several accounts. It enables them to perform aright their own part in the business of education, in which the parental is of more consequence than any other, as its influence is most constant and powerful. A parent who is aware of the evils resulting from a failure of his duty, in this respect, and the benefits which flow from a contrary course, will confer permanent good on his offspring, of which those who are ignorant on the subject have no conception. Such knowledge also enables parents to select the best accessible teachers for their children, whereas ignorance often hands them over to educational quacks, who profess and promise whatever will increase their notoriety and gains, the sole objects of their labors. Thus it has sometimes happened that children who were badly educated at home, were consigned to worse teachers abroad: and when they afterwards followed evil ways, the disciples of darkness held them up as a striking proof of the uselessness of education. Yet such results do not in the least disprove the great truth that a *good* education (not *any kind* of education) secures right conduct in after life.

CHAPTER XXVI.

OF PARTICULAR KNOWLEDGE.

§ 1. OF HISTORY.—Foundations and character of History.—Its Uses. — Cautions. — Empiricisms. — Inductions.—Limits of History.— Philosophy.—Prophecy.—Causes and Effects.—Selection of Events. — Best Historian.—First Merit of History.—Frequent Errors.— Two Classes of Historians.—Principal original Authorities.—Caution. — Common Defect. — Observations on Study.—Traditional Narratives.

HISTORY is founded on personal observation, the testimony of others, and documents relating to the events recorded. The character of its statements is to be ascertained by the ordinary criterions of testimony, as if we were inquiring into the truth of a narrative regarding a recent occurrence, in our own neighbourhood. As the events recorded are neither necessary nor general, its truths are all contingent and particular.

History not only gratifies a natural curiosity regarding the actions of mankind, and thus affords a refined pleasure, to persons of all ages, but it also furnishes materials for various sciences and arts: for what has happened, in certain circumstances, must always happen where the determining agencies are the same, although there be several minor differences. Hence we can often determine the results of certain agencies, before they have become matters of History. Here we also learn various lessons regarding the influence of particular institutions, principles, and opinions: we are aided in ascertaining the character of man, what circumstances are favorable to the due development of his faculties and to national prosperity, and whence spring public distress and misery; and we are warned to exercise proper caution in our intercourse with others.

Scientific and artistic histories furnish many suggestions and warnings to those who cultivate the different branches, as has been already remarked.

History is likewise a powerful instrument of moral discipline and enjoyment. For virtuous feelings are excited, strengthened, and gratified by tracing the career of the good, while vicious passions are checked by the exhibitions which are made of their repulsive nature, and the consequences to which, sooner or later, they uniformly lead. Conception enables us, in a great measure, to view the events of History like spectators of the scenes and events described, while sympathy with laudable actions conveys to us a part of the joy of those who performed them, and, at the same time, stimulates to emulation. On the other hand, the indignation excited by a vivid contemplation of base or vile actions, strengthens our resolutions to beware of them, and pursue a different course. Thus the achievements of a great and good man become the common inheritance of mankind, while the course of a contrary character furnishes a perpetual warning.

In order to secure such advantages, the events must be fairly and accurately depicted, and we must beware of receiving the partial and distorted narratives of unprincipled men as faithful delineations. If an historical work is not substantially accurate, and does not exhibit men and events nearly, if not precisely, as they were, it will only mislead and corrupt, instead of enlightening and improving.

Empirical generalizations are properly used in History, to render its statements more concise and forcible: but we should beware of the fallacy of irrelevant empiricism, which abounds in various historical works.

Inductions have frequently been attempted to be introduced into History: but they are quite alien to its nature and objects: and although the truths of History are properly employed in establishing inductions, in those branches of knowledge to which they properly belong, yet the attempt to introduce them here produces only incongruity, distraction of attention, and error. The duty of the historian is confined to relating past occurrences, with their immediate causes and effects, so far as these can be certainly ascertained, leaving every man to make such applications of his narrative as he deems proper.

With regard to futurity, it evidently lies wholly beyond the province of History; and consequently the historian wanders altogether out of his way when he speculates about the future: he has quite as little to do with prophecy as with philosophy.

A knowledge of causes and effects forms one of the most important parts of History: but we should beware of receiving as true, statements made on this subject, where it does not appear that they are anything more than conjectural opinions. Fallacies of causation abound in many histories; and actions are often attributed to motives which had no effect in their production, while their real causes are not set forth. Motives must generally be learned from the conduct of the parties, their statements being frequently mere pretences, or the results of self-deception. The real motives were not unfrequently very different, even from what the parties themselves believed: for men wish to stand fair with their own consciences, and thus frequently mislead themselves regarding the true character of their motives.

Remote causes and consequences are not easily traced, with any degree of certainty; and they run to an indefinite extent. The consideration of them, therefore, forms no part of the historian's duty, as it would produce doubt and distraction, rather than instruction or entertainment.

Actual occurrences are far too numerous to be mentioned in History, while most of them are too trivial or unimportant to be worth recording: and few things are a better test of an historian's judgement than the mode in which he makes his selection. Some weary us with commonplace details and gossip, neither interesting nor instructive, while the most important parts of the subject are either omitted or stated vaguely. Others are equally uninviting and uninstructive on account of their extreme conciseness, and the consequent indefiniteness and obscurity of their statements.

The best historian is he who avoids prolixity, and yet sets the events vividly before us, as they really were, so that, after reading his narrative, we may have such a notion of them as if we had actually witnessed them. While he totally excludes what is not worth recording, and devotes little space to ordinary occurrences, which we could accurately conjecture, he relates all that is requisite to a clear and accurate exhibition of the events and incidents, as he seizes and faithfully depicts the characteristic traits.

The first merit of a history is, truth, without which it does not deserve the name. Hence, in selecting works for study, those which are scrupulously accurate and impartial, should have a preference over all others. These excellences are frequently wanting in histories otherwise written with much ability. Some historians exaggerate or distort for the sake of effect; not a few misrepresent from prejudice or malicious motives; and others frame their statements so as to support their own views, theories or secret purposes, while many err from hurry or carelessness.

The statements are, in many instances, literally true; and yet they convey a false impression, because some important points are omitted, or not related with sufficient fullness. Consequently the reader is apt to be misled, and to form an erroneous opinion of the people or the occurrences. In other cases, general statements are made, which conflict with facts as numerous as those on which they are based. Fallacious appeals to the passions are not unfrequent, especially in narratives written by partisans. Opprobrious epithets are freely applied to their opponents, while they are equally profuse in commendations of their friends, set off by sneers, laudatory maxims, or feeble expressions of praise or censure, where truth required much stronger language.

History is vitiated by painting men in colors which are either brighter or darker than the reality: for we are thus misled regarding the characters delineated and the influence of certain agencies on communities. It is dangerous to believe that mankind are more honest and faithful than they really are, while we wrong them, and disturb our own peace of mind by attributing to them vices of which they are innocent. Nations and individuals have often suffered severely, owing to misplaced confidence, arising from their ignorance of the dark side of human nature. On the other hand, the generous and benevolent feelings of the mind are deadened, and intercourse with others is rendered unnecessarily difficult, by our being led to believe that all mankind are knaves and liars. Hence the propriety of selecting those historians who neither blacken nor whitewash, but represent both the lights and the shades precisely as they were.

Historians may be divided into two great classes—*original authorities*, who record facts never before written—and *compilers*, who draw their materials exclusively from previously written narratives or compositions. Not unfrequently an author belongs to both classes.

Of original authorities, some write from hearsay only, and others record their personal observations. The former are generally entitled to much less credit; and when the events related were remote from their own time, they are seldom implicitly reliable. Sometimes it is difficult to determine to which head a statement belongs; and, in such cases, we must be guided by a knowledge of the author's character and circumstances.

Among original authorities, the narratives of personal witnesses, public records, statutes, treaties, inscriptions, official documents, and information derived immediately from eyewitnesses, are generally the best. But many of these are by no means unexceptionable; and it sometimes happens that a compilation is more accurate than any single authority from which it is composed.

In Ancient History, it is requisite to ascertain the genuineness of the composition, as several are spurious; and the same remark applies to some modern narratives also.

Several narratives relate chiefly military and political transactions, while they profess to give the entire history of the country. Wars are always effects of certain opinions: and unless these are delineated, a military history is like a picture of a battle-field, which exhibits chiefly fury and carnage. No history of a nation or community deserves that title, unless it gives a view of the progress and changes of religious opinions, morals, law, science, art, literature, and domestic life and manners. These are the most important subjects of History; and an accurate delineation of them forms the best exhibition of a nation's real state and progress.

In studying History, we should first obtain a general view of the whole course of events, from the earliest times, including the rise and fall of states, the intercourse of one country with another, and the general condition of mankind, during the various periods. This may be acquired from a good set of historical charts, and corresponding outlines. We can afterwards study to advantage the history of any particular country, which we cannot well do till we learn the outlines of General History.

The most advisable course then is, to take the best accessible general compilation, as a text-book, and to read and compare original narratives and documents as the importance of the particular subjects or periods, and our circumstances, require or admit. Original narratives generally cover only a small portion of a nation's history; and, therefore, it is desirable to have a good compilation as a general guide. Yet the former are usually more full and graphic, and frequently more accurate in details.

After acquiring the outlines of General History, and a good knowledge of the events of Holy Writ, our own country should engage our attention, unless our circumstances permit, and our tastes lead, us to study the history of the principal nations of ancient and modern times. In that case, it is best to follow the course of events, beginning with the earliest, and coming down gradually to the present day.

Where we cannot study the history of every celebrated nation, we should select those which are most interesting, either on account of the mental development and general intelligence of the people, or of its connection with our own.

Before commencing the study of an author, we should ascertain his relation to the facts he professes to relate, and the nature of the testimony on which his statements are based. This will sometimes cause a little delay and additional labor: but these are well repaid, by the checks which they furnish against receiving erroneous statements for historical truths, which will frequently be the result of neglecting those precautions.

We should always endeavor to form an accurate and lively conception of the events and scenes, so that they may be represented to our minds as nearly like the originals as possible. For this is requisite, in order to avoid erroneous views, and to feel that degree of interest in the narrative which is requisite to agreeable or profitable study. It is impracticable to secure the principal advantages of this study without an accurate conception of the events, and of the characters and circumstances of the community whose history we peruse.

In regard to the earliest history of nations, the student will readily find that, although it may have long existed in a written form, yet it is generally based on traditional accounts, without deriving much light from contemporary written narratives. These accounts should neither be utterly rejected without examination, nor implicitly received. On the one hand, traditional narratives of remarkable events have been sometimes handed down, without any material variation, through many generations. On the other hand, they are generally liable to be greatly corrupted, during their transmission, even where they were originally correct; and, in many instances, they are totally spurious, being the result of knavery and fraud, operating on ignorant credulity.

It is true that pure fiction cannot be directly imposed on mankind for authentic history: yet this may be done indirectly; and the circumstances of the case, compared with the character of the tradition, and well authenticated facts, derived from other sources, will generally enable us to ascertain the actual truth. In the absence of any corroboration derived from present facts or the testimony of History or Philology, traditional narratives are seldom entitled to any weight: and where they are evidently absurd, or contradicted by authentic evidence, they should be unhesitatingly rejected. On the other hand, where they derive corroboration from other credible sources, and are neither absurd nor at variance with known truths, they may be safely adopted as substantially true, save that they are very rarely reliable as to numbers, dates, and places.

§ 2. OF CHRONOLOGY. — Use of Chronology. — Epochs. — Means of determining Dates.

A knowledge of the time when events occurred is necessary in order to understand their progress and connections: for otherwise they present themselves to our memory as a confused and unconnected mass of facts. Hence Chronology forms an important auxiliary to the right understanding and remembering of History.

In order to determine the time when the various events occurred, some epoch or fixed period is chosen, from which they are reckoned, either backward or forward, and the number of years which intervene between it and an event, determines the time of the occurrence.

Various epochs have been chosen by different nations: but as the Christian states have obtained the control of the world, and far outstripped the rest of mankind in every important respect, the epoch of our Savior's birth, or the commencement of the Christian era, which they all employ, is now the only one of much consequence.

The interval between an occurrence and the epoch is ascertained either from the direct statements of persons who knew, or from public records, or from computing the length of the various periods which form the interval, from the testimonies of historians, monumental inscriptions, coins, astronomical phenomena, &c.

When the interval between two epochs has been ascertained, dates reckoned from one can be converted into those reckoned from another, by a simple process of addition or subtraction.(32)

§ 3. OF BIOGRAPHY.—Relations of Biography to History.—Its Uses. — Its Advantages and Disadvantages, compared with History.— Common Faults, and how they may be obviated in Study.—Characteristics of good Biographies.—Autobiographies.

Owing to its nature, Biography is much narrower than History in its range: but it is more particular in its views, and gives a fuller insight into individual character. History gives an account of communities collectively, while Biography gives an account of the comparatively small number of individuals whose lives present something remarkable, and worthy of remembrance. The exact line which separates them, is not very clearly marked: yet the general distinction is obvious; and indeed the two cannot properly be amalgamated. Accounts of occurrences in which individuals alone were concerned, and which did not directly affect any community or class, are out of place in History; and narratives of public transactions are equally irrelevant in Biography. A professed history which consists, in reality, of a string of biographies, is apt to mislead us, as the characters of the individuals whose lives are recorded, may have differed widely from that of the community with which they were connected.

Biography so closely resembles History that most of the remarks made in the preceding section are equally applicable here. Biography, however, states many minute particulars which History overlooks; and thus it often enables us to form more definite conceptions than the more general statements of History admit; and it is more available in enabling us to ascertain how an individual of a certain character will act, in given circumstances. At the same time it is of comparatively little use, in enabling us to determine the effects of laws, opinions and institutions, because individual peculiarities frequently control or prevent the ordinary results of those agencies, which become conspicuous only when we view their effects on nations or communities.

As biographies are generally written by friends or admirers, they are still more affected by prejudices than histories. The characters and motives of the principal personage and his friends are often represented in colors too favorable, while those of their opponents or rivals suffer proportionally. Hence due allowances should be made. Where the biographer is unfriendly, this state of things is reversed; and we should accordingly lighten the picture, as the results of proper investigation require.

Prolixity is a common fault of biographies; and this is not unfrequently carried to such an extent that the works are unworthy of general perusal: and they should be used only for occasional reference, when we wish to learn something interesting which is not found elsewhere, or to compare the statements with other authorities.

A good biography gives a correct and vivid representation of the life and character of its object, of the various agencies that directly influenced him, and of the immediate effects of his words or actions upon others. It avoids the extremes of adulation and detraction, dull prolixity and vague generalities.

In reading autobiographies, we should guard particularly against paralogisms of testimony: for the writers often impose on themselves unconsciously, and consequently are very apt to mislead others. A man is often greatly mistaken in his estimate of his own character, opinions, motives and conduct: and, therefore, if he writes his own life, the work will be very apt to abound with errors, of many of which he may have no conception or suspicion.

CHAPTER XXVII.

OF THE KNOWLEDGE OF FUTURITY.

§ 1. OF THE KNOWLEDGE OF FUTURITY IN GENERAL.—Common Desires regarding the Future.—Why not gratified.—Future Things which can be known, and advantages of knowing them.—Probability often sufficient.

A KNOWLEDGE of the past is of no value except as it affects the present or the future: and as the present moment is incessantly running into the past, while the future will never terminate, some knowledge regarding it is both important and desirable. Although such knowledge is narrower than what we may acquire regarding the past, yet it may be so extensive as to furnish no reason to regret its smaller extent.

The anxiety of mankind to know the principal events of their future lives, appears from the general and long-continued prevalence of many superstitious arts, which professed to accomplish this purpose, but which are only delusions, resulting from thoughtless credulity, prejudice and knavery, as appears both from experience and reason. The Most High has wisely and graciously covered the particular events of our future lives with an impenetrable veil. If evil awaits us which we cannot obviate, it is enough when it comes; and if we receive some unexpected good, it is the more agreeable for not having been anticipated.

While some things are thus concealed, and all attempts to foreknow them are fruitless, there are many future events that we may foresee, and which it is of the utmost consequence that we should. The farmer must know that summer and winter will return; otherwise he would not be justified in toiling long and hard, and incurring great expense, to prepare and sow his fields: and we must know how persons of a certain character generally act, when placed in given circumstances; otherwise we could not safely trust others with property, or employ them to perform important business. So it is always of great importance that we should know the general results of certain lines of conduct; otherwise we should be strongly disposed to follow that course which is most agreeable at the present time, without paying any proper regard to the distant future. And indeed this is one of the most common and fatal errors that men ever commit.

If all the unprincipled persons in the world would employ the means within their reach, to discover the certain and unavoidable results of their present conduct, they would see a picture which would tend very powerfully to lead them into better courses. And if many parents would examine into the inevitable results of the evil training which they give their children, and the good effects of a proper course, they would discard the former and adopt the latter. But a foolish curiosity to know particular future events, has frequently been combined with a supine indifference regarding the individual's future destiny.

In many cases, certainty regarding the future is unattainable; but we can arrive at a high degree of probability, by reasoning from the circumstances of the case, the effects of agencies known to be operative, or present signs of futurity: and, in all cases of this kind, probability suffices, for practical purposes. Thus, in regard to the way in which individuals will act, in given circumstances, we are sometimes disappointed: but if we use due caution, such disappointments will be very rare. So, although nothing is more uncertain than the life of an individual, yet the average duration of human life, in a large community, varies little, from year to year. Even the variations caused by a change in the determining agencies, such as famines or pestilences, can generally be predicted with considerable accuracy.

§ 2. SOURCES OF OUR KNOWLEDGE OF FUTURITY.—Necessary and Hypothetical Truths independent of Time.—Two Classes of Contingent Truths.—Source of our Knowledge of the future Permanence of Natural Laws.—Proofs that the Termination of the present Course of Nature is very remote.—(1) Astronomical Argument.—(2) Geological Argument.—(3) Historical Argument.—(4) Argument from Prophecy.—Inferences.—Means of knowing particular future Events.—Presentiments.—Extent of our Knowledge of Futurity.—Erroneous Dogma.

Necessary and hypothetical truths are, from their very nature, independent of time and place, and, therefore, as true for the past and future as for the present. The propositions of Mathematics, for instance, will be as true at any future time as they are to-day; and if a certain consequence is necessarily implied in a supposition now, it will be equally implied forever.

With respect to contingent truths, the case is different. These consist of two classes, general and particular. The former comprises most of scientific cognitions, exclusive of those of Mathematics, which are all necessary truths. The latter express only particular occurrences, and therefore apply neither to previous nor to subsequent times. When we say—"Alexander the Great conquered the Persian Empire," we express only a particular event, which will never occur again: but when we say—"all vertebrate animals possess a brain," we express a truth which held true in past times, and will do so, as long as the present system of the world is upheld. So, the law of gravitation, and the properties of heat, light, and electricity, are equally permanent; and the structure of every species of organic beings will be what it now is, as long as the species exists.

A knowledge of the operation of a constant and unchanging agent informs us of its future effects, when we have learned what they are in one case: and thus we know an indefinite number of future truths, including the greater portion of all general contingent truths. We know that the laws of nature will be the same hereafter as they are now, because the same agencies will operate, and in the same circumstances. Nothing short of a miraculous operation of the Deity would alter those agencies or circumstances: and we know that he will not so interfere. Temporary miraculous interference may possibly take place, for special purposes: but the supposition that this will extend to a permanent alteration of the laws of nature, is absurd, since these are the results of boundless benevolence, unlimited power, and omniscient wisdom; and, therefore, they admit of no improvement.

The laws of inorganic nature have been unchangeable, so far as we can trace them: and although many old species of organic beings have disappeared, and others succeeded in their places, no species was ever altered, so far as we can find. Now as the same Eternal and Immutable Creator, who takes no second thought, and all whose works are perfect of their kind, will forever continue to rule nature, we must conclude that there will be no change of the natural laws, either of matter or of mind, at least until the system shall be dissolved by some great upbreaking, to be followed by another, and probably similar, system.

That the present system of the world will continue as we now behold it, for an immensely long period, may be proved by various arguments, of which the following are the principal.

1. Astronomers have shown that the solar system would go on forever, as it now does, if there were no resisting medium, and God did not interfere miraculously. The resisting medium, however, appears to exist: but there are proofs that its influence on the planets will produce no important change, for many millions of years. It is further found that the Sun revolves round some distant center, at a rate which will require a very long series of ages to perform a single revolution.

Astronomy, therefore, leads to the conclusion that the Earth and the other planets will continue to move, and receive light and heat from the Sun, as they now do, for many millions of years. The immense magnitude of the Sun, and the extensive chemical and electric action which must be incessantly occurring among its elements, obviate any fears of its heat or light failing, during all that period: and the effect of the resisting medium may possibly be, to move the planets very slowly nearer the Sun, as these gradually diminish, so as to counterbalance the deficiency, until the time of the final catastrophe has arrived.

2. Geology shows that the Earth has been the residence of animals incomparably inferior to man, for countless ages before he existed: and it is a manifest absurdity to suppose that it was to be miraculously destroyed, a few thousand years after it became gradually fitted to be his abode, while it continued to be as suitable as ever for his residence.

3. The preceding argument is corroborated by observing the gradual improvement of mankind, from early times, and the absurdity of supposing that the Earth should be destroyed when they had become enlightened and happy, instead of being the reverse, as they were in early times. When we closely observe the condition of those nations who stood highest, since the dawn of His-

tory, during the successive periods, there is a marked advance, at every step, since those times when the most enlightened nations offered human sacrifices, and formally worshiped such revolting characters as Moloch and Jupiter. The nature of truth and error, and the character of the Most High, who favors the advancement of knowledge and virtue, imply that the course of improvement will be progressive, until mankind shall have become what they were designed to be, and the future shall vastly excel the present.

4. If we consult the prophecies of Revelation, we are led to the same conclusion. There he who foresees the end from the beginning, foretells a time when wars will cease, knowledge be universally diffused, and all men be truly religious, moral and happy.* Although we are not distinctly told when this state of things will begin, or how long it will last, it is clearly implied that the time of its continuance will be very long. Thus, the time during which the pagans were to continue in ignorance, is called a moment; and it is added that they should be visited with great mercies, and everlasting salvation. So it is said that the glorious time predicted would shortly arrive, whence it follows that the interval is of no account, compared with the time of its continuance. Now when it is observed that this interval extends over some thousands of years, we must infer, not only that the future of man's history will be brighter than the past, but that it will be incomparably longer.

Thus we are led, by different lines of argument, to the conclusion that what is science now, will continue to be so, for a vast period of time. The Almighty has endless duration before him: and he has planned the present arrangements of the universe on as great a scale with respect to their continuance, as they are in regard to their extent.

Besides the statements just adverted to, Prophecy foretells various particular events, of the utmost consequence. Of these the general judgement of all mankind, to be followed by a state of endless weal or endless woe, is conspicuous for its paramount importance, and forms a subject which every one should examine with a corresponding degree of attention and care.

* Isaiah, Chapter ii., 1-4; xi., 1-9; xxxv., liv.; Jeremiah, xxxi., 31-34; Micah, iv., 1-4; Revelation of John, xxi., xxii., 1-6.

Particular future events may be frequently foreseen by knowing that causes operate which will certainly produce them, or by observing present indications of them. But many cases of this kind contain some uncertain elements, which take away certainty, and render the future only highly probable. The degree of probability may be determined by means of the principles already stated; and it will often be found to be so high that we may safely act upon it as if it were a certainty.

With regard to *presentiments*, or pretended previous knowledge of future events, based on something undefined, they are nothing but probable inferences from present indications or feelings, which are drawn so rapidly and easily that the process is overlooked or forgotten, as is habitually done in the case of other guesses. These frequently turn out to be correct; and such instances are then noted, and adduced as proofs of the truth of presentiments. In the equally, and possibly much more, numerous class of cases in which they turn out to be false, they are either overlooked or ridiculed. They are not essentially different from the conjectures which we are incessantly forming regarding futurity, with a distinct discernment of their nature. The only difference of any consequence is, that presentiments are generally less reliable, because they are more frequently based on delusive indications, such as dreams, reveries or omens.

A review of the nature and extent of our knowledge of futurity shows that it is adequate to our wants, while it falsifies the dogma that knowledge may be dangerous. God never does anything which he desires to conceal. On the contrary, he desires that we should acquaint ourselves with his works, all of which exhibit the high and attractive attributes of his character; and none but those who read them aright, fulfil the great end of their being. What God does desire to conceal, there is not the least danger that any human effort will ever unfold.

PART V.

OF THE RETENTION OF KNOWLEDGE.

CHAPTER XXVIII.

OF THE RETENTION OF KNOWLEDGE BY SIMPLE REMEMBRANCE.

§ 1. GENERAL LAWS AND RULES OF REMEMBRANCE.—Proper application of remembrance.—(1) Law of Attention.—How Attention may be excited.—Three Rules regarding it.—Pernicious Error, and how it may be avoided.—Why an Adept often remembers better than a Tyro.—(2) Law of Comprehension.—Importance of Apprehensions.—(3) Law of Generalization.—(4) Law of repeated Examinations and Reviews.—Requisites.—When mental Reviews preferable to Apprehensional.—(5) Law of Intervals.—Important Aids.—(6) Law of Re-comprehension.—Its Applications.—(7) Law of System and Arrangement.—Why leading Principles should be first attended to.—Advantages of proper Classification.—Principles of Arrangement.—(8) Hygienic Law.—Agencies which injure the Memory.—Why it is particularly affected by Dissipation.—Practical Inference.—(9) Law of the Relation of Thoughts.—(10) Law of Recollection.—Where it applies, and how its Application may be facilitated.

THE power of remembrance varies greatly in different persons, independently of culture: but the power of remembering permanently, vividly, and without confusion, depends greatly on the way in which the faculty is applied; and we shall consider how this may be done to the best advantage.

1. The influence of *attention* on remembrance is so striking that it has been noticed from the earliest times. Things which we have seen or heard, are generally soon forgotten, if they excited little attention; and whatever excites strong attention, is generally remembered permanently. This holds true of such things as excite strong emotion, since these powerfully attract attention. A scene of overwhelming grief or ecstatic joy, for instance, is rarely forgotten. We may therefore lay it down as a law of remembrance, that *whatever strongly excites the attention is generally remembered, and whatever excites little or no attention is generally forgotten*. It is owing chiefly to the various directions of attention that several persons, who have all witnessed the very same scenes, remember different things; every one remembers what forcibly excited his attention, and forgets the rest.

In order to excite attention, one or other of two things is necessary. (1) The subject must spontaneously excite a strong curiosity, without any effort on our parts, as happens when we view a thing that appears very strange and striking, or read a very remarkable and affecting narrative. (2) If it be not intrinsically interesting, it must appear to have a bearing on something which we highly value, and forgetfulness or ignorance on the subject must be thought to involve serious consequences. These facts suggest the three following rules:

(1) *Endeavor to obtain a knowledge of the elements of a subject, and fix them firmly in the memory, during the first study:* for then it possesses the charm of novelty, which is favorable to close attention. If we follow a different course, and skim over the surface at first, it becomes more difficult to remember the subject afterwards, since the attention is much more apt to wander from the point under consideration.

(2) *If the subject possesses no intrinsic novelty, we should carefully examine its bearings, until we clearly see its importance, and its connection with the future.* When we distinctly perceive that much depends on our understanding and remembering a subject, there is little danger that the attention will flag, however uninviting it may be in itself. This is proved by the alacrity with which mankind do many things, much more disagreeable than to study attentively any subject which is worth studying, solely for the sake of earning a little money, or perhaps gratifying some favorite desire. The rewards of the explorer of truth are richer and surer; and, therefore, it is only necessary that he should understand their value, in order to work with attentive eagerness in the pursuit.

(3) *Overlook trifling details, or irrelevant matter; and attend only to what is worth remembering, and capable of being remembered permanently.* For the attention is distracted, and remembrance consequently weakened, by being drawn to many objects simultaneously. We should guard against the pernicious error of burdening and confusing the Memory by attending to many things at once, or by cramming it with a multitude of facts which are not worth remembering, or which are sure to be speedily forgotten. The cognitions embraced in all departments of knowledge are far too numerous to be retained

SEC. 1.]   GENERAL LAWS AND RULES.   415

by any Memory; and there are many truths the recollection of which tends rather to perplex and mislead the Judgement than to answer any good purpose. The consideration of them leads us to confound matters of no consequence with what is material, and to withdraw the attention from the essential parts of the subject.

In order to avoid such results, we should beware of paying any attention to things which were better overlooked, for the time being, and of attempting to commit to memory what is not worth remembering, or what cannot be remembered for any length of time. Where some important object calls our attention to anything of this kind, we should pay no more attention to it than the object requires.

We may now understand why a person already familiar with a subject remembers new discoveries regarding it so much better than others. His previous knowledge excites strong attention, on account of the interest he feels in it, while it guides his attention to those things that are most important. Hence, when we have once mastered the elements of a science, our future attainments in it are acquired and retained with unusual facility.

2. Memory is aided by clearness and distinctness of the original comprehensions. As ideas are copies of their prototypes, any obscurity or vagueness in the latter necessarily attaches to the former, and leads to confusion and forgetfulness. Hence *clearness and distinctness of comprehension are favorable to lasting remembrance.* Thus close attention to comprehensions aids remembrance as well as a right understanding of the subject at first.

As we can seldom form, from mere descriptions, conceptions so vivid as our own comprehensions of the objects, Memory is aided by subjecting to our apprehension either the very things to be remembered or accurate representations of them. Thus, we can remember a scene which we have witnessed much better than one of which we merely read a description; and we remember the positions of objects in a country much better by surveying a map of it, than by simply reading a description of those positions.

3. As a general proposition includes many particular ones, and yet is remembered as easily as one of the lat-

ter, *Memory is aided by the extension of generalization.* As inductions are wider than empiricisms, and at the same time more free from details, they assist memory as much as they extend knowledge; and this is probably one reason why mankind have been so prone, in all ages, to assume mere empiricisms for scientific generalizations.

4. *Remembrance is rendered more distinct and permanent by repeated examinations and reviews.* This truth is learned at a very early period of life; and the practice which it suggests is one of the most common means of committing to memory pieces of composition. Yet the value of mere repetition is generally overestimated. In order to render it effectual, it must be done deliberately and attentively: otherwise the whole is apt to be soon forgotten, as when we learn a string of words in an unknown language. It is further observable that a close mental review of a subject, soon after the original comprehension, without appealing to any external object except when the Memory is at fault, generally aids permanent remembrance more than a re-apprehending of the objects. Thus, we remember the contents of a book much better by reviewing it mentally after perusal, and referring to the book only when we are at a loss, than if we were simply to read it over a second time.

The cause of this is probably the greater degree of attention required to recollect than to re-apprehend an object. In the latter case the mind may be, in a great measure, passive, and wander incessantly to other matters, whereas, in the former, it is necessarily active and attentive; otherwise the process could not be performed. But where a long time has elapsed since the original comprehension, the case may be otherwise, as the ideas may then have become indistinct, or been wholly lost.

5. *A thing is committed to memory more easily by repeating and reviewing at short intervals, than by the same number of exertions, at long intervals.* In the latter case, one impression is, in a great measure, lost before another succeeds, whereas, in the former case, the succeeding impression seems to strengthen the former; and if the process is properly repeated several times, it may produce permanent remembrance, while the same number of operations, repeated at long intervals, might wholly fail to do so.

SEC. 1.]   GENERAL LAWS AND RULES.   417

When a subject is so extensive that frequent reviews of it are impracticable, we may draw up summaries of the most important parts, showing their relations to each other, and confine our reviews to these. In such cases abridgements and tabular synopses are of great use, as well as conversations on the subject with persons who know it well, or feel much interested in it.

6. *Fading similitudes are rendered precise and vivid by re-comprehending their prototypes.* Thus, if we attempt to recall the idea of a friend, whom we have not seen for many years, we may find that it is vague and indistinct: but if he come in sight, we may possibly recognize the smallest peculiarity in his appearance, the idea being now rendered very clear and accurate. Although this effect is partly temporary, it is by no means wholly so: for the idea continues permanently more distinct and vivid than before, especially if we have viewed the prototype with close attention. By a proper application of this principle, the attainments of early life may be, in a great measure, preserved through all our later years.

7. *Memory is aided by a systematic course of learning, and a proper arrangement of our acquisitions.* In all subjects, there is a certain relation between one part and another, which may be made subservient to the remembrance of both. In order to this, we have only to trace and attentively mark these relations; and then the idea of the one will recall that of the other. But this will not generally happen, unless we examine the various parts of a subject, in regular succession, according to their relations: for if we pursue a different course, the ideas of the various parts will be confused in our minds, and the connecting bond will be wanting.

The remembrance of a leading principle will generally secure that of its subordinate truths or consequences, provided we first learn the former, then proceed to the latter, and mark the relations in which they stand to each other. A proper system of classification will enable us to do this with little difficulty, as it renders the relations of the various parts to each other easy to be discovered.

In arranging our acquisitions we should be guided by the ordinary principles of classification, already discussed; and when we have once made an arrangement, on particular principles, we should not change it afterwards,

S 2

unless for some urgent reason, as the change confuses the Memory.

8. *The Memory is much influenced by the state of the nervous system.* This is frequently observable in severe injuries of the head, and in several diseases affecting the brain. A wound in the head has sometimes led to a person's forgetting a great part of his language, and much of his knowledge of other things. So everything which produces a languid or diseased state of the blood, weakens the Memory, owing to the injurious influence exerted on the brain. That this is not owing merely to distracting the attention, appears from its existing where there is no such distraction, although, in many cases, this difficulty also is added.

Dissipated habits affect memory directly by their deleterious influence on the blood, and consequently on the brain, while they affect it indirectly by destroying the power of close and continued attention. Hence the regular and temperate habits which are requisite for general vigor of intellect, are equally necessary for faithful and permanent remembrance.

What was formerly said regarding the propriety of attending to the laws of health, during investigation, is equally applicable to this subject, because every serious violation of these laws injures the brain, more or less, and thus impairs the power of clear and vivid remembrance.

9. *Certain thoughts are so related to each other that thinking one leads us to think the other.* This law operates so incessantly that it can hardly escape the notice of anybody; and, as we have already seen, Memory consists wholly in the faculty of thus passing from one thought to another, whence it appears that *direct remembrance is wholly dependent on the relations of thoughts.*

10. *Recollection is exercised by recalling the thoughts to which the one sought is related.* It frequently happens that, although we have no direct remembrance of the thing sought, we know that it is related to something which is remembered directly. Thus, we may have no direct remembrance of the time when we last saw a particular friend; but we may remember it was at such a place; and we may then directly remember the time when we were there. So the time may often enable us to recollect the place where something happened.

In order to facilitate the application of this law, we should note to what familiar thought the thing to be recollected is related: for by this means the former will be apt to be remembered when we are trying to recollect the latter.

§ 2. OF THE RELATIONS OF THOUGHTS.—Two kinds of Relations.—Laws of Natural Relation.—(1) Law of Contiguity and Succession.—Usual Course of the Similitudes.—Rule based on this Law.—Modifications.—How the Relation may be strengthened.—(2) Law of Resemblance.—Analogies.—Requisites to render it available.—How aided.—(3) Law of Emotions.—Its Operation.—Rule based on it.—Advantage attending it.—Important Difference.—(4) Law of Contraries.—Why less reliable than the preceding.—Arbitrary Relations.—Mnemotechny.—Harmony of Laws.—How new Acquisitions strengthen Remembrance.—Different Effects of Exercise on Memory.—Evils of Cramming.—Important Rule.—Means of widening the Range of Relations.—Why Sight generally affects us more than mere Description.—Disadvantages of visible Representations.—Effects of Desires.—Why some things are well remembered and others readily forgotten.

Some thoughts are so related or connected that one excites the other without any effort. Thus, the sight of one thing leads us to think of another which looks very like it; and when we think of a remarkable scene that we witnessed, we think of what occurred. Such connections may be termed *natural.* In other cases, a desire to remember a particular thing leads us to search for some familiar thought with which we may connect it, by a voluntary effort. Thus, if we know when Alexander the Great lived, we may remember when the emperor Constantine lived, by noting that he was just as long after the birth of Christ as Alexander was before it. Such relations may be termed *arbitrary* or *artificial,* as the thoughts attempted to be connected are not naturally related.

The following are the principal laws of the natural relations of thoughts.

1. The law of *contiguity and succession,* which may be expressed thus: *the similitudes of contemporaneous thoughts arise simultaneously, and, if not interfered with, in the order of their prototypes.* If we think, for example, of a certain place which we have lately visited, the ideas of the things seen and heard there arise at once, with the utmost rapidity, and also the similitudes of all our thoughts at that time, so that we know what emo-

tions we felt, what intellectual processes we performed, and what conclusions we deduced. Then the similitudes of subsequent thoughts arise, in the order of time, down to the present moment, if no volitions or other laws interfere. But a slight effort of the will may alter, or even reverse, the process, and lead us backward, through previous thoughts, or some resemblance or contrast between something in the series and some other thought, suggested either by some external object or by our own feelings, may lead the Memory into a totally different channel, so that, in the course of a few seconds, our thoughts may possibly have roved over the whole creation.

This law suggests the rule that *our thoughts should be concentrated on a few objects, where we desire that the remembrance of them should be clear and permanent.* For, if we act otherwise, so many similitudes arise afterwards, when we think of them, that the mind is apt to be confused, and led off from the path which we desire that it should hold. Hence many studies at one time are unfavorable to remembrance, independently of the bad effects of distracted attention while we are learning.

The law of *antecedent and consequent* is only a particular case of that of contiguity and succession, these always standing to each other in that relation. A special modification of this case is that of *cause* and *effect*, which are always related as antecedent and consequent. Another frequent modification is, the law of *premise* and *inference*, which are always contiguous in thought, although they may be separated in expression.

These relations are strengthened by other agencies, such as the suggestions of the Judgement, influenced by the desire of securing one object or avoiding another, and the easy transition from premise to inference, on account of the self-evident connection. Thus, when we witness a disastrous effect, the desire of removing it leads us to think of the cause, and so recalls anything which we may have formerly observed, bearing on this point. So, when we witness a powerful cause in operation, we are reminded of what we may have formerly observed regarding the effects of the same or similar agencies. Hence things connected as cause and effect recall each other more readily than mere antecedent and consequent. In the same way, when we think of premises, the inferences are apt to come into view, independently of previous reasoning; and hence these are readily brought to remembrance.

To the same general law belong various other laws, which are evidently nothing but modifications of it, such as that of *means* and *end*, *name* and *object*, *whole* and *part*, to all of which the remarks made in the preceding paragraph are applicable.

2. The law of *resemblance*, which may be expressed thus: *one thought recalls similar thoughts.* Thus, when we see an object which looks very like one that is familiar to us, we immediately think of the latter. So when we hear a musical air, like one which we admire, the idea of the latter immediately arises; and if a foreign field exhales a fragrance like those of our native place, we immediately think of the latter, and the scenes of childhood. One of the most common exercises of this law is in recognition, where the comprehension of the prototype recalls the similitude, as formerly stated, and thus we recognize and identify many objects formerly apprehended. Although apprehensions recall ideas more vividly than similitudes do, yet the connection extends equally to all.

This law extends to analogies, as well as to direct resemblances; and some of the most important applications of it are based on this property, as the success of a discoverer or inventor frequently depends on some analogy, suggested by what is already well known.

In order to render the law of resemblance sufficiently available, the points of similarity must be perceived; and the more numerous these appear, or, in other words, the more complete we observe the likeness to be, the more readily and surely will the one object recall the other. As the closest resemblances are not always apparent at first sight, the operation of this law will be facilitated and extended by our knowing the recondite, as well as the manifest, points of similarity between two objects; and hence remembrance is aided by marking the former as well as the latter.

3. The law of *emotions*, which may be expressed thus: *things directly connected with powerful emotions, are recalled with unusual facility.* Let a person, for example, attend to his ordinary business, just after hearing of the death of a beloved friend: his thoughts may, for the moment, be wholly occupied with his business; but, speed-

ily, and without any external cause, he thinks of the deceased: again he banishes the painful thought, and again it soon intrudes. On the other hand, a person who has just heard a very joyful piece of news, ever and anon returns to the agreeable theme.

Owing to the operation of this law, which acts with great force and constancy, things which have once excited strong emotions are apt to be easily recalled ever afterwards, independently of the greater degree of attention which they may have excited in the first instance. Hence *we should, if possible, place those things which we desire to remember, in such a light that they will powerfully affect our sensibilities.* If we do so, there is little danger that they will be afterwards forgotten.

A great advantage attending the operation of this law is, that those subjects which are most important excite the deepest emotions, when things are well understood and seen in their true bearings. The strong emotions requisite to permanent remembrance are by no means unfavorable to the acquisition of truth: for they rather secure than distract attention, and generally become strongest after the investigation has been concluded, as the very uncertainty of a proposition tends to moderate emotion, until its real nature has been ascertained.

Those violent and transient emotions termed *passions*, differ widely from the deep and permanent feelings which arise from taking a calm and extensive survey of an important subject. The former always spring from exaggerated, narrow, or one-sided views of a subject; and they are liable to occur only when we consider a subject of real or supposed importance, which we have never rightly understood or seen in its true light. They are decidedly unfavorable to the acquisition or retention of truth, since their violence concentrates the attention too much on certain points, and thus leads to others being viewed hastily or altogether overlooked. But those permanent emotions are not so violent as to produce any such results; and they generally arise from views essentially correct.

4. The law of *contraries*, according to which *thinking of a thing recalls its contrary.* When we are oppressed with the heat of summer, we are reminded of the cold of winter; and when we see a desert waste, we are apt to think of a fertile land. Here, however, the connection

does not seem to be primary and immediate; but some emotion or sensation suggests its contrary, by some intermediate steps. The uneasy sensation of heat, for example, leads us to desire cold; and the sight of the desert leads us to think of the cause of its barrenness, whence we pass, by a natural transition, to the fertile region.

Owing to its dependent nature, this law is less constant and reliable than the others, and of comparatively little value in aiding memory.

Arbitrary relations are frequently useful for temporary purposes: but they are seldom of much use in aiding permanent remembrance, as the relation selected is very apt to be forgotten. Rules for the formation of such relations are of no value whatever, since that which first offers is generally the best for effecting the temporary recollection for which alone these relations are of any use, as it is the one which will most readily suggest itself when required. Systems of Mnemotechny are applicable chiefly to dates and numbers: and even these can generally be remembered more effectually by other means.

Those relations which are aided by extraneous suggestions, like that of cause and effect, or premise and inference, are more reliable than such as depend solely upon the intrinsic power of relation, since the suggestions cooperate in producing the desired result. Hence relations based on some real likeness or natural connection of the things related, are more effectual than such as depend on fancied resemblances or casual juxtaposition; and, therefore, we should attend chiefly to the former. Thus, a public speaker may remember the different parts of his discourse much better by giving it a logical form, and marking the mutual connection and dependence of its several parts, than by attempting to connect them with the various rooms of a house, as was frequently done by some of the ancient orators.

In noting resemblances, we should attend primarily to the most important, or those which determine the general character of the objects, and attend to minor points of similarity afterwards, since the former are most easily remembered, and suggest the latter. Hence the laws of remembrance here harmonize with those of original acquisition.

Since every observed relation of one thing to another forms a connection, remembrance is strengthened as the number of such relations increases: and hence, as long as new acquisitions multiply such connections, the more we know, the better we remember. Such additions to our knowledge multiply the bands that unite its various parts in the Memory, while they are themselves associated with so many of the old elements that they are in little danger of being forgotten. A new discovery often unites into one whole many elements of knowledge hitherto apparently unconnected. It is thus that Memory is apparently strengthened by exercise: for if we merely load it with unconnected or unrelated facts, it becomes confused and weakened, instead of being strengthened. We should, therefore, never attempt to burden it with unimportant details, which are not worth remembering.

In order to strengthen the Memory, *we should mark the various relations which the thing to be remembered bears to several others that are well known, and avoid unimportant and unconnected details.* By this means it will become associated with all these, so that it may be recollected by thinking of any of them. Consequently the practice of observing the relations of new acquisitions to our previous attainments, is as favorable to remembrance as it is to discovery and invention.

A good means of widening the range of known relations is, to observe a thing in as many different ways as circumstances will permit: for it may thus become associated with the ideas of the different senses; and if one fail or hesitate, the others may still avail. Thus, in the case of chemical or mineral specimens, we may see, feel, smell and taste the substance whose properties we desire to remember.

Owing to the comparative feebleness of the power of Conception, in the great majority of mankind, things addressed to the eye produce a more permanent impression than mere descriptions: but the narrow limits of ocular representations render the latter generally indispensable; and, in many of the most important subjects, the former are unavailable, as no accurate representation can be given. Thus, the best representations of the solar system are so defective that the notions which they convey are extremely erroneous, unless they are corrected by the aid of description and conception, while, in the case

Sec. 1.]. GENERAL OBSERVATIONS. 425

of purely mental objects, such representations are often worse than useless.

Although desires and volitions are incessantly changing the previous currents of our thoughts, and substituting others, yet every train of thought strictly obeys the laws of relation: and hence these ever-varying currents may recall any former act or thought of our lives which is connected with others. Things much valued are well remembered, not only on account of the attention which they excite when they are thought of, but also because we voluntarily search for them, and dwell on them when they are presented to view, whence they become strongly connected with many other things. For similar reasons, things little valued are apt to be totally forgotten, except where they have severely pained our feelings.

CHAPTER XXIX.
OF THE RETENTION OF KNOWLEDGE BY MEANS OF EXTERNAL SIGNS.

§ 1. OF EXTERNAL SIGNS IN GENERAL. — Advantages of External Signs.—Various Modes of their Operation.—Monuments and Commemorations.—Advantages and Disadvantages of direct Likenesses, Symbols, and Phonetic Representations.—Best Course.

THE extent of human attainments is such that, after employing all the aids of Memory discussed in the preceding chapter, external signs are requisite, in order to secure the largest and most important portions of knowledge. For, although similitudes represent the things to be remembered, without any external sign, yet they are liable to be forgotten; and as they pass away with the individual, they cannot make knowledge permanent in a community.

External signs operate in various ways. When a person ties a string on his little finger, or a knot on his handkerchief, the sight of the unusual object recalls its origin, evidently in virtue of the law of cause and effect. Of the same kind are *monuments* and *periodic commemorations*. These preserve a knowledge of occurrences, because they excite curiosity and inquiries regarding their origin, so that every generation learn the cause from their predecessors, and afterwards communicate it to those who succeed. So the boundaries of land are frequently known from posts or stones fixed in the ground. But such devices, at the best, preserve only the remembrance of the principal facts or events: and, in order to retain a knowledge of details, we must have recourse to more varied and unequivocal signs.

Direct likenesses possess over all other mnemonic signs the advantages of representing the very things to be perpetuated, in a lively and striking manner, placing a scene before the eye without calling in the aid of Conception, and exhibiting some peculiarities which description cannot so well convey. They give us more accurate notions, in many instances, than can possibly be obtained from mere descriptions, while they render similitudes more precise and vivid, like re-comprehending their prototypes, so that, when very accurate and complete, their effects are not much inferior, in these respects, to the latter process.

Symbolic representations are employed chiefly where direct likenesses are inadmissible, owing to the thing which it is designed to perpetuate being invisible, or incapable of direct pictorial representation: and they sometimes set this forth more forcibly and comprehensibly than verbal descriptions, just as figurative expressions sometimes portray an occurrence more clearly and strikingly than those which are literally true.

Yet much is still requisite in which those devices wholly fail. Direct likenesses can generally represent only a small portion of what is visible, while the invisible, which is usually the most important part of the subject, cannot be thus represented at all. The range of symbolic likenesses, again, is very narrow, unless their number is multiplied so as to overburden the Memory, and render their import very liable to be misunderstood or forgotten. Hence the value of *writing* or *printing*, which, by means of characters denoting its elementary sounds, perfectly expresses language, so that we can accurately and easily retain whatever can be spoken.

Writing furnishes the means of expressing all human thought with great precision and the utmost generality, while the number of characters required is so small that the Memory need never be burdened in remembering them. It also removes the vagueness and uncertainty which frequently attach to all other methods, when we attempt to express details, or a continuous chain of events or thoughts; and it prevents the numerous mistakes and total loss of knowledge incident to relying on simple remembrance. Moreover, as the composition can be rewritten or reprinted without limit, it can be handed down unchanged through countless ages. Thus writing preserves the knowledge of past ages, and enables those of succeeding times to use or improve upon the attainments of their predecessors.

As every class of signs possesses advantages and disadvantages, the best method is that which combines the advantages of all, as the subject may require. The expressiveness and directness of pictures and solid representations may be combined with the brevity and generality of symbols, and the simplicity, comprehensiveness, ease and exactness of phonetic writing. We are thus furnished with the means of retaining all our thoughts with force, precision, perspicuity and accuracy, transmitting them to the most distant parts of the world or remotest posterity, and studying the very words of others as deliberately and frequently as we please.

§ 2. OF THE RETENTION OF KNOWLEDGE BY WRITING.—Requisites to render what is written available.—Things to be attended to, in writing for our own use.—How to be secured.—Abstracts.—Copying.—Various Courses. — Extracts. — Tables. — Caution. — Definitions.

When we have properly written down anything which we wish to remember, all that is requisite, in order to its being available afterwards, is, that the writing should be preserved, and that we should remember where to find it, and what is denoted by the characters and words employed. This we can generally do by means of the methods discussed in the preceding chapter.

In writing solely with a view to the retention of knowledge, all that we require to do, is, to express ourselves with perfect clearness and precision. But this is not so easily effected as we might suppose. While we distinctly remember what we have written, we may think that all our expressions are quite unobjectionable; but they may present a different aspect by the time we have forgotten what we wrote.

In order to determine whether our expressions are sufficiently perspicuous and exact, we should view them like one who knows nothing of the subject, except what he derives from the writing; and a little practice will enable us to do so with little difficulty. We shall seldom err if we write as we should do, in addressing a person to whom we desired to communicate clearly everything which is to be remembered, in such a manner that there would be no danger of his mistaking or hesitating, regarding our meaning.

When all that we require to remember has been properly written already, it will be unnecessary to copy it, except where we have not easy and constant access to the works in which it is found. But it is frequently advisable to make an abstract of a treatise, as this will fix the attention more closely than mere reading, and lead to our obtaining a clearer view of the subject. Mere copying is generally of little value, as the process is so purely mechanical that we may be thinking more intently of something else during the operation.

Facts which ought to be remembered, and yet are not contained in any composition within constant and easy reach, may be minuted in a common-place book. We may either have several books of this kind, for various subjects, or write our entries continuously in one, and, when it is full, take another volume. Wherever our minutes are voluminous on each of several subjects, it is better to adopt the former plan, as the different subjects will thus be kept distinct.

Where we desire to note something in a composition to which we have constant access, it will generally be sufficient to mark the passage, and give a reference to the book and page in a general index. But if the matter worth remembering is blended with much that is not, a minute in the common-place book may save much time when we afterwards come to refer to it.

In carrying out such plans, young learners are apt to collect a good deal of dross, along with valuable matter. But experience generally corrects this error, after a few years: and those who follow no such plan, will often find that they formerly read much which they would like to retain permanently, but which has now either gone beyond their reach, or lies they know not where.

In minuting matters which abound with details, much benefit may often be derived from forming synoptical tables, which may enable us to take a bird's eye view of the principal facts, and thus greatly assist remembrance, while they facilitate reference. Running the eye repeatedly and attentively over their contents, and then reviewing them, will generally fix them in the Memory better than all the machinery of Mnemotechny; and the remembrance of what is thus acquired will greatly aid us in recollecting the substance of the whole subject. Such aids, however, should be used only as auxiliaries, and by no means to the exclusion of regular narratives or expositions; for they necessarily present nothing but a skeleton of leading facts or events, and exclude the most interesting parts of the subject.

Wherever we have occasion to use a word regarding whose exact meaning we may possibly experience some difficulty afterwards, we should either properly define it, or refer to some accessible definition which we follow. Definitions are required only where a term is unknown, obscure, equivocal, or liable to be forgotten, as otherwise it can cause us no difficulty.

CHAPTER XXX.

OF THE MEANS OF POSSESSING A READY COMMAND OF OUR KNOWLEDGE.

§ 1. REQUISITES FOR POSSESSING A READY COMMAND OF OUR KNOWLEDGE.—Importance of this Subject.—Eight Requisites, with Remarks.

THE methods already discussed will generally secure knowledge so that it is not lost beyond the power of recollection or recall. Many things which never occurred to us for years, and which appeared to be completely forgotten, are often vividly recalled afterwards, by means of some relation or external object. But this degree of retention is quite insufficient for many purposes. For if we form our decisions, and act, upon partial views of a subject, while we overlook or forget important facts, the subsequent recollection of these will generally be too late. Knowledge forgotten or overlooked when it is wanted, is of little avail. The evils arising from this source are of frequent occurrence, and, in many cases, extremely serious. In order to avoid these, and to have our knowledge sufficiently at our command whenever it is needed, the following things are requisite.

1. The things to be remembered must have been attentively considered and well understood, when they were originally learned, or at least at some former time: for we do not well remember what we never rightly considered or understood.

2. They must have been repeatedly reviewed with attention, and their relations carefully observed, so that they are strongly connected with things which we are not in danger of overlooking or forgetting.

3. Our knowledge must be classified and arranged according to its natural connections: otherwise the preceding requisite cannot be secured; for if all our knowledge lies in a confused mass, it is impossible to discover the principal relations, and we must rely chiefly on casual associations, which rather lead us astray than guide us to what we ought to remember.

4. We must be systematic in our habits: for if we are accustomed to run from one subject to another, entirely unconnected with it, then back to the first, and so on, our knowledge of them will be so confused that the same evils will result which are mentioned in the preceding paragraph.

5. We must cultivate equanimity, and suppress anxious and violent emotions. These are almost as injurious in their influence on memory as they are in the original acquisition of knowledge: for they concentrate the attention on other objects than those with which it should be occupied; and hence we cannot call up our knowledge when required, whereas, if our minds were calm and unruffled, the requisite cognitions would readily arise.

6. We should not attempt to recall our knowledge while we are under the influence of strong sensations, whether pleasant or painful, for the same reasons as those just mentioned.

7. Attention must be paid to bodily health. Disease not only distracts the attention, by the painful sensations which accompany it, but it directly affects the Memory by its influence on the nervous system. Hence the advantage of a close adherence to the laws of health.

8. Where our knowledge is partly in writing, we must have some accurate notion of the nature of what is written, and where it is to be found: otherwise we cannot avail ourselves of it when required.

§ 2. MEANS OF ACQUIRING AND EMPLOYING THE PRECEDING REQUISITES.—Systematic Habits.—When a Change of Plan is desirable.—Common Error of Young Persons.—Evils of Hurry.—Difference between Reading and Study.—Means of securing Equanimity, and moderating Passion.—Proper mode of dealing with Sensations.—Power of bad Habits.—Means of securing Health.—Common Error regarding Writing.—Means of rendering Writing available.—Indexes.—Influence of the Judgement on Memory.

The modes in which the three first requisites mentioned in the preceding section are to be acquired, have sufficiently appeared; and therefore any further consideration of them is unnecessary.

In order to secure systematic habits, we must form some plan of proceeding, and adhere to it, as closely as circumstances will permit. Young persons are apt to abandon all system, as soon as they find that the one planned cannot be carried out, owing possibly to its being too rigid, and too minute in its details. Yet if we modify and simplify our original scheme, as experience shows to be desirable, we shall find that occasional interruptions and deviations will have little influence on the benefits derivable from systematic habits. If we find that we are frequently interrupted, during the time allotted to a particular pursuit, we may possibly alter the time to advantage. The circumstances of individuals differ so much that every one should form his plans to suit himself.

Young persons are apt to err in expecting to accomplish too much within a given time: and they often allot a few months to what will require as many years to effect properly. To have important results well accomplished, generally requires time: and to hurry through our studies and investigations, is a very bad course, because we thus generally learn little and that superficially, while even that is apt to be mingled with much error, and to be mostly forgotten, within a short time. It also leads to a habit of careless study and investigation, which may cause innumerable errors and misconceptions. Hurry, and the confusion and oversights that uniformly attend it, are as injurious to remembrance as they are to right understanding; and therefore we should avoid it, with the utmost care.

We should also guard against the common practice of cramming our memories with the mere statements of others, without ascertaining either their exact import or their character and bearings. As it is not everything which is swallowed that furnishes bodily aliment, but only food, which is properly digested and assimilated; so it is not what we read or hear that instructs the mind, but only truth, rightly understood and permanently remembered.

In order to preserve an even tenor of mind, and avoid the extremes of stormy passions, and the apathy which inevitably succeeds, we should note the blinding effects of the former on the intellect, and how completely they vitiate every investigation performed while we are under their influence or during the succeeding prostration. The habit of taking wide and close views of the subject, will also conduce greatly to the same result. By taking narrow and one-sided views of a subject, we are very liable to work ourselves into a passion about things which are easily seen to be quite insignificant, when considered closely and from the true point of view. Our feelings may be both deep and permanent, without ever rising into gusts of passion, which can always be subdued by extending our view beyond their exciting objects, or by observing these in their true bearings; and habit will make this course comparatively easy.

As to sensations, it is generally easy to abstain from important investigations or decisions while we are affected by any which materially interferes with attention. Every one will see the impropriety of raking the Memory, and deciding important questions, while he suffers acute pain, and the same objection exists in the case of all violent feelings, whether pleasant or painful. Many sensations are much influenced by our habits; and those formerly recommended will place the most dangerous of this class, such as the appetites, sufficiently under the control of the Judgement.

To secure health, we must know what its conditions are, and rigidly observe them: for if we once fall into the habit of neglecting them, we are apt to go on from bad to worse, one violation leading to a deeper, and the strengthening habits rendering a return to better courses more and more difficult. People often flatter themselves that they can easily abandon a habit when, in fact, they do not possess sufficient resolution and energy to do so. As a man who floats with the stream knows not the difficulty of stemming the current, so the slave of bad habits knows not their strength till he has wholly subdued them.

In ascertaining what are the laws of health, we must be guided by the disclosures of science regarding the structure and wants of our bodily organization, and the influence exerted on it by the various agents within and around it. To follow the opinions and practices of the ignorant, as good guides, is like adopting the views of a cheat as a correct representation of honesty.

With respect to writing, we must avoid the common error of thinking that when once we have a thing written down in manuscript, or in a printed volume, it is henceforth at our service, whenever it may be required: for it is liable to be overlooked altogether, or we may be unable to find it in time, or it may never have been properly understood. To obviate such difficulties, we should well understand what we read or write, and have matters which we require to refer to afterwards, entered in an index. We should also occasionally review our acquisitions, so that we shall have a correct idea of their nature, and neither search for what is not there, nor neglect what is, when there is occasion for it. If we have separate books for different subjects, each should have its own index, in which may be entered both what we have ourselves written, and what we have read and desire to note.

An index should be so constructed that it will enable us to find any passage to which it refers, without difficulty. In order to this, the initial word should be that which we are most likely to refer to; and, where there is room for doubt, there should be several, referring to the passage, so that we cannot fail to find one or other of them, without much trouble. An index can be made by ruling a blank book, and writing proper headings, or it may be purchased ready made from a stationer.

In forming a judgement on any important subject, we should carefully run over the various parts of it, till we have ascertained whether we have distinctly before us everything requisite to obtain a correct view of it: and, if we have any doubt on this point, we should search and think, till it is fairly removed. Such exercises strengthen the Memory, by concentrating the attention on particular things; and thus they not only bring up all our knowledge on that occasion, but they render it more serviceable for future use.

T

CHAPTER XXXI.

TABULAR VIEW OF THE MEANS OF RETAINING KNOWLEDGE.

KNOWLEDGE is retained by means of

I. Simple Remembrance or Recollection, dependent on
- 1. Attention.
- 2. Clearness and distinctness of comprehension.
- 3. Extension of generalization.
- 4. Repeated examinations and reviews.
- 5. System and arrangement.
- 6. Soundness of the nervous system.
- 7. Relations of thoughts, which are
  - (1.) artificial, and
  - (2.) natural, including laws of
    - (*a.*) Contiguity and succession,
    - (*b.*) Resemblance,
    - (*c.*) Emotions, and
    - (*d.*) Contraries.

II. External Signs, including
- 1. Commemorative signs.
- 2. Likenesses.
- 3. Writing and printing, rendered available by
  - (1.) Perspicacity and precision of expression,
  - (2.) Proper definitions,
  - (3.) Abstracts and common-place books, and
  - (4.) Tables and indexes.

NOTES.

NOTE 1, PAGE 13.

Some have defined Logic as "the science of the laws of thought." But this definition is far too wide: for it comprises a great portion of Psychology, even if we limit the word "thought" to *intellectual* thought, a limitation not justified by the long established usage of the language.

NOTE 2, PAGE 38.

The Aristotelians divide a proposition into three parts, the subject, the predicate and the *copula*, or a word connecting the two. Thus, in the proposition "man is mortal," *is* is the copula. I have not adopted this division, because it is based on a particular mode of expression, different from what is generally used. In ordinary language, no copula appears, except where the substantive verb *to be* is used, in some of its forms.

NOTE 3, PAGE 44.

What is loosely termed a "moral certainty," seems to be nothing but such a high degree of probability as will induce belief, in ordinary circumstances. The phrase is objectionable, since it leads us to confound certainty with a high degree of probability, things which are essentially different. The former does not admit of degrees; the latter does: the former is wholly unaffected by future discoveries; the latter may be reversed or destroyed by them.

NOTE 4, PAGE 50.

Reasoning has been frequently defined as a comparison of one thing with another, and observing whether they agree or disagree. But this definition is very faulty: for it misrepresents the nature of reasoning, and does not express its peculiar characteristic. There is in reasoning a comparison of one thing with another; but this is done in order to ascertain whether the one necessarily implies the other, while such comparisons as those expressed in the definition are made for a different purpose, and form no part of reasoning. Thus, if I look at two crows, perched side by side, and see that both are of the same color, here is a comparison of one thing with another, and a discernment of agreement, but no reasoning whatever, since there is no discernment of necessary implication or connection. So, if I see a crow on a snow-drift, and observe that the former is black and the latter white, there is a comparison of two things, and a discernment of disagreement, but no reasoning. In both cases, there is simply an apprehension of two things, observed, by means of abstraction, to be either like or unlike. On the other hand, when I see the crow before me, I know intuitively, and without any discernment of agreement or disagreement, that he is nowhere else at that instant. Reasoning is an application of Intuition; and, therefore, where the latter is not employed, the former cannot exist.

Note 5, Page 51.

The Aristotelians represent reasoning as a comparison of two premises, in order to evolve the inference, whereas it consists of comparing the premise with the inference, in order to ascertain whether the latter is necessarily implied in the former. Their view of it is substantially the same as that mentioned in the preceding note. Hence they represent a syllogism as consisting of four, instead of three, different parts, two premises, a connective, and an inference, which they term the conclusion. They maintain that it embraces three separate objects, or notions, two of which are successively compared with the third, in the two premises, and then pronounced, in the conclusion, to agree or disagree with each other.

The subject of the conclusion is called the *minor*, and its predicate, the *major term:* the premise in which the latter occurs is called the *major*, and the other, the *minor* premise. The subject of the major premise, which forms the predicate of the minor, is denominated the *middle term.* Let us take the following syllogism for an example:

    Every man is mortal; (major premise)
    John is a man: (minor premise)
    Therefore John is mortal. (conclusion)

Here *man* is the middle, *mortal,* the major, and *John,* the minor term.

This view misrepresents the real process of reasoning in such cases, which may be variously expressed as follows:
(1) John is one of a class individually mortal;   (premise)
    But whatever belongs to a class individually, belongs to every one of that class: (connective)
    Therefore John is mortal.   (inference)
(2) Every proposition which is true universally is true of every case included in it: (connective)
    Now it is true universally that every man is mortal; (premise)
    Therefore the man called John is mortal.   (inference)

The Aristotelian syllogism is unobjectionable only in those instances where the major premise is a self-evident truth which shows that the minor premise necessarily implies the conclusion, in which cases it evidently becomes equivalent to what I have termed the connective. The two kinds of premise are alike in form; and they never looked any farther: yet they are intrinsically different. Inductive propositions cannot form logical connectives, even when they are universally true, because they are not self-evident, and much less where they are only generally true, as in the instance just given.

Their vague and erroneous view of reasoning and syllogism have led the Aristotelians astray to such an extent that a great part of their Logic is worthless, or even worse. "'Terms" seem to constitute the corner-stone of their fabric, and they evidently attended chiefly to expressions, without sufficiently analysing the processes of thought. Like their master, they often lost themselves in words, and discussed merely different forms of expression, while they professed to unfold what these denote. The whole machinery of their "moods," "figures," and "rules for reduction," are useless as an intellectual exercise, and positively detrimental in the actual pursuit of truth, since they only cloud and clog the investigator.

NOTE 6, PAGE 54.

Some affect to doubt whether extension is infinite; but if they seriously attempted to determine its boundaries, they would probably think otherwise. Boundaries to extension are evidently as impossible as a termination to duration. If we think of any part of extension, we know there is space all round it; and if we think of any part of duration, we know that part preceded and part succeeds. Hence the former is infinite, and the latter eternal.

Another absurd dogma is, that duration and extension are only conceptions of the mind, and have no existence beyond it. There is nothing more self-evident than that these two things necessarily exist, and that they are immutable, and independent of every other thing. The dogma probably arose from confounding their nature with the faculty by which we discern it, as if one should maintain that the quality of hardness exists only in the points of our fingers, because it is through these chiefly that we learn its nature. Instead of duration and extension existing only in the mind, the latter and every other real being exist only in duration and extension.

NOTE 7, PAGE 61.

The essential nature of *change*, *cause*, and *power* are known intuitively; and, therefore, the terms expressing them neither require nor admit of any real definition, any more than such words as *blue*, *sweet*, *sour*, *pleasure*, *pain*, &c., but their precise signification is learned by simply considering what they are.

NOTE 8, PAGE 61.

It has been said that the 11th principle is not even true, and much less self-evident, for that the Sun attracts the Earth without any medium. This objection is like declaring that a man may be in two places at the same instant, for that we often see ourselves on the north side of a room when we are actually on the south side, or that the three angles of a triangle are not exactly equal to two right angles; for that, if we actually measure them, we shall always find them either a little greater or a little less. That the Sun really attracts or *draws* the Earth, or any other planet, is an absolute impossibility, because it has no hold on them, and consequently it can no more draw them than it can draw empty space. As action is only a kind of motion, the principle is only another form of saying that a being cannot be in motion where it is not, which is as evident as that a man cannot be where he is not.

NOTE 9, PAGE 71.

The Aristotelians seem to confound *judgements*, or conclusions from premises, which have been investigated and are believed to be true, with *propositions*, without adverting to the fact that the latter may be wholly mental. Some of them seem to have been aware that we reason in forming judgements, yet to have erred in supposing that this is different from ordinary reasoning, and also in thinking that we reason whenever we discern the truth of a proposition, which is by no means the case. All discernments are known directly, without any reasoning whatever.

Kant's *analytic* judgements are simply truisms, while his *synthetic* judgements embrace both intuitions and other propositions, so that his division only increases confusion. In order to determine the logical character of a proposition, we must evidently analyse it, so as to understand what it is; and then, in order to determine whether it is an intuition or not, we must compare the subject with the predicate. Consequently his analytic judgements are equally synthetic, and conversely, while neither class can properly be termed judgements at all. The truth of truisms is discerned precisely like that of other intuitions, and the distinctions drawn by Kant are nugatory. Thus we know that "a man cannot be in two different places at the same instant," just as one knows that "every Englishman is a man."

Note 10, Page 72.

One of the most remarkable cases of spectral illusions of which I have ever learned, is that related by Dr. R. Patterson in the *Edinburgh Medical and Surgical Journal*, for January, 1843. In that instance, a man saw the figure of a deceased friend, heard him speak, and also felt him pinch his arm. The specter was so distinct that he could perceive the color of the clothes: yet he adds that it was dim and imperfect throughout, and that it could not for a moment be considered a real object. The illusion regarding his arm, he attributed to cramp of the triceps muscle.

Note 11, Page 77.

If any person should think that he is immediately conscious of producing the changes consequent on his volitions, I answer that a careful consideration will show we are conscious of nothing but the volition and simultaneous comprehensions. We are no more conscious of moving our arms than we are of inhaling the air which is forced into the lungs by the pressure of the atmosphere, when we expand the chest.

Note 12, Page 98.

Direct likenesses appear to have been the first signs employed for perpetuating knowledge, as they are more obvious and expressive than any other; and some races, like the Aboriginal Americans, never advanced beyond this method. It was probably followed by symbolic writing, as being the next link in the series. This method has been most extensively used by the Chinese, who never went beyond it.

Phonetic writing was probably invented later than the symbolic, because, although the simplest and most complete, it is the least obvious and most recondite; yet it is so ancient that its origin and early history are, in a great measure, lost in the mists of antiquity. So far as we can learn, however, the invention appears to be due to the ancient Egyptians, who certainly practised it at a very remote period. Theory would lead us to assume that the earliest writing of this kind was *syllabic*, like the Cherokee alphabet of George Guess, and that of the Veh people in Africa. But of this there is no clear proof; and we find characters representing letters, and not syllables, in the earliest extant specimens of phonetic writing.

From Egypt letters passed to the Hebrews, Phœnicians, Assyrians, Babylonians, and Indians. The Phœnicians introduced them into Greece, whence they spread over Europe; and thence they passed into America.

The first letters were complete pictures of visible objects, the power of the letter being the initial sound or articulation of the object's name. Thus, an ant, an apple, or an axe might stand for A; a book, a bee, or a box, for B, and so on. Such are the letters found on the Egyptian monuments, the hieroglyphics being chiefly inscriptions in such letters, often blended with symbolic and pictorial figures. Even in several of the Roman characters, the original forms of the objects are still apparent. Thus, A was an ox's head, and D, a door, two of the angles having been rounded off for convenience of writing; O was an eye, T, a cross, and U, a hook. The Hebrew names of the letters still indicate the original objects; for those of the letters just mentioned signify *ox, door, eye, cross*, and *hook*.

The primitive literal system was afterwards improved by simplifying the forms of the letters, and employing only one form to denote one power, whereas originally several objects were employed, whose names began with the letter. The Greeks perfected the art of writing, by expressing all the vowel sounds, which does not appear to have been previously done by any nation. It was not till upwards of fifteen hundred years afterwards, that the Syro-Arabian races attempted to supply this defect, by the clumsy invention of the vowel points, after the old pronunciation had been lost, while the ancient Hindoos did not even make the attempt.

NOTE 13, PAGE 107.

An analysis of the inductive processes shows the futility of the distinction that some have drawn between what they term *deductive* reasoning, *syllogism* or *ratiocination* and *inductive* reasoning or *induction*.

We may also see the absurdity of maintaining that a new kind of Logic was invented in the seventeenth century, which they distinguish as the *Inductive* Logic, accompanied with a new method of investigation. Induction indicates only the quantitative relation of the premises to the conclusion, not the nature of the reasoning by which this is established; and every kind of inductive process was practiced in ancient times. The more rapid progress of knowledge, in modern times, has been owing, not to any new method of investigation, but to several other causes, some of which are quite obvious.

Another similar error is, dividing all reasoning into *deductive* and *inductive*. The former term is applied to those syllogisms where we infer that a particular case of a general proposition is true, provided the latter is true; and, consequently, when valid, it is confined to necessary truth. By *inductive* reasoning they understand syllogisms in which it is inferred that a general proposition is true, provided every particular proposition embraced in it is true, or has been previously found to be so, which is evidently nothing but empirical generalization, and which does not, in fact, include any real process of induction. Moreover most of our reasonings are quite different from either of these processes, since both the premises and inferences are particular. This is usually the case in the ordinary affairs of life, and very frequently in scientific and historical investigations.

NOTE 14, PAGE 128.

Some authors have attempted to go beyond Consciousness, and prove its faithfulness: but they are obliged to take this for granted while they attempt to prove it; and, therefore, they reason in a circle. Thus, Kant attempted to prove the possibility of intuitions; but he was obliged to assume at the outset, not only their possible, but their actual truth, so that his proceeding was as illogical as it was preposterous.

So when Fichte says that the *Me* puts forth a spontaneous effort, and, meeting an impediment in something external to itself, Consciousness results, he overlooks the fact that we can know nothing of a spontaneous effort of the *Me* without Consciousness, and that such an effort without Consciousness is a self-evident impossibility. Again, when he says "A is equal to A," I ask how does he know this? The only rational answer is, that it is self-evident, and requires no proof. I reply, it is self-evident that every other proposition which has the same amount and kind of evidence is equally certain, and equally unsusceptible of any proof which can add to its certainty; and of this class is every proposition expressing an intuition or any present comprehension.

Similar remarks apply to Hegel's process, when he begins with pure nothing, and tells us that something added to nothing, makes something, which is only a particular case of the general intuition that a thing is what it is.

The doctrine of the Pyrrhonists or universal sceptics, labors under the same difficulty, and is liable to the same objection. Every thinking being necessarily believes some proposition as certain, if it be only the reality of his present thought: and, therefore, the Pyrrhonist, in expressing himself sceptical concerning his doubts, only flies from one certainty to plunge into another, which is not a whit more certain than any other discernment.

NOTE 15, PAGE 170.

The law of gravitation is often expressed by saying that "matter attracts matter, directly as the mass, and inversely as the square of the distance:" but it would be much more correct to say that "ponderable bodies are urged towards each other, by a force which varies directly as their mass, and inversely as the square of their distance," or, that "every tangible substance tends to move towards every other, with a force which varies directly as their mass, or quantity of solid matter, and inversely as the square of their distance." The proofs given of this law are quite fallacious, so far as they attempt to show that there is any real attraction, their authors having overlooked the fact that all the phenomena may result from a compulsive, instead of an attractive force.

The heavenly bodies have no hold on each other; and without this it is manifestly as impossible for them to *attract* or *draw* each other, as it is to lift a stone from the ground without having anything attached to it by which it can be lifted. Attraction without connection is a manifest impossibility. Another difficulty, in the way of attraction, is, that the bodies are inanimate; and therefore it is evidently as impossible for them to move either themselves or other bodies as it is for a rock to move itself from one mountain to another.

Note 16, Page 200.

The word *interpretation* was formerly employed in a wider sense, so as to include what we now term *translation:* but as the latter differs essentially from what is now generally understood by the former term, I have used this in its restricted signification. The old use of the word has evidently misled some logicians, regarding the true nature of interpretation. They say that the interpreter should be thoroughly acquainted with the language, and familiar with the subject of which the writing to be interpreted treats. These remarks are applicable to translation; but to apply them to interpretation reminds us of the old advice that we should not go into the water till we have learned to swim: for it is only by numerous exercises of interpretation that a thorough acquaintance with the language can be acquired; and we frequently study a work which treats of a subject regarding which we know little or nothing, and which we expect to learn from it.

Note 17, Page 244.

The ordinary names of the various divisions of organic nature are mostly Latin; and, even when they are derived from some other source, they are usually given in a Latin form. Attempts have been made, by several naturalists, to substitute terms in their own vernacular languages: but none of these have yet been generally adopted; and it is to be hoped that they never will. For the Latin terms possess the great advantage of being familiar to naturalists throughout the world, while they are free from the peculiar difficult or repulsive sounds that occur in others. It is evidently very desirable that scientific terms should be common to all mankind, without any change of spelling or pronunciation; and, therefore, where such terms exist, they should be retained, till they can be superseded by better, of which there is no immediate prospect.

The ordinary mode of naming genus and species is, to take the Latin name of some well-known genus for the generic name, and that of the other species, in the same language, for the specific designation. Thus, *bos*, the generic name of the ox tribe, is the Latin for the common ox, and *bubalus* for the common buffalo. The species whence the generic name is taken, is distinguished by some peculiar Latin epithet. Thus, the common ox is *bos taurus*, the latter word being the Latin for the common bull. So the cat is distinguished by adding to the generic term *felis* (which is the Latin for the common cat) the specific term *catus*, a Latinized form of *cat*, a word which is found in several of the Aryan languages. The specific name for the common cat is, therefore, *felis catus*, the lion being similarly denominated *felis leo*, the tiger, *felis tigris*, the puma, cougar or American lion, *felis concolor*,—the panther, *felis pardus*, and so forth. So the dog is termed *canis familiaris*, canis being the Latin for dog: *canis lupus* is the wolf, and *canis vulpes*, the fox.

If the species or genus has no Latin name, naturalists adopt the native name, or one which indicates a striking peculiarity of the division, or the name of its discoverer, or of some of his friends or favorites. Thus, *bos arnee* is the specific name of the arnee, or Indian buffalo—*bos grunniens*, that of the yak, or grunting ox of Central Asia—and *bos caffer*, the South African buffalo. So the gigantic genus of conifer recently discovered in California, is termed by some *Washingtonia*, and by others *Wellingtonia*, while a species of pine lately discovered in the same country, is termed *pinus Jeffreyi*, from Jeffrey, its discoverer.

It is to be regretted that, in naming newly discovered groups, the principle has not been universally followed of adopting a name which indicates the most remarkable peculiarity of the division, since other methods furnish no direct information regarding its nature.

NOTE 18, PAGE 252.

The *categories* of Aristotle consisted of a few heads, such as *time, place, quality*, &c., to one or other of which it was supposed every question might easily be referred; and they were apparently designed to assist in finding and keeping in view the point at issue in an investigation. But all such attempts are futile. For either the heads will be so general as to be good for nothing, or the enumeration will run into a mass of details too burdensome for the memory, and possibly inaccurate or incomplete after all.

A careful consideration of the subject will always show the nature of the question much better than any categories: for many questions are of a complex nature, and the inquiry branches into several heads, so that the categories would mislead, rather than guide.

NOTE 19, PAGE 334.

I use the singular, instead of the plural, form of the names of several sciences, not merely because all nations except those who speak English, do so, but because the sciences are one, and, therefore, the singular is the proper form. It is, in reality, as absurd to talk of *Mechanics, Optics* and *Ethics*, as to speak of *Logics, Rhetorics* and *Astronomies*, although our ears are more familiar with the former than with the latter terms.

NOTE 20, PAGE 340.

Ethnography differs so widely in its nature and subjects from what is properly termed *Geography* that they ought to be distinguished by different names. The latter is purely scientific, while the former is only partially so. Yet its truths possess so much generality that they cannot properly be classed with History or Biography, which discusses only particular facts and occurrences.

NOTE 21, PAGE 347.

As the general uses of knowledge have been discussed in the Introduction, those pointed out in Part IV. are only the special uses of the various branches.

NOTE 22, PAGE 347.

Various methods have been employed in establishing the fundamental principles of the Higher Analysis, or, as it is often termed, the Infinitesimal Calculus. But the method of *limits*, or of *prime and ul-*

timate ratios, which was first employed by Newton, is that which is now generally adopted; and it appears to be the best, as it leads to the required principles by the clearest and easiest steps. The principal difficulty attending it is, that we cannot comprehend the infinity of changes or variations which it assumes; and hence we are apt to conclude that they cannot be. But the following simple theorem enables us to surmount the difficulty.

A quantity which gradually diminishes, so that it becomes less than any assignable quantity, vanishes, or becomes nothing.

If the quantity does not become absolutely nothing, let its least value be x: then, since it becomes less than any assignable quantity, it becomes less than $\frac{x}{1000}$: that is, a quantity becomes less than the thousandth part of itself, which is absurd. Therefore the diminishing quantity becomes nothing.

It follows, as a corollary, from this theorem, that when a quantity approaches indefinitely near another, it *ultimately* coincides with it, or, as it is otherwise expressed, it becomes equal to it *at the limit*.

Another, and more general, corollary is, that a quantity which becomes less than any assignable quantity, may be rejected in a final expression, without changing the value of the expression. The former corollary is only a particular case of this one. For, let C be a constant quantity, x a variable which approaches indefinitely near it, and $d\,x$ the quantity by which x differs from C: then

$$x + d\,x = C;$$

and, therefore, by the second corollary, when $d\,x$ becomes indefinitely small, or less than any assignable quantity,

$$x = C,$$

which is the first corollary, analytically expressed.

These theorem and corollaries also furnish the best foundation for the doctrines of proportion, as they enable us to treat incommensurables like commensurables.

Note 23, Page 352.

To the inductive laws of motion usually given by writers on Mechanic, should be added the following:

The momentum of a body is proportional to its mass multiplied by its velocity. This law is sometimes introduced as a definition: but it is evidently a theorem; and it is established like other inductive laws of motion.

The *momentum* of a body means its moving force, or its power to move, penetrate, break, tear, or crush, another body with which it comes in contact, or which it otherwise affects. The *mass* of a body means its quantity of solid matter, which is generally measured by its weight. The *velocity* of a body is either *actual* or *virtual*. The former is that with which it actually moves: the latter is that with which it would move, if some counteracting force were withdrawn, and is that meant in the proposition.

Note 24, Page 358.

Attention has now been so extensively directed to every branch of Astronomy that there is little probability of any entirely new field being discovered; yet the immense subject of the fixed stars still offers many problems for future observations, which it will require many generations to solve; and even the field of the solar system is by no means exhausted.

NOTE 25, PAGE 362.

The identity of the law of intensity of light, heat and electricity with that of the force of gravitation, is an indication that the phenomena of gravitation result from undulations of ether, passing incessantly, at very short intervals, through every point, in all directions. The fact that we can clearly see any one point in a room from any other, proves that this is the case with the waves of light; and it may, therefore, hold true of those of gravitation.

If we suppose these undulations so small that they penetrate through all ponderable bodies, and impinge against the atoms which compose them, those phenomena will be a necessary consequence. A single atom would be apparently unaffected, since the forces acting on it in all directions would be equal: but when there were two atoms, the waves on the outer sides would force them together, because those flowing in the opposite direction were stopped by the intervening body. The nearer the atoms, the more interfering waves would be stopped; and the force would follow the law of the inverse square, for the same reason that a person at two yards from a fire receives only one fourth as much heat as when he is one yard distant. As every atom would be similarly affected, gravity would vary directly as the mass, or number of atoms.

According to this view, gravitation is a *compulsive*, and not an *attractive* force, as it is constantly termed; or, in other words, it is a *pushing*, and not a *pulling* force. It would also follow that gravitating bodies do not in any way affect each other, except where they are in contact. But these results can form no real objection to the doctrine: for the demonstrations given of the law of gravitation wholly fail to prove that there is any real attraction, or that the bodies really affect each other. All they prove is, that the bodies move towards each other; and this is explained by the one theory quite as well as by the other.

We may suppose that the waves of light, heat and electricity originate in the gravitation waves being disturbed, and thus generating them, owing to the peculiar form, position or motion of the atoms of ponderable bodies, although these positions and motions might have been produced, in the first instance, by the gravitation waves themselves.

If we farther suppose that the gravitation waves sometimes impinge against substances which they cannot permeate without causing their parts to adhere closely to each other, either directly or by means of new waves, we should have an explanation of cohesion and chemical affinity.

The elasticity of bodies might be explained by supposing that wherever some of the atoms were pulled a little apart while others were forced nearer than usual, new waves were generated which exerted a repellent action on the latter, and a compulsive force on the former.

The peculiar phenomena of heat and electricity are easily explained on this hypothesis. Thus, the melting and expanding power of heat, and its influence on chemical action, would result from its strong waves simply counteracting the cohesive force of those of gravitation, as in fluids; or they might even produce a repulsive contrary force, as in gases. In both cases we might expect the distance of the atoms from each other to be increased, which would account for the expansion that usually takes place, although the form of these atoms might be such that they would become more compactly arranged, and consequently occupy less space, on being heated, as in the case of melted iron, and water near the freezing point.

The effects of heat and electricity on chemical action would result from their loosening the particles, and thus favoring a combination of the elements, in some cases, while the increased expansibility of a gas would cause it to fly off, in other cases, as in the common process of burning limestone, or what is technically termed *roasting* metallic ores.

The attractive and repulsive phenomena of electricity would be produced by its waves interfering with each other, according to the various directions in which they flowed. So, latent heat would result from the waves being destroyed by resistance in one case, like those of light falling on a black body, and being again re-generated from the rapid motion of the particles, in assuming their former state.

Thus the gravitation waves, and the peculiar forms of the various ponderable substances, may account for a great portion of the phenomena of the physical creation, as their immediate causes. But we must look to a presiding Intelligence, not only to form and arrange those substances originally, but also to sustain the motions of the ethereal particles, by continued and most powerful acts. This is possibly effected by undulations propagated from a center, as light proceeds from the Sun, and reflected from the various points of a solid sphere, surrounding the visible creation.

If it be objected that this arrangement would render gravitation of unequal force, I answer that we have no proof it is not so; we do not know, for instance, that the force of gravitation is the same at the nearest, and much less at the remotest, fixed star, as it is at the Sun; and the space occupied by the solar system is only a very small part of the universe. It may be further answered that the particles of ether may be so arranged as to compensate for the unequal action of the central force, in different parts of space.

Note 26, Page 364.

Organic specimens have generally been preserved by drying them, or putting them in alcohol; but they could be preserved in a vacuum better, probably, than by any other means, although this method has been little used, if at all.

Note 27, Page 373.

The grand outlines of Geology have already been clearly traced: but the science is very extensive, and much remains to be done, in filling up details, modifying propositions which may have been too loosely or generally expressed, and settling doubtful or disputed points. Several of the terms, also, should be superseded by others, of a more general and scientific character.

Note 28, Page 376.

Logic properly embraces the few truths relating to the general properties of beings, which have been attempted to be formed into a separate science, under the name of *Ontology*. This is variously defined as the science of "being in general," and "that which investigates the nature and properties of being or reality, as distinguished from phenomena or appearances." Our knowledge of such properties is wholly intuitive, and far too scanty to form a science.

What some term the science of *Æsthetics* is only a part of that division of Psychology which treats of the emotions. Discussions regarding the modes of producing æsthetical emotions belong to art, and not to science.

To Logic and Psychology belong most of the subjects discussed under the vague names of *Pneumatology* and *Metaphysics*, and all the rest properly belong to other branches of knowledge, so that those divisions should be discarded.

NOTE 29, PAGE 385.

The subjects discussed under the name of *Political Economy* belong partly to Morality, partly to Jurisprudence, and partly to Technology. The combination forms a compound of incongruous elements which were much better discussed under the heads to which they respectively belong. What relates to laws, is best discussed in Jurisprudence; the duties of an individual, in regard to his vocation, belong to Morality; and the mode in which an art or profession should be carried on or exercised, in order to make it most useful or profitable, rightly belongs to Technology, and has nothing to do with political science.

NOTE 30, PAGE 390.

From misunderstanding some passages in the eleventh chapter of Genesis, many have supposed that the original language of mankind was miraculously formed into several at Babel. But there is no real foundation in Scripture for any such opinion. The literal rendering of the first verse of the chapter referred to, is—"And the whole Earth was of one lip and of one words." The last term certainly means language, as it does not admit of any other interpretation in this place: and, consequently "lip" must mean something else; otherwise the expression would be absurdly tautologous: and it is observable that the expression "words" is not repeated, in any part of the narrative. The literal rendering of the last clause of the seventh verse is—"that a man will not hear the lip of his neighbour," which is well rendered in the old Greek and Latin versions—"that one will not hear the voice of his neighbour." The word rendered "hear" (which is of very frequent occurrence in the Old Testament), properly signifies *hear* in Hebrew as much as this does in English. It is sometimes employed figuratively to denote *understand:* but this is unusual. Here it evidently means *listen to* or *regard;* and the confusion spoken of was simply dissension, arising from differences of opinion, the instrument, by a common figure of speech, being put for that which is expressed.

The more closely and extensively languages are examined, the more irresistible appears the evidence of their common origin. The old philologists were often mistaken in their views regarding the derivation of languages: for when they found several that closely resembled each other, they inferred that all the rest must have sprung from the one which they thought the oldest, a process like inferring that the oldest-looking of several sisters must be the mother of all the rest. Yet the affinities on which they argued were mostly real.

The theory which attributes the affinities of language to the similarity of the organs of speech, in the various races of men, is refuted by several well-known facts. Thus, the Turks resemble the Germans in physical structure, much more than do the Hindoos; and yet the languages of the latter resemble the German much more closely than the Turkish does. Again, words which appear to be derived from imitating natural sounds, widely differ, in various languages, while words which have no such origin, are alike. Thus, the words for *weep* are entirely different, even in several of the kindred Aryan languages, while the word *sack* is found, with the same signification, in several distinct classes of languages, along the whole length of the old world.

If we request several unconnected persons to imitate some natural sound, such as the note of a bird, they will generally pronounce it very differently, while the similarities of language are apparent in words whose particular form must have been casual, and also in the structure of language, as well as in the vocabularies.

To compare human speech, with its myriads of words and its complexity of structure, to the few instinctive and inarticulate cries of the lower animals, could proceed only from persons who never properly examined the subject. To those who have, the comparison will appear absurd, as it really is.

It is observable, however, that the facts just mentioned do not warrant Müller's theory, that language has sprung from "phonetic types, produced by a power inherent in human nature." For words have been so much changed that the onomatopoetic origin of many will have disappeared in that way, while, in other cases, it will have vanished by mere epithets or learned terms having usurped the original words. Of this, the *whip-poor-will* furnishes a recent instance. Although the bird is generally known by no other name throughout the United States, yet its scientific name is totally different, being *caprimulgus vociferus*.

That language originated partly in onomatopœia and ejaculation, admits of no doubt; and when we consider the power of man to form compounds, and the natural tendency to contraction and alteration, during successive ages, we shall see that those two sources, taken in connection with man's ordinary intellect, are amply sufficient to account for all the phenomena of speech. The supposed "inherent power" is, therefore, destitute of a tittle of evidence that it ever existed: that it does not exist now, is admitted. The slightest application of the doctrine of permutations and combinations, will show that onomatopoetic and instinctive ejaculations would readily produce many more words than the total number of primitives found in any language.

The more complex structure of Sanscrit and Greek, compared with English or French, has been applied as an argument in support of the original superiority and divine origin of language. But, besides the facts already stated, there are several others, which show the worthlessness of this argument. Many of the aboriginal languages of Africa and America are much more complex than either Sanscrit or Greek; and yet they are found ill adapted for conveying thought on abstract subjects, with either force or precision.

The copiousness of inflection, which has been frequently adduced as a striking proof of the superiority of the ancient classic languages, originated chiefly in colloquial blunders, or in confounding distinctions which ought to have been preserved. The personal inflections of verbs, for instance, arose from confounding the personal pronouns with the verbs of which they were nominatives.

As a simple machine, which performs well all the requisite functions, is superior to one which contains many useless parts; so a language which contains no useless inflections, is, so far, superior to one which exhibits many such cumbrous appendages. Thus, in the instance just mentioned, it is better to indicate the person by a separate word, as in English, than to blend nominative and verb, as is done in Latin and Greek. The natural consequence of the latter practice has been, that the meaning of the affix was lost; and hence the pronoun came to be repeated, or used superfluously. *Ego am-o* is—I love-I; *illi am-ant* is—they love-they; and *homines dic-unt* is—men say-they.

The wide room for variety in the collocation of words, allowed by the classic languages, tended to produce confusion in the speaker or writer, and impeded a right understanding of his expressions, on the part of his hearers or readers.

As every ancient language must have undergone numerous alterations from time to time, long prior to the invention of writing, it follows that the original language of mankind can nowhere be found. Nor can we say, with any degree of certainty, what language most resembles it. But as it must have been a very rude and scanty speech, the question is one of little interest, and of no practical importance.

NOTE 31, PAGE 394.

Although much has been already accomplished, yet the application of scientific principles to the improvement of the arts still presents a wide field for invention. For many of those who attended to this subject either failed to see the most pressing wants of art, or mistook the best modes of supplying them. Hence their inventions were either unimportant or inefficient.

NOTE 32, PAGE 403.

Besides the Christian, the following are the principal epochs used in History:

1. The *Creation of Man*, 3760 years before Christ, according to the Hebrew reckoning, or 5509 according to the Septuagint. This epoch is used by the Jews, and often by Christians, in treating of events prior to the Christian era.

2. The *Olympic* epoch of the Greeks, 776 years B.C., from which they reckoned by olympiads, or periods of four years.

3. The *Building of Rome*, 752 years B.C. This epoch was long used by the Romans.

4. The epoch of Budda, 544 B.C., used by the Buddists throughout Southern and Eastern Asia.

5. The *Samvat*, or era of *Vicramaditya*, beginning 56 years B.C., used by the Northern Hindoos.

6. The *Saca*, or era of *Sulwanah*, commencing A.D. 78, used in Southern and Western India.

7. The *Hejira*, or flight of Mohammed from Mecca, 16th July, A.D. 622, generally used by Mohammedans.

INDEX.

A.

Aberrancies, definition of, 250.
" illustration of, 251.
" of confusion, 306–316.
" of appeals to authority, 316–320.
Aberrancies, of appeals to desires, 320–324.
Aberrancies, table of, 327.
Abridgements, how distinguished, 233.
Absolute proposition, definition of, 39.
Abstract quantity, principles regarding, 54–58.
Abstraction, definition of, 35.
" two kinds of, 35.
Abstracts, uses of, 138, 417.
Accumulating probabilities, sophism of, 299.
Acoustic, definition of, 335.
" foundations and uses of, 354.
Adopting a mean, paralogism of, 280, 281.
Æsthetics, its nature, Note 28, 446.
Ætiology, definition of, 336.
Affirmative proposition, definition of, 30.
Agencies (see Causes and Determining Conditions).
Algebra, definition and divisions of, 333.
Algebra, uses of, 346, 347.
Alphabets, origin of, Note 12, 438, 439.
Altering propositions, sophism of, 288, 289.
Alternative proposition, definition of, 39.
Ambiguity, latent, 207.
Ambiguous expression, paralogism of, 276.
Ambiguous expression, modifications of, 276, 277.
Analogy, definition of, 158.
" uses of, 158, 159, 178.
" abuse of, 159.
Analysis, Mathematical, definition of, 333.
Analysis, Mathematical, characteristics of, 345.
Analysis, Mathematical, uses of, 346, 347.
Analysis, mental, remarks on, 160.
Analytical Geometry, definition and divisions of, 333, 334.
Analytical Geometry, its relation to Analysis, 345, 346.
Analytical Geometry, uses of, 350.
Anatomy, definition and divisions of, 336.
Anatomy, foundations and uses of, 365.
Angles, principles regarding, 55.
Animate substances, definition of, 58.
Animate substances, principles regarding, 58–60.
Antecedent, how distinguished from cause, 61.
Antecedent and consequent, mnemonic law of, 420.
Antiquated significations, when to be adopted, 202, 203.
Antiquities, definition of, 340.
Appeals to authority, aberrancies of, 316–320.
Appeals to desires, aberrancies of, 320–324.
Apprehensions, nature of, 32.
" inferences from, 33, 74–85.
Apprehensions, reality of, 45, 46, 73.
Apprehensions, requisite to, 46.

Apprehensions, how distinguished from ideas, 46, 47, 72, 73.
Apprehensions, distinct from their causes, 73, 74.
Apprehensions, subsidiaries of, 86–91.
Apprehensions, admit of no proof, 128.
Apprehensions, safe assumptions regarding, 130.
Apprehensions, errors regarding, 131, 132.
Approximation, method of, 88.
Archæology, definition of, 340.
Arguments, nature of, 68, 69.
" modes of testing, 68–70, 137–139, 146, 147, 264, 265.
Arguments, illustration of, 69.
" by what invalidated, 251.
Aristotelians, their division of a proposition, Note 2, 435.
Aristotelians, their doctrine of reasoning and the syllogism, Note 5, 436, 437.
Aristotle's Categories, Note 18, 442.
Arithmetic, definition of, 333.
" its uses, 346.
Art, definition of, 339.
" various kinds of, 340.
" characteristics and foundations of, 392, 393.
Art, requisites to proficiency in, 393.
Art, how related to Theory, 393, 394.
Art, field for improvements in, Note 31, 448.
Association of Thoughts (see Thoughts).
Assuming conditions, paralogism of, 277.
Assuming the question, paralogism of, 267, 268.
Astronomy, definition and divisions of, 335.
Astronomy, character and uses of, 354.
Astronomy, modes of establishing its truths, 354–358.
Astronomy, future discoveries in, Note 24, 443, 444.
Attention, nature of, 34.

Attention, its importance, 46–48, 121, 122, 214.
Attention, how secured, 253, 254, 414.
Attention, mnemonic law and rules of, 413–415.
Attraction of gravitation, remarks on, Notes 8 and 15, 437, 440.
Attributes, extrinsic, 79.
" intrinsic, 79–85.
Authority, fallacies of appeals to, 316–320.
Authorship, how ascertained, 229–232.
Axioms, mathematical, what they are, 345.

B.

Begging the question, paralogism of, 266–267.
Beings, definition of, 58.
" various kinds of, 58.
" principles regarding, 58–60.
Beings, how existence and properties of some known, 75–85.
Belief, definition of, 49.
" how distinguished from knowledge, 30, 49.
Belief, common error, 315.
Bias, its sources, and influence on testimony, 214, 215, 217.
Bias, how its presence may be ascertained, 215, 216.
Biblical Theology, definition of, 338.
Biblical Theology, foundations and uses of, 379.
Biblical Theology, study of, 379–381.
Biography, definition and divisions of, 341, 343.
Biography, how related to History, 403.
Biography, its uses, 403, 404.
" common defects in, 404.
Body, definition of, 62.
" natural tendency of, 170, 179.
Books, uses of, 141, 142, 148.
" selection of, 149, 150.
" study of, 150, 151.
" means of ascertaining their origin, 233.

INDEX. 451

Botany, definition and divisions of, 336.
Botany, sources and uses of, 364.

C.

Calculus, Infinitesimal, remarks on, 347. Note 22, 442, 443.
Categories of Aristotle, Note 18, 442.
Causation, principles of, 61–63, 175–177, 180.
Causation, sophisms of, 292–298.
Cause and effect, mnemonic law of, 420.
Causes, definition of, 61.
" how their nature known, Note 7, 437.
Causes, principles regarding, 61–63, 175–177, 180.
Causes, necessary and contingent, 165.
Causes, inadequate, 166.
" uses of a knowledge of, 166–169.
Causes, important distinction, 169.
" efficient and conditional, 169, 170.
Causes, immediate, mediate, and ultimate, 170.
Causes, only ultimate causes, 170, 180.
Causes, frequent error, 170.
" modes of determining, 171–175.
Causes, criterions where these fail, 175–177.
Causes, sole and joint causes, 177, 178.
Causes, uses of analogy, 178.
" " of experiments, 178, 179.
Causes, new agencies, 179.
" chain of causes, 179, 180.
" reactive agencies, 181.
" cautions, 181, 297.
" fallacies regarding causes, 292–298.
Causes, distinct from laws, 295.
Certainty, foundations of, 45–48.
" how distinguished from probability, 66, 301.
Chain of reasoning, definition of, 65.
Chain of reasoning, illustration of, 66.

Chain of reasoning, requisites to validity of, 66.
Chain of reasoning, how related to arguments, 68, 69.
Change, nature of, 69.
" conditions of, 60.
" principles regarding, 60–63, 437.
Characteristic marks, definition of, 64.
Chemistry, definition and divisions of, 337.
Chemistry, foundations and uses of, 371, 372.
Chemistry, important distinction, 371.
Children's testimony, remarks on, 217, 218.
Chronology, definition of, 341.
" uses and foundations of, 402, 403.
Circumstantial evidence, 211, 212.
Classes, organic, formation and naming of, 241–243, 441, 442.
Classification, definition of, 238.
" mental and physical, 238, 239.
Classification, how distinguished from generalization, 238.
Classification, main objects of mental, 239, 240, 417.
Classification, principles of, 240.
" chief rules of, 240, 241.
Classification, of organisms, 241–244.
Classification, application of principles, 244, 245.
Classification, influence of prejudices, 245.
Cognitions, definition of, 30.
" test of, 48, 244–246.
Combination, definition and use of, 69.
Commemorations, mnemonic effects of, 425, 426.
Common-place books, use of, 428.
Comparison, definition and nature of, 101.
Complex proposition, definition of, 39.
Compositions, modes of ascertaining their origin and character, 229–237.

Compound proposition, definition of, 39.
Comprehension of terms, definition of, 191.
Comprehension, mnemonic law of, 415.
Comprehensions, definition of, 29.
" what learned by, 29, 33, 34.
Comprehensions, require no proof, 128.
Comprehensions, paralogisms of, 269-271.
Conceptions, definition and nature of, 35, 36.
Conceptions, always particular, 98, 99.
Conceptions, their reality, how known, 128.
Conclusions, definition of, 65.
" requisites to validity of, 66, 67.
Conclusions, modes of testing, 67, 68, 136-139, 264, 265.
Conditional causes, definition of, 169, 170.
Conditional proposition, definition of, 39.
Conflicting opinions, aberrancy of, 317.
Confounding cause and effect, sophism of, 293, 294.
Confounding different senses, paralogism of, 283.
Confounding means and end, aberrancy of, 316.
Confusion, aberrancies of, 307-316.
" sophisms of, 287-289.
Connective, definition of, 51.
" various forms of, 68.
" necessary character of, Note 5, 436.
Consciousness, definition of, 30.
" futile attempts regarding, Note 14, 440.
Contiguity, mnemonic law of, 419, 420.
Contingent connective, sophism of, 300.
Contingent knowledge, nature of, of, 31.
Contingent knowledge, primary modes of acquiring, 71-93, 96, 97.

Contingent knowledge, primary modes of retaining, 93-98.
Contradictory propositions, definition of, 38.
Contraries, mnemonic law of, 422, 423.
Contrary proposition, definition of, 38.
Controverted subjects, remarks on, 142, 143.
Conversation, its logical character, 142.
Converse of a proposition, definition of, 38.
Converse of a proposition, when true, 288.
Copying, why of little mnemonic value, 428.
Corruptions of written testimony, 232, 233.
Counting witnesses, paralogism of, 281.
Cramming, evils of, 145, 414, 415, 431, 432.
Credulity, paralogism of, 281, 282.
Criterion of intuitions, 267.
" of reasoning, 65-71.
" of truth, 48.
Criterions of testimony, 213-219.
Crystalography, definition of, 338.
Curves, particular use of, 91, 92.

D.

Dates, how ascertained, 402, 403.
Dead languages, how learned, 385-389.
Deciding by appearances, aberrancy of, 311, 312.
Deciding by character, aberrancy of, 309.
Deciding by consequences, aberrancy of, 309, 310.
Deciding by motives, aberrancy of, 310, 311.
Deduction, definition of, 130.
" error regarding, Note 13, 439, 440.
Deduction, tests of, 136, 137.
Definitions, verbal and real, 192.
" rules of verbal, 192, 193.
" characteristics of good, 193.
Definitions, when required and when not, 429.

INDEX. 453

Desires, principles regarding, 62, 63.
Desires, aberrancies of appeals to, 320-324.
Desultory habits, evils of, 122, 123.
Determining conditions, definition of, 60.
Determining conditions, principles regarding, 60-63.
Deviations, evils of, 145.
Dictionaries, sources and uses of, 385-389.
Difficulties, art of surmounting, 144-146.
Diminishing improbability, sophism of, 305.
Diminishing probability, sophism of, 305.
Direct discovery, nature of, 154.
"   "   observations on, 155, 156.
Direct proof, definition of, 134, 135.
Discernments, definition of, 30.
"   require no proof, 128.
Discordant opinion, sophism of, 301, 302.
Discovery, direct, 154-156.
"   indirect, 154, 156-162.
"   "   usual course of, 159.
Discovery, indirect, chief difficulty, 160.
Discrepancies in testimony, 220-232.
Disjunctive proposition, definition of, 39.
Disposition proper for investigation, 119-121.
Division, nature of, 239, 240.
Dogmatism, character and origin of, 119, 120.
Duration, principles regarding, 53, 54.
Duration, errors regarding, Note 6, 437.
Dynamic, definition of, 334.

E.

Education, definition of, 340.
"   its importance, 395, 396.
Effects, definition of, 61.

Effects, how distinguished from consequents, 61.
Effects, principles regarding, 61-63.
"   importance of knowing, 166-169.
Effects, how sometimes discovered, 167, 168.
Effects, peculiar use of, 169.
"   modes of tracing, 171-181.
"   important principle, 180.
"   reactive effects, 181.
"   cautions, 181, 297.
Efficient causes, definition of, 169, 170.
Electric, definition and divisions of, 335, 336.
Electric, foundations and uses of, 361.
Electricity, probable nature of, 362.
Electrodynamic, definition of, 335, 336.
Electrodynamic, importance of, 361.
Electromechanism, definition of, 336.
Electromechanism, uses of, 361.
Electrostatic, definition of, 335.
"   importance of, 361.
Emotional arts, nature of, 340.
Emotions, definition of, 34.
"   what known by, 34.
"   mnemonic law of, 421, 422.
Emotions, important distinction, 422, 432.
Empiricisms, definition of, 101.
"   how formed, 100, 101, 312.
Empiricisms, uses of, 108.
"   not laws of nature, 109.
Enemies' opinions, sophism of, 303.
Epochs, principal used, Note 32, 448.
Equanimity, advantages of, 119, 122, 430.
Equanimity, its influence on remembrance, 430.
Equanimity, how to be secured, 432.
Equivocation, paralogism of, 276.
Eras (see Epochs).
Errors, sources of, 45-48, 131, 132.

Errors, means of avoiding, 46–49, 71–75, 261–265.
Errors, why powerful, 260.
Essential properties, definition of, 64, 242.
Ether, nature and probable effects of, 335.
Ethereal sciences, definition and divisions of, 335, 336, 342.
Ethereal sciences, nature and uses of, 358.
Ethereal sciences, probable origin of their phenomena, 362; Note 25, 444, 445.
Ethic, definition of, 338.
" foundations and uses of, 381, 382.
Ethic, study of, 382, 383.
Ethnology, definition of, 336–366.
Ethnography, definition and divisions of, 339, 340, 343.
Ethnography, foundations and uses of, 391, 392.
Ethnography, why distinguished from Geography, Note 20, 442.
Etymologies, paralogism of following, 284.
Evidence, definition of, 129.
" general principles of, 209–212.
Evidence, signs, 129, 209, 210.
" testimony, 129, 210.
" concurring, 210, 211.
" circumstantial, 211, 212.
" criterions of testimony, 212–219.
Evidence, concurring testimonies, 219.
Evidence, discrepancies, 220–222.
" probable testimony, 222–224.
Evidence, influence of prejudices, 223, 224.
Evidence, futile distinctions, 225, 226.
Evidence, various kinds of testimony, 226–228.
Evidence, written testimony, 227–237.
(See Signs, and Testimony.)
Exaggerating improbability, sophism of, 305.
Exaggerating probability, sophism of, 304, 305.

Excluding causes, sophism of, 296.
" effects, sophism of, 297.
Exclusion, principles of, 63–65.
Existence of self, how known, 75.
Experiments, definition of, 90.
" uses and objects of, 90, 91.
Experiments, two kinds of, 90.
" rules regarding, 161.
Explicit testimony, 226.
Expressions, importance of attending to, 143, 144.
Expressions, rules of proper, 190, 191.
Expressions, five classes of, 200, 201.
Extension, method of, 88, 89.
" or Space, principles relating to, 53, 54.
Extension or Space, errors regarding, Note 6, 437.
Extension of terms, definition of, 191.
External signs, 97, 98.
Extracts, uses of, 428, 429.
" how used in remembrance, 425–427.
Extrinsic probability, remarks on, 224.
Extrinsic properties, 79.

F.

Faculty, definition of, 32.
Fallacies, definition of, 249.
" operation of, 249.
" evils of, 249, 250.
" three classes of, 250.
" universal defect in, 250.
" independent of each other, 251.
Fallacies, illustration of, 251.
" sources of, 252–254.
" effects of prejudices on, 254–260.
Fallacies, means of guarding against, 261–265, 324.
Fallacies, common error, 266.
" table of, 325–327.
(See Paralogisms, Sophisms, and Aberrancies.)
Fallacious implication, paralogism of, 286.
Fallacious propriety, paralogism of, 286.

INDEX. 455

False association, paralogism of, 274.
False cause, sophism of, 292, 293.
" effect, sophism of, 293
Falsehood, paralogism of, 278.
Families, organic, formation and naming of, 243, 441, 442.
Fiction, how to be detected, 237, 402.
Figurative expressions, abuse of, 196.
Figurative interpretation, when proper, 201, 202.
Figures, Aristotelian, remarks on, Note 5, 437.
Flattery, why powerful, 259, 260.
Following etymologies, paralogism of, 284.
Force, definition of, 59.
" principles regarding, 59, 61–63.
Forgetfulness, paralogism of, 275.
Forgetting, nature of, 97.
Fraud, chief source and support of, 21, 22.
Friends' opinions, sophism of, 299.
Futurity, why often undervalued, 257, 321.
Futurity, extent and importance of our knowledge of, 405, 406.
Futurity, sources of such knowledge, 34, 406–410.

G.

Galvanism, definition of, 336.
" importance of, 361.
Gases, characteristic of, 335–354.
Genera, formation and naming of organic, 241–243, 441, 442.
General belief, aberrancy of, 317.
" Geography, definition of, 337.
General Geography, sources and uses of, 368–370.
General Grammar, definition of, 339.
General Grammar, uses of, 391.
" proposition, definition of, 39.
General terms, uses of, 36, 190.
" " what they denote, 99.
Generalization, definition and kinds of, 36, 98.

Generalization, uses of, 99.
" various processes of, 99–107.
Generalization, superior and subordinate laws of, 107.
Generalization, uses of empirical, 108.
Generalization, advantages of extending, 108, 109.
Generalization, how distinguished from classification, 238.
Generalization, sophisms of, 289–292.
Generalization, its influence on remembrance, 415, 416.
Geognosy, definition of, 337.
" sources and uses of, 370.
Geographical sciences, definition of, 334.
Geographical sciences, divisions of, 337, 338, 342.
Geographical sciences, character and study of, 367, 368.
Geography, definition and divisions of, 337.
Geography, sources and uses of, 368–371.
Geology, definition and divisions of, 337.
Geology, foundations and uses of, 372, 373.
Geology, defects of, Note 27, 445.
Geometry, definition and divisions of, 333, 334.
Geometry, how connected with Analysis, 345, 346.
Geometry, uses of, 346, 347.
God (see Theology).
Grammar, sources and uses of, 385–388.
Gravitation, law of, 170.
" objectionable views of, Notes 8 and 15, 437, 440.
Gravitation, how established, 356.
" inferences from, 356–358.
Gravitation, probable origin of its phenomena, Note 25, 444, 445.

H.

Habits, influence of, 121, 418.
" various kinds of, 121–126.
" two important laws of, 125.

Habits, how good to be formed, 127, 431.
Harmonizing conclusions, sophism of, 300.
Health, its importance, 148, 153, 418, 430.
Health, how to be secured, 148, 365, 366, 432, 433.
Health, its influence on remembrance, 418, 430.
Hearing, how aided, 90.
Hearsay evidence, remarks on, 216, 226, 227.
Heat, importance of knowing its laws, 360.
Heat, probable nature of, 362, 444, 445.
History, definition and divisions of, 341-343.
History, foundations and uses of, 396, 397, 400.
History, boundaries of, 397, 398.
  " first merit of, 399.
  " frequent imperfections in, 399, 400.
History, two kinds of, 399, 400.
  " study of, 400, 401.
  " traditional, 401, 402.
Homonymous expressions, aberrancy of, 314.
Hurry, evils of, 431.
Hydric, definition and divisions of, 334, 335.
Hydric, foundations and uses of, 353.
Hydrodynamic, definition of, 335.
  " uses of,
Hydrology, definition of, 337.
  " sources and uses of, 371.
Hydromechanism, definition of, 335.
  " uses of, 353.
Hydrostatic, definition of, 335.
  " uses of, 353.
Hygiene, definition of, 336.
  " importance of, 366.
Hypotheses, definition of, 109.
  " uses of, 110-112, 159.
  " why often undervalued, 110.
Hypotheses, abuse of, 112, 113.
  " modes of testing, 113-115.
Hypotheses, peculiarities of a certain class of, 113.

Hypotheses, phenomenal, 113-115.
  " refutation and confirmation of these, 114.
Hypotheses, preferable, 115, 159.
  " common errors, 115.
Hypothetical causes, sophism of, 294.
Hypothetical proposition, definition of, 39.
Hypothetical truths, nature of, 31.

I.

Ideas, definition of, 33.
  " distinctions of, 46, 47.
  " trains of, how traceable, 419-421.
Identical proposition, definition of, 38.
Identity, personal, remarks on, 93, 94, 132.
Idioms, definition of, 195.
  " interpretation of, 204, 205.
Ignorance, evils of, 15-25.
Ignorant interpretation, paralogism of, 285.
Illusions, nature of spectral, 72.
Illusive contradiction, aberrancy of, 315.
Illusive sign, paralogism of, 271-273.
Imaginary absurdity, sophism of, 304.
Imaginary apprehension, paralogism of, 274.
Imaginary cause, sophism of, 296.
  " effect, sophism of, 297.
  " quantities, nature of, 350.
Imaginary universality, sophism of, 292.
Imagination, nature of, 35, 36.
  " misapplications of, 255, 305.
Imitations, how distinguished, 131, 132.
Immaterial substances, definition of, 58.
Immediate cause, definition of, 170.
Immediate knowledge, definition of, 30.
Immediate testimony, definition of, 226.
Implications, difficulty regarding, 207, 208.

INDEX.   457

Implicit testimony, 226.
Imponderable substances, definition of, 334.
Imponderable substances, probable nature of, 362.
Impossible quantities, nature of, 350.
Impostors, on what chiefly dependent, 21, 22.
Inanimate substances, remarks on, 58, 59.
Inattention, how the immediate source of error, 48, 49, 252, 253.
Inattention, evils of habitual, 122.
" causes of, 253, 254.
Inclusion, principles of, 63-65.
Incomprehensible connective, sophism of, 300, 301.
Incomprehension, paralogism of, 270, 271.
Inconclusive investigation, sophism of, 301.
Indefinite proposition, definition of, 39.
Indefinite terms, aberrancy of, 307, 308.
Index, uses of a general, 428, 433.
Indications (see Signs).
Indirect discovery, nature of, 154.
" various observations on, 157-162.
Indirect proof, definition and character of, 134, 135.
Indiscrimination, paralogism of, 282.
Individual proposition, definition of, 39.
Indolence, evils of, 125, 126.
Induction, definition of, 101.
" requisites to, 101.
" various processes of, 101-106.
Induction, on what based, 104, 105.
" how established, 106, 107.
Induction, uses of, 108, 393.
" advantages of extending, 108, 109.
Induction, aided by classification, 239.
Induction, errors regarding, Note 13, 439, 440.
Inductive Logic, remarks on, Note 13, 439, 440.

Inertia, remarks on, 83.
Inferences, definition of, 29, 50, 51.
" requisites to validity of, 47.
Inferences, how distinguished, 48.
" from comprehensions, 129-134.
Inferences, from probabilities, 136.
" from testimony, 216.
Inferring hypotheses, sophism of, 298, 299.
Inferring the agreeable, aberrancy of, 321.
Inferring the converse, sophism of, 288.
Inferring the probable, sophisms of, 298-301.
Infinitesimal Calculus, remarks on, 347, Note 22, 442, 443.
Information, best sources of, 147.
Inorganical Sciences, definition of, 334.
Intellect, definition of, 71.
Intentional sense, when to be adopted, 203.
Internal signs, 97.
Interpretation, nature and use of, 200.
Interpretation, objectionable view of, Note 16, 441.
Interpretation, what expressions require, 200, 201.
Interpretation, various rules of, 201-207.
Interpretation, frequent sources of difficulty, 207, 208.
Interpretation, use of translations, 208.
Interpretation, influence of prejudices, 208.
Interpretation, fallacies of, 278-280.
Interrogation, paralogism of, 277.
Intervals, mnemonic law of, 416.
Intrinsic probability, remarks on, 224.
Intrinsic qualities, 79-85.
Intuitions, definition and nature of, 29, 64.
Intuitions, knowledge obtained by, 49.
Intuitions, admit of no proof, 127, 128.
Intuitions, paralogisms of, 266, 267.

U

458 INDEX.

Intuitions, criterion of, 267.
Intuitional assumption, paralogism of, 266, 267.
Intuitional rejection, paralogism of, 266, 267.
Invention, nature of, 154.
" how related to indirect discovery, 162.
Invention, two kinds of, 162.
" principal fields of, 162, 163.
Invention, requisites to success, 163, 164.
Invention, various aids in, 164.
Investigation, dispositions affecting, 119-121.
Investigation, habits affecting, 121-127.
Investigation, requisites to success, 120, 121.
Investigation, evils of superficial, 124. (See Proof, Study, and Original Investigation.)
Irrelevant admission, aberrancy of, 319, 320.
Irrelevant analogies, aberrancy of, 308, 309.
Irrelevant empiricism, aberrancy of, 312.
Irrelevant illustration, aberrancy of, 307.
Irrelevant induction, aberrancy of, 312.
Irrelevant modification, aberrancy of, 314.
Irrelevant objection, aberrancy of, 313.

J.
Judgements, definition of, 71.
" other views of, Note 9, 437, 438.
Jurisprudence, definition and divisions of, 338, 339.
Jurisprudence, nature, foundations and uses of, 383, 384.
Jurisprudence, study of, 385.

K.
Knowable, boundaries of the, 30.
Knowledge, definition of, 30.
" advantages of, 15-25.
" requisites in its pursuit, 16, 120, 121, 318.

Knowledge, its limits, 29, 30.
" its threefold division, 29.
Knowledge, how distinguished from belief, 30, 265.
Knowledge, immediate and mediate, 30, 31.
Knowledge, necessary, contingent, and hypothetical, 31.
Knowledge, of real beings, on what founded, 71.
Knowledge, mental processes for acquiring, 75-85.
Knowledge, external processes for acquiring, 86-93.
Knowledge, primary modes of retaining and perpetuating, 93-98.
Knowledge, primary and secondary, 139, 140.
Knowledge, best sources of, 147.
" no royal road to, 149.
" test of its amount, 191.
" table of the means of acquiring, 246.
Knowledge, classification of, 331.
" scientific, 331-339.
" mixed, 339, 340.
" particular, 340, 341.
" table of, 341-343.
" of futurity, 405-410.
" retention of, by simple remembrance, 413-425.
Knowledge, retention of, by external signs, 425-429.
Knowledge, requisites to a ready command of, 429, 430.
Knowledge, means of acquiring and employing these, 431-433.
Knowledge, table of the means of retaining, 434.
Known, boundaries of the, 30.

L.
Language, how related to reasoning, 70, 71.
Language, vernacular, how learned, 85.
Language, importance of understanding, 144, 315.
Language, origin of, 182, 183.
" progress of, 183-188.
" compounding terms, whence, 185, 186.
Language, not of divine origin, 188.

INDEX. 459

Language, uses of, 189.
" natural, 189.
" advantages of speech, 189, 190.
Language, three rules of proper expression, 190, 191.
Language, definitions, 192, 193.
" new terms, when required, 193, 194.
Language, new terms, when to be avoided, 194.
Language, what it represents, 194.
" imperfections of, 195, 196.
Language, abuses of, 196–199.
" interpretation of, 200–208. (See Interpretation.)
Language, misinterpretation of, 283–287.
Language, modes of learning, 385–389. (See Philology.)
Law (see Jurisprudence).
Laws of habits, 125.
Laws of health, advantages of attending to, 148, 153.
Laws of motion, 59, 443.
" proposed addition to, Note 23, 443.
Laws of nature, how established, 101–107.
Laws of nature, superior and subordinate, 107.
Laws of nature, advantages of extending, 108, 109.
Laws of nature, distinguished from empiricisms, 109.
Laws of nature, what they are, 109, 170.
Laws of nature, importance of knowing, 166, 168, 173, 174.
Laws of nature, not efficient causes, 295.
Laws of nature, their uniformity and permanence, 407–409.
Laws of remembrance, 413–425.
Lectures, characteristics of, 142.
Letters, origin and spread of, Note 12, 438, 439.
Life, criterion of, 364.
Light, probable nature of, 359, 360.
Likenesses, different kinds and uses of, 97, 98, 426, 438.
Likenesses, how subservient to remembrance, 93–95, 96–98, 426.

Lines, mathematical, property of, 54.
Liquids, definitions of, 334, 353.
Literal interpretation, when proper, and when not, 201–203.
Logic, definition of, 13.
" objectionable definition of, Note 1, 435.
Logic, nature and foundations of, 13.
Logic, its proper limits, 14.
" its objects and uses, 14, 15.
" study of, 15.
" remarks on Aristotelian, Note 5, 436, 437.
Logic, error regarding, Note 13, 439, 440.
Logic, how distinguished from Psychology, 376.

M.

Magnetism, definition of, 336.
" foundations and uses of, 361.
Magnitudes, principles regarding, 54–58.
Mannerism, remarks on, 205.
Manuscripts, means of ascertaining their origin, 231, 233.
Manuscripts, tests of their authenticity, 232, 233.
Manuscripts, sources of errors in, 234.
Manuscripts, means of removing sources of errors in, 235–238.
Many arguments, aberrancy of, 319.
Mass, definition of, Note 23, 443.
Mathematical Geography, definition of, 337.
Mathematical Geography, foundations and uses of, 368–370.
Mathematics, definition and divisions of, 333, 334, 341.
Mathematics, peculiarities of, 344–346.
Mathematics, errors regarding, 344.
" uses of, 346, 347.
" study of, 347–350.
" effects of exclusive study of, 350, 351.
Matter, definition of, 58.
" principles regarding, 58–60.
" its natural tendency, 170, 179.

Matter, two kinds of, 334.
Measurement, various modes of, 87, 89.
Measures, standard of, 87.
Mechanic, definition and divisions of, 334.
Mechanic, foundations and uses of, 352, 353.
Mechanical properties, definition of, 334.
Mechanical sciences, definition and divisions of, 334, 335, 342.
Mechanical sciences, foundations, uses and study of, 352-358.
Mechanism, definition of, 334.
Mediate causes, definition of, 170.
" knowledge, definition of, 30.
Mediate knowledge, how established, 34, 93-98, 129.
Mediate testimony, 226, 227.
Medical arts, rational bases of, 366, 367.
Memory, definition of, 33.
" uses of, 71, 93, 96.
" reliability of, 93-95.
" primary processes of, 93-95, 96-98.
Memory, means of avoiding its primary errors, 95, 96.
Memory, recognition, 95, 96.
" use of similitudes, 97.
" safe assumptions regarding, 132.
Memory, sources and safeguards of error, 132, 133, 216, 217.
Memory, how aided by classification, 239, 417.
Memory, paralogisms of, 274, 275. (See Remembrance.)
Mendacity, paralogisms of, 278-280.
Mental discipline, advantages of, 23, 24, 374, 375.
Mental sciences, definition and divisions of, 338, 339, 342, 343.
Mental sciences, characteristics of, 374.
Mental sciences, foundations and uses of, 374, 375.
Mental sciences, study of, 375, 376.
Metaphysics, remark on, Note 28, 446.

Meteorology, definition of, 337.
" sources and uses of, 371.
Method (see System).
Methodical habits, advantages of, 122, 123.
Microscope, uses of, 90, 362, 363.
Mind, usual tendency of, 265, 266.
Mineralography, definition of, 338.
Mineralogy, definition and divisions of, 338.
Mineralogy, uses and sources of, 372.
Miscomprehension, paralogism of, 269, 270.
Misconception, paralogism of, 285, 286.
Misconstruction, paralogism of, 285.
Misinterpretation, sources of, 207, 208.
Misinterpretation, paralogisms of, 283-287.
Misinterpreting ambiguities, paralogism of, 283.
Misinterpreting technicalities, paralogism of, 283.
Misplacing the accent, paralogism of, 284, 285.
Misrepresentation, paralogisms of, 278, 279.
Misrepresenting comprehensions, paralogism of, 279.
Misrepresenting testimony, paralogism of, 279, 280.
Mistaking allusions, paralogism of, 286.
Mistaking expressions, paralogism of, 283.
Mistaking ideas, paralogism of, 274, 275.
Mistaking the chief cause, sophism of, 294.
Mistaking the chief effect, sophism of, 294.
Mistaking the ultimate cause, sophism of, 295.
Mistaking the style, paralogism of, 284.
Misunderstanding archaisms, paralogism of, 283.
Mixed knowledge, definition and divisions of, 339, 340-343. (See Philology, Ethnography, and Technology.)

Mnemotechny, character of, 423.
Modern opinious, aberrancy of, 317.
Momentum, definition of, Note 23, 443.
Monuments, mnemonic use of, 425, 426.
Moods, remarks on Aristotelian, Note 5, 437.
Moral certainty, nature of, Note 3, 435.
Morality, definition of, 338.
" foundations and importance of, 381, 382.
Morality, study of, 382, 383.
Mortifying proofs, sophism of, 303.
Motion, nature and laws of, 59, Note 23, 443.

N.

Natural History (see Zoology).
" Theology, definition of, 338.
Natural Theology, foundation and importance of, 378, 379.
Natural Theology, study of, 379.
Nature, laws of, how established, 101-107.
Nature, laws of, superior and subordinate, 107.
Nature, laws of, advantages of extending, 108, 109.
Nature, laws of, distinguishable from empiricisms, 109.
Nature, laws of, importance of knowing, 166, 168, 173, 174.
Nature, laws of, not efficient causes, 295.
Nature, hypothesis regarding, Note 25, 444, 445.
Nature, uniformity and permanence of, 407-409.
Necessary implication, definition of, 29.
Necessary qualities, principles regarding, 58-60.
Necessary truths, nature of, 29, 31.
Negative proposition, definition of, 39.
Negative quantities, nature of, 348, 349.
Nervous system, its influence on remembrance, 418.

New words, remarks on, 193, 194.
Non-interpretation of signs, paralogism of, 273, 274.
Nosology, definition of, 336.
Notions, definition of, 33.
Numbers, principles regarding, 54-58.
Numbers, abstract and concrete, 349.

O.

Objection, paralogism of irrelevant, 313.
Obscure expression, paralogism of, 277.
Obscurity, means of avoiding, 191.
" sources of, 196-198.
One-sided arguments, sophism of, 299, 300.
Ontology, why not a science, Note 28, 445, 446.
Opinion, definition of, 49.
Optic, definition of, 335.
" foundations and uses of, 358-360.
Oral testimony, observations on, 227, 228.
Orders, formation of organic, 242, 243.
Orders, naming of organic, 243, 441, 442.
Ordinary significations of words, when to be adopted, and when not, 201-203.
Organic species and genera, properties of, 105, 106.
Organic species and genera, naming of, 243, 441, 442.
Organic specimens, means of preserving, Note 26, 445.
Organical sciences, definition of, 334.
Organical sciences, divisions of, 336, 337, 342.
Organical sciences, characteristics and study of, 362, 363.
Organical sciences, guiding principles in, 363, 364.
Organisms, classification of, 241-244.
Original investigation, 151-164.
" " the general character of, 151.
Original investigation, uses of, 152.

Original investigation, selection of subjects, 153.
Original investigation, prerequisites to, 153, 154.
Original investigation, methods of, 154, 155.
Original investigation, the principal rules of, 155.
Original investigation, direct discovery, 155, 156.
Original investigation, indirect discovery, 156-162.
Original investigation, invention, 162-164.
Overlooking conditions, paralogism of, 277.
Overlooking testimony, 282.
" the alternative, sophism of, 302.
Overlooking the idiom, paralogism of, 284.

P.

Palæontology, definition of, 337.
Paralogisms, definition of, 250.
" illustration of, 251.
" of assuming what is attempted to be proved, 267-269.
Paralogisms, of comprehension, 269-271.
Paralogisms, of signs, 271-274.
" of memory, 274, 275.
" of testimony (intrinsic), 275-280.
Paralogisms, of testimony (extrinsic), 280-282.
Paralogisms, of misinterpretation, 283-287.
Paralogisms, table of, 325.
Particular knowledge, nature of, 331.
Particular knowledge, divisions of, 340, 341, 343.
(See History, Chronology, and Biography.)
Particular proposition, definition of, 39.
Passions, evil effects of, 119, 422, 430, 432.
Passions, distinction, 422, 432.
" means of moderating, 432.
Past, how known, 34, 93-98.

Pathology, definition and divisions, 336, 337.
Pathology, uses of, 366, 367.
" two important principles, 357.
Pathology, foundations of, 357.
Peculiar marks, definition of, 64.
Perception, definition of, 32.
Perseverance, advantages of, 125.
Personal identity, nature and proof of, 93, 94, 132.
Personal observations, advantages of, 152.
Perspicuity, how to be secured, 191.
Petrology, definition of, 337.
Phantasm, definition of, 33.
Philology, definition and divisions of, 339, 343.
Philology, modes of learning a language, 385-389.
Philology, principles and results of comparative, 389, 390.
Philology, errors regarding, Note 30, 446, 447.
Philology, general, 390, 391.
" uses of, 391.
Phonetic writing, nature of, 97.
" " origin and spread of, Note 12, 438, 439.
Phonetic writing, advantages and disadvantages of, 426, 427.
Physical sciences, definition and divisions of, 334-338.
Physical sciences, characteristics and study of, 351.
Physical sciences, uses of, 352.
" welfare, requisite to, 200.
Physiology, definition and divisions of, 336.
Physiology, foundations and uses of, 365, 366.
Phytology, definition of, 336.
Pictures, uses of, 91, 92, 426.
" disadvantages of, 424, 425, 426.
Pneumatic, definition of, 335.
" foundations and uses of, 353, 354.
Pneumatology, remark on, Note 28, 446.
Point, property of mathematical, 54.
Political Economy, remarks on, Note 29, 446.

INDEX. 463

Ponderable matter, definition of, 334.
Power, how its nature known, Note 7, 437.
Predicate, definition of, 37.
Prejudices, evils of, 126, 208.
" rules regarding, 143.
" influence of, on testimony, 223, 224.
Prejudices, influence of, on classification, 245.
Prejudices, nature and general operation of, 254–256.
Prejudices, causes of power of, 255, 256, 260.
Prejudices, several kinds of, 256–260.
Prejudices, combination of, 260.
" means of guarding against, 261, 262, 265.
Premise and inference, mnemonic law of, 420.
Premises, definition of, 50.
" inaccurate use of, 160.
Primary premises, definition of, 65.
" " criterions of, 66, 135–138.
(See Reasoning, Evidence, and Testimony.)
Principles of reasoning, 50–65.
" " general principle, 50.
Principles of reasoning, special principles, 52–65.
Principles of classification, 240–242.
Probability, definition of, 40.
" different kinds of, 40–43.
Probability, general character of, 43, 44, 223, 224.
Probability, uses of, 44, 45.
" resultant, 45.
" how distinguished from certainty, 66, 301.
Probability, reasoning from, 136.
" circumstantial, 211, 212.
Probability, effects of, on testimony, 222–225.
Probability, extrinsic and intrinsic, 223, 224.
Probability, futile distinctions, 225, 226.

Probability, sophisms of, 296–306.
Probable reasoning, nature of, 136.
Proof, definition of, 127, 129.
" what truths require none, 127–129.
Proof, twofold division of, 129.
" what may be admitted as proved, 129–134.
Proof, what propositions require, 134.
Proof, two kinds of, 134, 135.
" general modes of testing, 134–139.
(See Deduction, Evidence, and Testimony.)
Properties, extrinsic, 79.
" intrinsic, 79–85.
" of organic beings, 105, 106.
Properties, essential and non-essential, 242.
Propositions, definition of, 37.
" parts of, 37, 38.
" expression of, 38.
" Aristotelian view of, Note 2, 435.
Propositions, various kinds of, 38, 39.
Propositions, ambiguities in, 39.
" various forms of, 40.
" combinations of, 40.
Prototype, definition of, 38.
Psychology, definition of, 338.
" boundaries of, 376, 377.
Psychology, foundations and importance of, 377.
Psychology, study of, 377.
Pyrrhonism, futility of, Note 14, 440.

Q.

Qualities, extrinsic, 79.
" intrinsic, 79–85.
Quantities, principles regarding, 54–58.
Quantities, means of accurately determining, 86–91.
Quantities, standards of, 87.
" negative, 348, 349.
Quantities, imaginary or impossible, 350.
Quantities, unit of, 87, 346, 355.
(See Mathematics.)

R.

Ratiocination, error regarding, Note 13, 439, 440.
Real definitions, nature of, 192.
Reason, definition of, 33, 49, 50.
Reasoning, definition of, 33, 49, 50.
" objectionable views of, Notes 4, 5, and 13, 435, 436, 439, 440.
Reasoning, general principle of, 50.
" identity of, in all cases, 50.
Reasoning, expression of, 50, 51.
" special principles of, 52-65.
Reasoning, processes of, 65-70.
" chain of, 65, 66.
" requisites to validity of, 66, 67.
Reasoning, ultimate foundations of, 67, 130.
Reasoning, modes of testing, 67, 68, 136-139.
Reasoning, arguments, 68.
" combination, 69.
" why unimpugnable, 70.
" how related to language, 70, 71.
Reasoning, aids of, 71.
" from probabilities, 136.
" in a circle, paralogism of, 268, 269.
Recognition, nature of, 95, 421.
Recollection, definition of, 97.
" operation of, 418, 419.
Re-comprehension, mnemonic law of, 417.
Recreation, advantages of, 148.
Reductio ad absurdum, nature of, 134, 135.
Reduction of syllogisms, remark on, Note 5, 437.
Rejecting the disagreeable, aberrancy of, 321, 322.
Rejecting the improbable, sophisms of, 301-304.
Rejecting theories, sophism of, 302.
Relations of thoughts, nature and general law of, 418.
Relations of thoughts, two kinds of, 419.
Relations of thoughts, natural, 419-425.

Relations of thoughts, remarks on artificial, 423. (See Remembrance, and Thoughts.)
Religion, proper foundation and importance of, 378.
Religious knowledge, only reliable sources of, 381.
Remembrance, definition of, 33.
" nature of, 33.
" importance of, 34, 93, 96.
Remembrance, reliability of, 93-95.
" recognition, 95, 96.
" similitudes, 96, 97.
" primary processes of, 93-95, 96-98.
Remembrance, safe assumptions regarding, 132.
Remembrance, sources and safeguards of error, 132, 133, 216, 217.
Remembrance, general laws and rules of, 413-419.
Remembrance, laws of the relations of thoughts, 419-425.
Remembrance, two kinds of, 419.
" contiguity and succession, 419-421.
Remembrance, resemblance, 421.
" emotions, 421, 422.
" contraries, 422, 423.
" various observations on, 423-425.
Remembrance, external signs, 425-427.
Remembrance, writing, 427-429.
" requisites to readiness of, 429, 430.
Remembrance, how these to be secured, 431-433.
Repetition, its influence on remembrance, 416.
Repetition, method of, 87, 88.
" combination of, 89.
Representations, uses of visible, 91, 92, 426.
Representations, uses of tangible, 92, 93.
Resemblance, mnemonic law of, 421.
Results, how tested, in certain cases, 90.
Retention of knowledge, primary processes of, 96-98.

INDEX. 465

Retention of knowledge, by simple remembrance, 413-425.
Retention of knowledge, relations of thoughts, 419-425.
Retention of knowledge, external signs, 425-429.
Retention of knowledge, writing, 427-429.
Revelation, evidences of, 379.
" importance of, 379.
" study of, 379-381.
Reviewing, its importance in study, 148.
Reviewing, mnemonic laws, 416, 417.
Rhetoric, definition of, 340.
" uses of, 394, 395.
Ridicule, its logical character, 322.
Rumor, general character of, 226, 227.
Rumor, when important, 227.

S.

Sages' opinions, aberrancy of, 317, 318.
Scepticism, character and origin of, 119, 120.
Scepticism, paralogism, 282.
Sciences, which dependent on experience, 137.
Sciences, definition of, 331.
" requisites to, 331, 332.
" boundaries of, 332, 333.
" three classes of, 333.
" remarks on some names of, Note 19, 442.
Sciences, table of, 341-343.
" their relations to Art, 392, 393.
Sciences, permanence of (see Mathematics, Physical Sciences, and Mental Sciences).
Self-control, advantages of, 126, 127.
Self-indulgence, evils of, 126, 127.
Semeiology, definition of, 336, 337.
Sensations, definition of, 32.
" proper mode of dealing with, 432.
Senses, credibility of, 71-75.
" origin of errors attributed to, 71, 72, 132.
Senses, how to avoid these, 72-75.

Senses, subsidiaries of, 86-93.
Severing probabilities, sophism of, 302, 303.
Sight, how aided, 90.
" cause of its superior power, 424, 425.
Signs, definition of, 129.
" internal and external, 97, 98, 129.
Signs, general means of testing, 135.
Signs, conclusive, 209.
" how these ascertainable, 209.
" probable, 209, 210.
" value of, how determined, 210.
Signs, use of, in aiding testimony, 222, 223.
Signs, paralogisms of, 271-274.
" use of, in aiding remembrance, 425-429.
Similitudes, definition of, 33.
" distinctions, 46, 47, 72, 73, 96.
Similitudes, relation of, to remembrance, 97.
Similitudes, how fading rendered precise and vivid, 417.
Simple proposition, definition of, 39.
Singular proposition, 39.
Sneers, logical character of, 322.
Sobriety, advantages of, 126, 127, 418.
Solid bodies, nature of, 58.
Solidity, mathematical, 54, 333.
" resistive, 58.
" mechanical, 334.
Solids, properties of mathematical, 54.
Sophisms, definition of, 250.
" illustration of, 251.
" of confusion, 287-289.
" of generalization, 289-292.
Sophisms of causation, 292-298.
" of probability, 298-306.
" table of, 326.
Sophistical combination, sophism of, 292.
Sophistical connection, sophism of, 287, 288.
Sophistical contraction, sophism of, 291.

Sophistical distinction, sophism of, 304.
Sophistical exclusion, sophism of, 291.
Sophistical explanation, sophism of, 295.
Sophistical extension, sophism of, 290.
Sophistical inclusion, sophism of, 290, 291.
Sophistical induction, sophism of, 295.
Sophistical leap, sophism of, 301.
" proof, sophism of, 295, 296.
Sophistical relation, sophism of, 296.
Space, principles regarding, 53, 54.
Special interpretation, when to be adopted, 202.
Species, classification of organic, 241, 242.
Species, mode of naming, 242, Note 17, 441, 442.
Speech (see Language).
Specters, nature of, 72.
" remarkable case, Note 10, 438.
Spurious compositions, how distinguished, 230-232.
Static, definition of, 334.
Straight line, principles regarding, 56-58.
Study, what acquired by, 139.
" importance of, 139.
" general objects of, 139, 140.
" other advantages of, 140.
" order of, 140.
" selection of subjects, 140, 141.
Study, objects to be considered, 141.
Study, extent of, 141.
" three modes of, 141, 142.
" of controverted subjects, 142, 143.
Study, general rules of, 143-148.
" prejudices, 143.
" meaning of terms, 143, 144.
" language, 144.
" difficulties, 144-146.
" evils of deviations and cramming, 145.

Study, advantages of careful and thorough, 146.
Study, testing statements, 146, 147.
" sources of information, 147.
" simultaneous subjects, 147, 148.
Study, recreation, 148.
" laws of health, 148.
" final reviewing, 148.
" selection of books, 148, 149.
" of books, 150, 151.
" evil practices of, 150, 151.
" different from reading, 431, 432.
Style, requisites to a good, 190-194.
Subject of a proposition, definition of, 37.
Substances, necessary qualities of, 58-60.
Succession, mnemonic law of, 419, 420.
Superstition, its origin and remedy, 22, 23.
Suppressing truth, paralogism of, 278, 279.
Surfaces, properties of mathematical, 54.
Syllogism, definition of, 50.
" Aristotelian view of, Note 5, 436, 437.
Syllogism, parts of, 50, 51.
" various modes of stating, 51.
Syllogism, modes of testing, 66-70.
" error regarding, Note 13, 439, 440.
Symbolic writing, character of, 98.
Symbols, uses and kinds of, 91-98, 426.
Symptomatology, definition of, 336, 337.
Synopses, uses of, 264, 417.
Synthetical Geometry, definition of, 334.
Synthetical Geometry, uses of, 350.
System, general advantages of, 122, 123, 430.
System, mnemonic law of, 417, 418.
" how secured, 127, 431.

T.

Table of the means of acquiring knowledge. 246.

INDEX. 467

Table of fallacies, 325–327.
" of the principal branches of knowledge, 341–343.
Table of the means of retaining knowledge, 434.
Tables, uses of, 428, 429.
Tangible representations, use of, 92, 93.
Technical senses, when to be adopted, 202.
Technology, definition and divisions of, 339, 340, 343. (See Art.)
Temperance, advantages of, 126, 127, 418.
Terms, uses of general, 36, 190.
" what these denote, 99.
" importance of understanding, 143, 144.
Terms, understanding of, how to be effected, 196, 199, 202.
Terms, when new requisite, 194.
" whence these best derived, 194.
Terms, when new objectionable, 194.
Terms, five classes of, 200, 201.
Testimony, definition of, 129.
" safe assumptions regarding, 133.
Testimony, general principle of its credibility, 133, 134, 210.
Testimony, two requisites, 135, 136.
Testimony, influence of witness's moral character, 210.
Testimony, importance of, 212, 213.
Testimony, criterions of, 213–219.
" general and special, 213.
Testimony, concurring testimonies, 219.
Testimony, nature of the statements, 219, 220.
Testimony, discrepancies, 220–222.
" difficulties, how surmountable, 221, 222.
Testimony, probable testimony, 222–224.
Testimony, influence of prejudices, 223, 224.
Testimony, futile distinctions, 225, 226.
Testimony, explicit and implicit, 226.
Testimony, oral and written, 227, 228.
Testimony, effects of lapse of time, 228, 237.
Testimony, evidences of authorship, 229–232.
Testimony, sources of material corruptions, 232.
Testimony, rules regarding these, 232, 233.
Testimony, means of ascertaining origin of writing, 233.
Testimony, various readings, 234.
" rules regarding these, 235, 236.
Testimony, authentic and fictitious, 237.
Testimony, paralogisms of (intrinsic), 276–280.
Testimony, paralogisms of (extrinsic), 280–282.
Theology, definition and divisions of, 338.
Theology, character and foundations of, 377–379.
Theology, study of, 379–381.
Theory, definition of, 114.
" its relations to Art, 393, 394.
Therapeutic, definition of, 337.
" foundations of, 357.
Thermotic, definition of, 335.
" foundations and uses of, 360.
Thinking, six things necessary to, 36, 37.
Thinking, source of errors, 37.
Thoughts, relations of, 419–425.
" two kinds of, 419.
" four laws of natural, 419–423.
Thoughts, arbitrary relations of, 423.
Thoughts, mnemotechny, 423.
" means of widening the range of relations of, 424.
Time, principles relating to, 53, 54.
Torture, its effects on testimony, 218.
Toxicology, definition of, 336.
Tradition, remarks on, 401, 402.
Translations, particular use of, 208.

Truism, eight forms of it, 52, 53.
Truisms, definition of, 52.
" general expression of, 52.
Truth, general criterion of, 45–48.
" common error in investigating, 119.
Truth, two extremes, 119, 120.
" their common origin, 120.
" proper course, 120.
" requisites to discovery of, 120, 121, 261-266.
Truth, frequent causes of failure, 120.
Truth, proper and improper habits, 121-127.
Truth, how former to be secured, 127.
Truth, necessary, 29, 31, 32.
" contingent, 31, 32.
" hypothetical, 31, 32.
" universal, how known, 52.
" what requires no proof, 127-129.
Truth, prerequisite to its admission, 128.
Truth, what may be admitted as proved, 129-134.
Truth, what requires formal proof, 134.
Types, organic, 242, 243.

U.

Ultimate causes, definition of, 170.
" " what alone are, 170.
Ultimate causes, error regarding, 170.
Unconditional proposition, definition of, 39.
Uniformity of nature, how known, 101-107.
Units of measure, 87, 346, 355, 357.
Universal belief, aberrancy of, 316, 317.
Universal proposition, definition of, 39.
Universal truths, only means of knowing, 29, 31, 32.
Universal truths, various expressions of, 64.
Usages of language, how learned, 204.

Usages of language, how employed in interpretation, 203-206.
Usages of language, which the best, and which good, 386, 387.
Usual significations, when to be adopted, 201-203.

V.

Vagueness of expression, means of avoiding, 191-198.
Vagueness of expression, sources of, 196, 197.
Various readings, sources of, 234.
" " written and printed copies, 234.
Various readings, rules regarding, 232, 233, 235, 236.
Various readings, general character of, 236.
Varying probability, sophisms of, 304-306.
Velocity, principles regarding, 59.
" actual and virtual, Note 23, 443.
Verbal definitions, nature of, 192.
" " rules regarding, 192, 193.
Verbal illusion, aberrancy of, 314, 315.
Verification, various methods of, 90.
Vernacular language, how first learned, 85.
Vicious circle, paralogism of, 268, 269.
Visible representations, uses of, 91, 92, 426.
Volition, definition of, 62.
" principles regarding, 62, 63.
Volition, error regarding, Note 11, 438.
Volition, changes caused by, 75-79.

W.

Weights, standard of, 87.
Wisdom, requisites to, 16, 17.
Witnesses (see Testimony).
Words (see Expressions, Language, and Terms).
Writing, origin and spread of, Note 12, 438, 439.
Writing, uses of, 141, 142, 148, 228, 426, 427.

Writing, retention of knowledge by, 427–429.
Writing, common error regarding, 433.
Writing, means of rendering available, 433.
Written testimony, remarks on, 227–237. (See Testimony.)

Wrong expression, paralogism of, 277, 278.

Z.

Zoology, definition and divisions of, 336.
Zoology, foundations and uses of, 364, 365.

THE END.

www.ingramcontent.com/pod-product-compliance
Lightning Source LLC
Chambersburg PA
CBHW022101300426
44117CB00007B/547